International Association of Fire Chiefs

 International Society of Fire Service Instructors

Live Fire Training
Principles and Practice
First Edition Revised

JONES & BARTLETT LEARNING

Jones & Bartlett Learning
World Headquarters
5 Wall Street
Burlington, MA 01803
978-443-5000
info@jblearning.com
www.jblearning.com

National Fire Protection Association
1 Batterymarch Park
Quincy, MA 02169-7471
www.NFPA.org

International Association of Fire Chiefs
4025 Fair Ridge Drive
Fairfax, VA 22033
www.IAFC.org

International Society of Fire Service Instructors
2425 Highway 49 East
Pleasant View, TN 37146
www.isfsi.org

Jones & Bartlett Learning books and products are available through most bookstores and online booksellers. To contact Jones & Bartlett Learning directly, call 800-832-0034, fax 978-443-8000, or visit our website, www.jblearning.com.

Substantial discounts on bulk quantities of Jones & Bartlett Learning publications are available to corporations, professional associations, and other qualified organizations. For details and specific discount information, contact the special sales department at Jones & Bartlett Learning via the above contact information or send an email to specialsales@jblearning.com

Production Credits

Chairman, Board of Directors: Clayton Jones
Chief Executive Officer: Ty Field
President: James Homer
Sr. V.P., Chief Operating Officer: Don W. Jones, Jr.
V.P., Design and Production: Anne Spencer
Executive Publisher: Kimberly Brophy
Executive Acquisitions Editor—Fire: William Larkin
Associate Managing Editor: Amanda Green
Editor: Jennifer Kling
Associate Editor: Nick Cronin
Associate Production Editor: Jessica deMartin

V.P., Sales, Public Safety Group: Matthew Maniscalco
Marketing Manager: Brian Rooney
V.P., Manufacturing and Inventory Control: Therese Connell
Composition: diacriTech, Chennai, India
Assistant Photo Researcher: Rebecca Ritter
Cover Design: Kristin E. Parker
Cover Image: Courtesy of Dave Casey
Printing and Binding: RR Donnelley Companies
Cover Printing: RR Donnelley Companies

Copyright © 2016 by Jones & Bartlett Learning, LLC an Ascend Learning Company, and the National Fire Protection Association®

All rights reserved. No part of the material protected by this copyright may be reproduced or utilized in any form, electronic or mechanical, including photocopying, recording, or by any information storage and retrieval system, without written permission from the copyright owner.

The procedures and protocols in this book are based on the most current recommendations of responsible sources. The National Fire Protection Association (NFPA), International Association of Fire Chiefs (IAFC), International Society of Fire Service Instructors (ISFSI), and the publisher, however, make no guarantees as to, and assume no responsibility for, the correctness, sufficiency, or completeness of such information or recommendations. Other or additional safety measures may be required under particular circumstances.

Notice: The individuals described in the "You Are the Live Fire Training Instructor" and "Live Fire Training Instructor in Action" throughout the text are fictitious.

There may be images in this book that feature models; these models do not necessarily endorse, represent, or participate in the activities represented in the images. Any screenshots in this product are for educational and instructive purposes only. Any individuals and scenarios featured in the case studies throughout this product may be real or fictitious, but are used for instructional purposes only.

Additional illustration and photographic credits appear on page 252, which constitutes a continuation of the copyright page.

ISBN: 978-1-284-04123-1

Library of Congress Cataloging-in-Publication Data
Live fire training: principles and practice / International Association of Fire Chiefs [and] National Fire Protection Agency.
    p. cm.
   Includes bibliographical references and index.
   ISBN-13: 978-0-7637-8188-0
   ISBN-10: 0-7637-8188-6
   1. Fire fighters—Training of. 2. Fire extinction. I. International Association of Fire Chiefs. II. National Fire Protection Association.
   TH9120.L58 2012
   628.9'25071—dc22
                        2010041568
6048

Printed in the United States of America
16   10 9 8 7 6 5 4 3 2

Dedication

This textbook is dedicated to Elias "Buck" Tomlinson and all of the live fire training instructors who teach their students to perform safely in all conditions.

Elias "Buck" Tomlinson
1940–2009
2004 Florida Fire Service Instructor of the Year

Brief Contents

1 Introduction to Live Fire Training ... 2

2 Critical Incident Planning ... 24

3 Preparation and Training of Instructors ... 38

4 Fire Fighter Physiology ... 54

5 Planning for Live Fire Training ... 76

6 Acquired Structures ... 98

7 Gas-Fired and Non-Gas-Fired Structures ... 124

8 Nonstructural Training Props ... 152

9 Live Fire Training Evolutions ... 176

Appendix A: An Extract From: NFPA® 1403, *Standard on Live Fire Training Evolutions*, 2012 Edition ... 188
Appendix B: NFPA® 1403 Correlation Guide ... 197
Appendix C: Acquired Structure Live Fire Training Model SOP ... 200
Appendix D: Permanent Structure Live Fire Training Model SOP ... 228
Appendix E: Permanent Structure Live Fire Training Burn Plan for _____ Fire Training Center ... 235
Appendix F: NFPA 1403, Standard on Live Fire Training Evolutions, Crosswalk from 2007 Edition to 2012 Edition ... 243
Glossary ... 246
Index ... 248
Credits ... 252

Contents

1 Introduction to Live Fire Training..... 2
Introduction............................5
   The History of Live Fire Training Evolutions... 5
Review of Live Fire Evolution Incidents........6
The Standard and Legal Considerations.......6
   The Impact of NFPA 1403 on
      Live Fire Training...................... 7
   Referenced Standards................... 7
   Using NFPA 1403....................... 8
Student Prerequisites.....................9
   The Student and the Live Fire Training
      Evolution............................. 16
   Recommended Student Training.......... 17
The Last Session Before Live Fire Training
   Evolutions 19

2 Critical Incident Planning.......... 24
Introduction...........................26
Line-of-Duty Significant Injury or Death Plan..26
The Investigation 27
   NIOSH Investigations................... 27
   External Reviews 29
The News Media 29
The Aftermath 32
   Regulatory and Criminal Charges and
      Prosecution 32
   Internal Strife......................... 32
   News Media Scrutiny and
      Public Opinion 33
   Employment 33
   Civil Litigation 33
Avoidance 33

3 Preparation and Training of Instructors 38
Introduction...........................40
Prerequisites to Becoming a Live Fire Training
   Instructor............................40
Training for the Live Fire Training Instructor...43
   Fire Behavior and Structural Fire Dynamics ...43
   Student Psychology: Fire Fighter Style 45
   Training Evolutions 45
   Accountability, Safety, and Practice
      for When Things Go Wrong............ 48
Florida's Live Fire Training Laws: Instructor
   Requirements 49
Pennsylvania Suppression Instructor
   Development Program (ZFID) 50
ISFSI Live Fire Trainer Program 50
The National Wildfire Coordinating Group 50

4 Fire Fighter Physiology 54
Introduction...........................56
Cardiovascular and Thermal Strain of
   Firefighting 56
Factors Affecting Cardiovascular and
   Thermal Strain 58
   Environmental Conditions 58
   Work Performed....................... 58
   Personal Protective Equipment........... 58
   Individual Characteristics............... 59
   Medical Conditions.................... 59
   Fitness Level.......................... 59
   Hydration Status 60
Heat Emergencies.......................60
   Thermal Balance...................... 61
   Heat Illness 61
   Risk Factors for Heat Illness 63
   Prevention of Heat Illness............... 64
Cardiac Emergencies 69
   Risk Factors for Developing
      Cardiovascular Disease................ 69
   Prevention of Cardiac Emergencies......... 69
Incident Scene Rehabilitation 70
   Goals and Purpose of Incident
      Scene Rehabilitation.................. 70

5 Planning for Live Fire Training 76
Introduction...........................83
Initial Evaluation of the Site...............83
Developing the Preburn Plan...............83
   Learning Objectives 84
   Participants 84
   Water-Supply Needs.................... 84
   Apparatus Needs 84
   Building Plan 84
   Site Plan............................. 85
   Parking and Areas of Operations 85
   Emergency Plans 85
   Weather 87
   List of Training Evolutions 87
   Order of Operations.................... 87
   Emergency Medical Plan................ 88
   Communications Plan 88

Staffing and Organization................ 88
Safety Officer......................... 89
Staff and Participant Rotation............ 94
Personal Protective Equipment Use........ 94
Agency Notification Checklist........... 94
Demobilization Plan.................... 94
Using the Preburn Plan................... 94
Post Live Fire Training Tasks............. 95

6 Acquired Structures.............. 98
Introduction.........................102
Initial Evaluation......................102
Owner Responsibilities............... 103
Preburn Plan....................... 103
Emergency Plans.................... 103
Water Supply....................... 104
Initial Preparation.....................106
Access............................ 106
Equipment and Supplies for
  Preparing the Building.............. 106
Entry and Egress for the Structure........ 106
Exterior Preparation................. 108
Interior Preparation.................. 109
Preparing the Neighborhood........... 114
Live Fire Training Evolution Preparation..... 115
Equipment and Supplies.............. 115
Preparing for Ignition................ 115
Operations..........................116
Setup............................. 116
Preburn Briefing and Walk-through....... 116
Go/No-Go......................... 116
Igniting the Fire..................... 117
Fire Behavior Considerations........... 117
Final Controlled Burn.................. 121
Overhaul.......................... 121
Postevolution Debriefing................ 121

7 Gas-Fired and Non-Gas-Fired Structures................... 124
Introduction.........................130
Code Requirements 130
Making and Enforcing the Rules..........130
Types of Live Fire Training Structures.......131
Gas-Fired Live Fire Training Structures..... 133
Non-Gas-Fired...................... 134
Shipping Containers.................. 136
Features of Live Fire Training Structures.....137

Preburn Plan..........................141
Emergency Plans.................... 142
Spectators, Media, and Visitors.......... 143
Water Supply....................... 143
On-Site Facilities.................... 143
Management of Live Fire Training
  Structures........................ 144
Preparation..........................144
Drillground........................ 145
Preparation and Inspection............ 145
Operations..........................148
Staffing and Organization............. 148
Lighting the Fire.................... 148
Overhaul............................149

8 Nonstructural Training Props..... 152
Introduction.........................159
Types of Props.......................160
Class A Props...................... 160
Class B Props...................... 161
Preburn Plan........................165
Student Prerequisites................ 165
Environmental Concerns.............. 165
Neighbors......................... 165
Weather.......................... 166
Water Supply...................... 166
Parking Areas...................... 166
Operations Area.................... 166
Emergency Medical Services............ 166
Safety............................ 166
Maintenance....................... 170
Preparation.........................170
Operations.........................170
Preburn Briefing Session.............. 170
Lighting the Fire.................... 171
Safety During Evolutions.............. 172
Instructional Technique............... 172

9 Live Fire Training Evolutions...... 176
Introduction.........................178
Learning Objectives 178
Types of Students in the Live Fire
  Training Environment..................179
Recruit Students.................... 179
Experienced Students................ 182
Operations.........................183
Postevolution Debriefing...............183

Appendix A: An Extract From: NFPA® 1403, *Standard on Live Fire Training Evolutions*, 2012 Edition 188

Chapter 4: Acquired Structures 188
Chapter 5: Gas-Fired Live Fire Training Structures 192
Chapter 6: Non-Gas-Fired Live Fire Training Structures 196
Chapter 7: Exterior Props 200
Chapter 8: Exterior Class B Fires 204
Chapter 9: Reports and Records 208

Appendix B: NFPA® 1403 Correlation Guide 209

Chapter 4: Acquired Structures 209
Chapter 5: Gas-Fired Live Fire Training Structures 210
Chapter 6: Non-Gas-Fired Live Fire Training Structures 211
Chapter 7: Exterior Props 213
Chapter 8: Exterior Class B Fires 214
Chapter 9: Reports and Records 215

Appendix C: Acquired Structure Live Fire Training Model SOP 216

Purpose 216
Initial Evaluation 216
Procedure 216
Initial Preparation 217
Preburn Plan 217
   Fuel Materials 218
   Water Supply 218
   Command Structure 218
   Participant Safety 220
Live Burn Manual Forms 222
   Sample Letter to Property Owner 222
   Sample Acquired Structure Property Owner Contact Information Sheet 223
   Sample Insurance Certification Form 223
   Sample Asbestos Letter to Property Owner .. 224
   Sample Letter to Neighbors 225
Sample Checklist for Live Fire Evolutions in Acquired Structures 226
   General Information 226
   Student Prerequisites 226
   Structures and Facilities 227
   Fuel Materials 232

   Safety 233
   Instructors 237
Preburn Notification Contact Sheet 238
Sample Participant Duties and Assignments 239
   Instructor-in-Charge 239
   Safety Officer 239
   Ignition Officer 240
   Instructors 240
   Instructor Trainees 240
   Students 240
Sample Preburn Plan 240
   Location 240
Preburn Site Plan 242
Preburn Floor Plan 243

Appendix D: Permanent Structure Live Fire Training Model SOP 228

Appendix E: Permanent Structure Live Fire Training Burn Plan for _____ Fire Training Center 235

Appendix F: NFPA 1403, *Standard on Live Fire Training Evolutions*, Crosswalk from 2007 Edition to 2012 Edition

Glossary 246
Index 248
Credits 252

Resource Preview

The National Fire Protection Association (NFPA), the International Association of Fire Chiefs (IAFC), and the International Society of Fire Service Instructors (ISFSI) are thrilled to bring you *Live Fire Training: Principles and Practice*. Fire fighters need the safe and controlled "real-life" training offered by live fire training evolutions in order to be fully prepared for the hazards of the fireground. *Live Fire Training: Principles and Practice* provides a guide for live fire training instructors on how they can ensure safe and realistic live fire training for their students. Based on NFPA 1403, *Standard on Live Fire Training Evolutions*, this essential resource features a singular focus on fire fighter safety throughout the text.

Chapter Resources

Live Fire Training: Principles and Practice serves as the core of a highly effective teaching and learning system. Its features reinforce and expand on essential information to make information retrieval a snap. These features include:

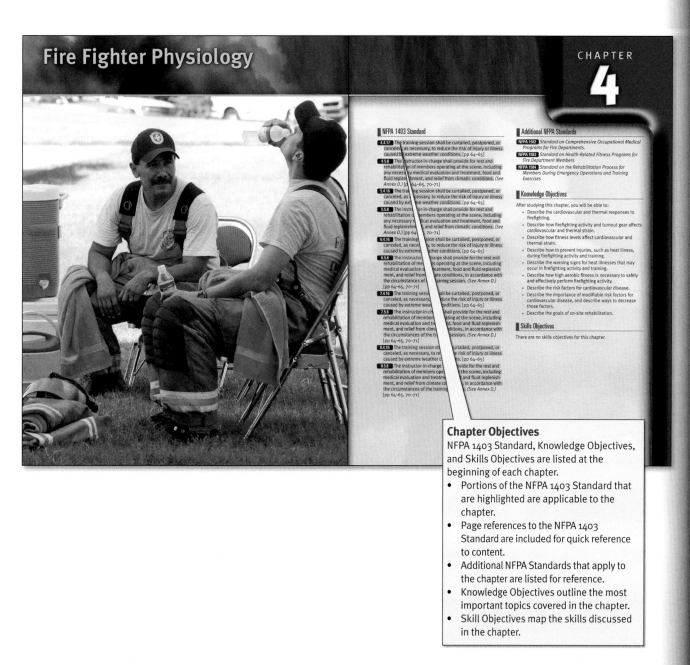

Chapter Objectives
NFPA 1403 Standard, Knowledge Objectives, and Skills Objectives are listed at the beginning of each chapter.
- Portions of the NFPA 1403 Standard that are highlighted are applicable to the chapter.
- Page references to the NFPA 1403 Standard are included for quick reference to content.
- Additional NFPA Standards that apply to the chapter are listed for reference.
- Knowledge Objectives outline the most important topics covered in the chapter.
- Skill Objectives map the skills discussed in the chapter.

Live Fire Training: Principles and Practice

You Are the Live Fire Training Instructor
Each chapter opens with a case study intended to stimulate classroom discussion, capture the students' attention, and provide an overview for the chapter.

Incident Report
Incident Reports of actual live fire training injuries and deaths are detailed in each chapter. Post-Incident Analysis summaries, including NFPA 1403 Standard compliance issues, are offered as lessons learned.

Resource Preview

Safety Tips
Safety Tips reinforce safety-related concerns.

> **Safety Tips**
> Every fire fighter, regardless of tenure, should be trained to constantly identify hazards and alternative escape routes during interior fire suppression operations, including training exercises. Live fire training in any structure should include directions for a secondary means of egress in case of an unexpected fire condition change. While it is not an NFPA 1403 requirement, it is a good idea for each fire fighter to identify two means of egress from each area, prior to any evolution.

Live Fire Tips
Live Fire Tips relate successful, common practices for live fire instructors and students.

> **Live Fire Tips**
> Lightning detectors are a useful tool to have on hand at live fire training events. They can provide advanced notice of lightning, give warning to shut down operations before lightning arrives, or alert the user to safe conditions after the threat of lightning has passed. Lightning detectors can vary greatly in price and in functionality, but a great tool to use is your head. Hearing thunder is an indication that lightning is close enough to shut down operations. Shelter should be sought immediately. Activities should not resume until the threat of lightning has completely passed.

Hot Terms
Hot Terms are easily identifiable within the chapter and define key concepts. A comprehensive glossary of Hot Terms appears in the end-of-chapter Wrap-Up.

Live Fire Training: Principles and Practice

Wrap-Up
End-of-chapter activities reinforce important concepts. Answers for all questions are available on the Instructor's ToolKit CD-ROM.

Chief Concepts
Chief Concepts highlight critical information from the chapter in a bulleted format to help in preparation for quizzes and exams.

Hot Terms
Hot Terms provide key terms and definitions from the chapter.

Live Fire Training Instructor in Action
This feature promotes critical thinking through the use of case studies and provides discussion points for the classroom presentation.

Instructor Resources

A complete teaching and learning system developed by educators with an intimate knowledge of the obstacles that instructors face each day supports *Live Fire Training: Principles and Practice*. These resources provide practical, hands-on, time-saving tools such as PowerPoint presentations, customizable lesson plans, test banks, and image/table banks to support instructors and students.

Instructor's ToolKit CD-ROM
ISBN: 978-1-4496-2223-7

Preparing for class is easy with the resources on this CD-ROM. The CD-ROM includes the following resources:
- **Adaptable PowerPoint Presentations:** Provide instructors with a powerful way to create presentations that are educational and engaging to their students. These slides can be modified and edited to meet instructors' specific needs.
- **Detailed Lesson Plans:** Keyed to the PowerPoint presentations, these complete, ready-to-use lecture outlines include all of the topics covered in the text. The lecture outlines can be modified and customized to fit any course.
- **Image and Table Bank:** Offers a selection of the most important images and tables found in the text. Instructors can use these graphics to incorporate more images into the PowerPoint presentations, make handouts, or enlarge a specific image for further discussion.
- **Electronic Test Bank:** Contains multiple-choice questions and allows instructors to create tailor-made classroom tests and quizzes quickly and easily by selecting, editing, organizing, and printing a test along with an answer key that includes page references to the text.

Navigate Course Manager
ISBN: 978-1-4496-2261-9

Combining our robust teaching and learning materials with an intuitive and customizable learning platform, *Navigate Course Manager* enables instructors to create an online course quickly and easily. The system allows instructors to complete the following tasks:
- Customize preloaded content or easily import new content
- Provide online testing
- Offer discussion forums, real-time chat, group projects, and assignments
- Organize course curricula and schedules
- Track student progress, generate reports, and manage training and compliance activities

To learn more about *Navigate Course Manager*, please contact your sales specialist at 1-800-832-0034.

Technology Resources

www.Fire.jbpub.com

This site has been specifically designed to complement *Live Fire Training: Principles and Practice* and is regularly updated. Resources include:

- Chapter Pretests that prepare students for training. Each chapter has a pretest and provides instant results, feedback on incorrect answers, and page references for further study.
- Interactivities that allow students to reinforce their understanding of the most important concepts in each chapter.
- Hot Term Explorer, a virtual dictionary, allows students to review key terms, test their knowledge of key terms through quizzes and flashcards, and complete crossword puzzles.

Acknowledgments

Editorial Board

Eddie Buchanan, President
International Society of Fire Service Instructors
Hanover, Virginia

Dave Casey
Director, Louisiana State University Fire & Emergency Training Institute

Doug Cline, Vice President
International Society of Fire Service Instructors
High Point, North Carolina

Shawn S. Kelley
Director of Strategic Services/GPSS
International Association of Fire Chiefs
Fairfax, Virginia

Steven Sawyer
National Fire Protection Association
Quincy, Massachusetts

Authors

Chapter 1
Dave Casey, MPA, EFO
Director, Louisiana State University Fire & Emergency Training Institute
Susan Schell, MBA
Training Programs Manager
Florida State Fire College
Ocala, Florida
Robert Brown
Retired Battalion Chief
Boynton Beach Fire Rescue
Boynton Beach, Florida

Chapter 2
Dave Casey, MPA, EFO
Director, Louisiana State University Fire & Emergency Training Institute

Chapter 3
Mike Kemp
Retired Battalion Chief
Palm Beach County Fire Rescue
Palm Beach, Florida

Dave Casey, MPA, EFO
Director, Louisiana State University Fire & Emergency Training Institute

Chapter 4
Denise L. Smith, PhD
Professor and Class of '61 Chair
Health and Exercise Sciences Department
Skidmore College
Saratoga Springs, New York
Research Scientist
Illinois Fire Service Institute
University of Illinois
Champaign, Illinois

Chapter 5
Mike Kemp
Retired Battalion Chief
Palm Beach County Fire Rescue
Palm Beach, Florida
Dave Casey, MPA, EFO
Director, Louisiana State University Fire & Emergency Training Institute

Chapter 6
Mike Kemp
Retired Battalion Chief
Palm Beach County Fire Rescue
Palm Beach, Florida
Dave Casey, MPA, EFO
Director, Louisiana State University Fire & Emergency Training Institute

Chapter 7
Dave Casey, MPA, EFO
Director, Louisiana State University Fire & Emergency Training Institute

Chapter 8
Susan Schell, MBA
Training Programs Manager
Florida State Fire College
Ocala, Florida

Chapter 9
Bryant Krizik
Deputy Fire Chief South Holland Fire Department South Holland, Illinois

Reviewers

Dave Belcher
Lieutenant, Fire Prevention Bureau
Violet Township Fire Department
Pinkerington, Ohio

Bill Blankenship
Captain
Town of Chapel Hill Fire Department
Chapel Hill, North Carolina

Dave Degnan
Captain
Sandusky Fire Department
Sandusky, Ohio

Dave Downey
Assistant Fire Chief
Miami-Dade Fire Rescue
Miami, Florida

Paul Duckworth
Lieutenant
Augusta Fire Department
Augusta, Georgia

Tim Farmer (retired)
City of Coon Rapids Fire Department
Coon Rapids, Minnesota

C. Larry Hansen
Battalion Chief of Operations
Oklahoma City Fire Department
Oklahoma City, Oklahoma

Ed Hines
Tennessee Fire and Codes Academy
Bell Buckle, Tennessee

Alan Joos, EFO, MS
Louisiana State University - Fire and Emergency Training Institute
Baton Rouge, Louisiana

Kevin McArthur
Fire/Rescue Training Specialist
North Carolina Officer of State
 Fire Marshal
Raleigh, North Carolina

Cliff McFarland
Training Chief
West Des Moines Fire Department
Des Moines, Iowa

Doug Orahood
State Fire Coordinator
Ohio Emergency Medical Services
Columbus Ohio

Mike Pannell
Captain
DeKalb County Fire Rescue
Tucker, Georgia

John Peltier
Massachusetts Fire Academy
Marlboro, Massachusetts

Jeff Pindelski
Deputy Chief
Downers Grove Fire Department
Downers Grove, Illinois

Tony Piontek
Training Division Captain
Green Bay Fire Department
Green Bay, Wisconsin

Forest Reader
Director of Training and Safety
Pleasantview Fire Department
LaGrange Highlands, Illinois

Paul Ricci
Assistant Chief of Operations
Sandusky Fire Department
Sandusky, Ohio

Rodney Straight
Training Officer
Bridgeport City Fire Department
Bridgeport, West Virginia

Chris Walker
Instructor
Fort Wayne Fire Department
Fort Wayne, Indiana

Foreword

When I began my career as a fire fighter in the early 1970's, saving property was just as important as saving lives. Today, the fire service understands that all life, especially the lives of fire fighters, is far more important than property. When my career began, if a fire fighter was hurt or killed in the line of duty, it was pretty much accepted as, "part of the job." The deaths of fire fighters were rarely questioned and there was little to no accountability for those deaths. When a death occurred, we would have a big funeral, build a memorial, and eventually go back to doing things the same way we always had.

Back then we didn't create a preburn plan. We didn't ensure that a backup water supply would be available. We didn't inspect the acquired structure to make sure it was free of hazards. EMS was not present on the scene. And probably the most hazardous practice of all, we often tossed gasoline directly onto the training area.

As I gained experience as a fire fighter and company officer, tragic accidents during fire fighter training began to gain attention. In 1982, Bill Duran and Scott Smith died when a live fire burn in Boulder, Colorado went wildly out of control in an abandoned structure. During the live fire training evolution, tires, motor oil, and tar paper were burned to generate smoke, causing very high heat conditions. The high heat produced black, thick, highly aggressive rolling smoke conditions. Despite these conditions, a crew of fire fighters entered the structure to begin training. The low-density fiberboard ceiling ignited behind them, trapping them and the live fire training instructor.

In addition to the two fire fighters who died in 1982, numerous other fire fighters have been senselessly injured or killed during training fires. With *Live Fire Training: Principles and Practice*, the authors, Dave Casey, Susan Schell, Mike Kemp, Dr. Denise Smith, Bob Brown, and Bryant Krizik share their knowledge and experience so that we, the readers, can minimize the chance of injury or death and maximize safety during live fire training events.

If you speak to the family of a fire fighter, they understand that injuries and deaths can occur during an emergency. However, to lose a life during a live fire training evolution is not something that is easily understood. The authors understand this and use each chapter to describe how live fire training instructors can ensure fire fighter safety while still providing excellent training opportunities. Dave Casey and Susan Schell have investigated line-of-duty deaths related to live fire training and have applied the hard lessons learned from those tragedies to this text. The authors present, in very simple and easy-to-follow terms, how to comply with NFPA 1403, *Standard on Live Fire Training Evolutions*, from preburn planning, to preparing live fire training structures for safe training evolutions.

Every once in a while, the fire service is fortunate enough to learn from qualified, experienced individuals, so that history is not repeated. We are very fortunate that Dave Casey brought this group together to share these critical lessons, so that we can best honor those who lost their lives. May they rest in peace knowing that we are improving in honor of their sacrifices.

Deputy Chief Billy Goldfeder, EFO
Loveland-Symmes Fire Department
Loveland, Ohio

Introduction to Live Fire Training

CHAPTER 1

NFPA 1403 Standard

4.1.1 Strict safety practices shall be applied to all structures selected for live fire training evolutions. [pp 8–9]

4.3.1* Prior to being permitted to participate in live fire training evolutions, the student shall have received training to meet the minimum job performance requirements for Fire Fighter I in NFPA 1001, *Standard for Fire Fighter Professional Qualifications*, related to the following subjects: [p 17]

  (1) Safety
  (2) Fire behavior
  (3) Portable extinguishers
  (4) Personal protective equipment
  (5) Ladders
  (6) Fire hose, appliances, and streams
  (7) Overhaul
  (8) Water supply
  (9) Ventilation
  (10) Forcible entry
  (11) Building construction

4.3.2* Students participating in a live fire training evolution who have received the required minimum training from other than the AHJ shall not be permitted to participate in any live fire training evolution without first presenting prior written evidence of having successfully completed the prescribed minimum training to the levels specified in 4.3.1. [pp 16–19]

4.4.2* All live fire training instructors and safety officers shall be trained on the application of the requirements contained in this standard. [p 7]

4.6.1 The instructor-in-charge shall have received training to meet the minimum job performance requirements for Fire Instructor I in NFPA 1041, *Standard for Fire Service Instructor Professional Qualifications*. [p 7]

4.6.2 The instructor-in-charge shall be responsible for full compliance with this standard. [pp 6–8]

4.6.6 All instructors shall be qualified by the AHJ to deliver live fire training. [p 7]

4.8.5* Where station or work uniforms are worn by any participant, the station or work uniform shall have been manufactured to meet the requirements of NFPA 1975, *Standard on Station/Work Uniforms for Emergency Services*. [p 8]

4.8.6 Personal alarm devices shall have been manufactured to meet the requirements of NFPA 1982, *Standard on Personal Alert Safety Systems (PASS)*. [p 8]

4.11.4 The minimum water supply and delivery for the live fire training evolutions shall meet the criteria identified in NFPA 1142, *Standard on Water Supplies for Suburban and Rural Fire Fighting*. [p 8]

4.12.1* 4.12.1* The fuels that are utilized in live fire training evolutions shall only be wood products. [p 7]

4.12.3 4.12.3 Flammable or combustible liquids, as defined in NFPA 30, *Flammable and Combustible Liquids Code*, shall not be used in live fire training evolutions. [p 7]

4.12.11* 4.12.11* The use of flammable gas, such as propane and natural gas, shall be permitted only in live fire training structures specifically designed for their use. [p 7]

4.12.11.1 4.12.11.1 Liquefied versions of the gases specified in 4.12.11 shall not be permitted inside the live fire training structure. [p 7]

Additional NFPA Standards

NFPA 30 *Flammable and Combustible Liquids Code*
NFPA 58 *Liquefied Petroleum Gas Code*
NFPA 59 *Utility LP-Gas Plant Code*
NFPA 1001 *Standard for Fire Fighter Professional Qualifications*
NFPA 1142 *Standard on Water Supplies for Suburban and Rural Fire Fighting*
NFPA 1406 *Standard on Outside Live Fire Training Evolutions*
NFPA 1971 *Standard on Protective Ensembles for Structural Fire Fighting and Proximity Fire Fighting*
NFPA 1975 *Standard on Station/Work Uniforms for Fire and Emergency Service*
NFPA 1981 *Standard on Open-Circuit Self-Contained Breathing Apparatus (SCBA) for Emergency Services*
NFPA 1982 *Standard on Personal Alert Safety Systems (PASS)*

Knowledge Objectives

After studying this chapter, you will be able to:

- Describe the reasons why NFPA 1403, *Standard on Live Fire Training* was developed.
- Describe the purpose of NFPA 1403.

- Describe the live fire training evolutions covered under NFPA 1403.
- Describe the additional NFPA standards that affect live fire training.
- Identify the legal requirements associated with live fire training.
- Describe the student prerequisites for participation in live fire training.

Skills Objectives

After studying this chapter, you will be able to:
- Develop an appropriate prerequisite training program for students who will be engaged in live fire evolutions.
- Develop a process to evaluate a student's mastery of the required knowledge and skills.

Additional Skills http://Fire.jbpub.com/LiveFire

Additional skills are available online. The following additional skills may be taught in conjunction with this chapter:
- Building construction
- Fire behavior
- Fire hose, appliances, and streams
- Forcible entry
- Ladders
- Overhaul
- Personal protective equipment
- Portable extinguishers
- Safety
- Ventilation
- Water supply

You Are the Live Fire Training Instructor

Your chief has asked you to set up a live fire training evolution using an acquired structure. The two closest fire departments that run mutual aid with you have been invited to participate. The acquired structure must be burnt within seven days. Your department is relatively small with 20 certified fire fighters and 2 engines. The acquired structure has been damaged by a previous fire, which your department contained to one room.

When you talk to the mutual aid companies, they are eager to participate, however some of their personnel are in the middle of training and are not yet certified. They are also not sure what apparatus they can provide. From previous experiences with the department, you have had concerns regarding their personal protective equipment not fitting properly and being worn out.

As you consider the possibility of conducting this live fire training evolution, there are many questions that must be answered. Some of these require knowledge of other NFPA standards.

1. Does the gear meet the standards for firefighting?
2. Will there be enough resources to provide an adequate water source to conduct the live fire training?
3. Do the participants have the amount of training needed to participate in the live fire training evolution?
4. Is there enough time to adequately prepare for this live fire training evolution?
5. Are there enough live fire training instructors and safety officers?

Introduction

Whether you are a career fire fighter, a volunteer fire fighter, a company officer, an instructor, a training officer, or a chief officer, as an instructor conducting live fire training exercises, you are responsible for the safety of all participants involved. Fire fighters and students learning to become fire fighters have died or have been severely injured during training evolutions. Achieving a balance between participant safety and training effectiveness can be a difficult task.

There are thoroughly written standards that relate the requirements of conducting live fire training evolutions. The National Fire Protection Association (NFPA) has published one of the most popular of these standards, NFPA 1403, *Standard on Live Fire Training Evolutions*. The purpose of NFPA 1403 is to provide a process for conducting live fire training evolutions to ensure that they are conducted in safe facilities and that the exposure to health and safety hazards for the fire fighters receiving the live fire training is minimized. NFPA 1403 was designed to set standards on what should be done to mitigate the inherently dangerous conditions of live fire training. The purpose of this text, however, is to go beyond those written standards and to focus on the how-to aspects of conducting valuable live fire training evolutions in a safe and compliant manner. This text is designed to show you how to meet the standards while preparing fire fighters through the experiences of live fire training, in both acquired structures and permanent live fire training structures. This text will also review the different types of facilities and props used for live fire training. Suggested practices, preparations, training aids, and evolutions are given and can be used to maximize the learning experience of the fire fighter and better prepare him or her for suppression duties.

The History of Live Fire Training Evolutions

From 1987 to 2001, America saw roughly a 30 percent decrease in the number of structure fires. Over the same time period, there was an approximate 15 percent *increase* in fire fighter injuries during training. Of those injuries, 10 percent were attributed to burns or smoke or gas inhalation; however there are no statistics to specifically indicate what injuries occurred during live fire training evolutions.

A special report by the United States Fire Administration, *Trends and Hazards in Fire Fighting Training*, emphasizes the importance and the need for realistic fire training. Because of the decrease in the number of structural fires, fire fighters are getting less fire suppression experience. Without realistic training, fire fighters will be less prepared in the field.

Over the years, the North American fire services have often not been too enthusiastic to embrace new standards and laws for fire fighter safety. Acceptance has varied geographically and sometimes even between neighboring fire departments. New NFPA standards and Occupational Safety and Health Administration (OSHA) standards are often not well-received in the fire service. Those who entered the fire service in the 1960s will remember the gradual transition from an open cab apparatus to covered cabs to a fully enclosed four-door apparatus. They may also remember the transition from fire fighters riding tailboards to open jump seats. The four-door apparatus finally became the norm in the 1990s, but departments like Cincinnati had them in 1930s. The acceptance of using a self-contained breathing apparatus (SCBA) has also been a 40-year battle.

Similarly, the acceptance of live fire training rules and standards by the fire services is proving to be a difficult battle. Following the deaths of two fire fighters in a "smoke training" evolution that erupted into an uncontrolled fire in 1982, the lead investigator noted, "It does not appear that there is a comprehensive recommended good practice on burning exercises available on a national basis." In 1986, NFPA 1403 *Standard on Live Fire Training Evolutions in Structures* was released. It was revised in 1992, and in 1997 NFPA 1406, *Standard on Outside Live Fire Training Evolutions* was merged with NFPA 1403 to create the *Standard on Live Fire Training Evolutions* that we know today. It was revised again in 2002 and 2007.

As a reminder, NFPA 1403 is a standard, and standards are voluntary. As a consensus standard, they do not have the power of law, unless they are specifically adopted by a governing entity such as a state or municipality. It should be noted that even when it is not specifically adopted, some state OSHA agencies have the power to cite to such standards in their investigations and can use such standards in court. Since the release of NFPA 1403, a number of states have adopted, in one form or another, different requirements for instructor qualifications, permit requirements, environmental concerns, etc. Since the release of NFPA 1403, from 1986 to 2007, there were 11 fatalities during live fire training in the United States. Two of these occurred in permanent training facilities designed for live fire training.

This chapter will review a few of these incidents, not to assign blame or to make an agency look bad, but so that lessons can be learned and mistakes can be avoided in the future. These documented case studies will also appear in each chapter throughout the text along with a brief analysis of what went wrong with each case.

Review of Live Fire Training Evolution Incidents

As stated earlier, there was not a national standard for live fire training before 1986. In 1982, a smoke training drill went horribly awry. As a result, two fire fighters lost their lives, garnering considerable public and professional attention. A memorial to the two fire fighters reads, "They gave their lives learning to save others," a tribute that they were there to hone their skills to save the public they serve **Figure 1-1**. In reality, better homage has been paid through the creation of NFPA 1403, with a more reaching and encompassing impact on the safety of fire fighters throughout North America.

Figure 1-1 Memorial marker for Boulder fire fighters that died in 1982 training fire.

The Standard and Legal Considerations

It is clear that *safety* should be the first and foremost concern with all live fire training evolutions. The live fire training environment is the one area in which the live fire training instructor can control the type of structure being used, the amount of fuel being used, and the evolutions being performed. NFPA 1403 provides guidelines that aid the instructor in assuring that training is performed in a safe environment, and requires that all live fire instructors and safety officers be trained in the application of the requirements of NFPA 1403. It should be noted that the current edition of NFPA 1403 must be followed and provides the *minimum* requirements for training in live fire conditions, and extra precautions may be needed. The **authority having jurisdiction (AHJ)** has the ability to make more stringent rules, but they should not be adopting practices or procedures that are below the minimum requirements. The authority having jurisdiction is an entity that is responsible for enforcing a code or standard.

Whether training a new recruit or working with seasoned fire fighters, the instructor must assure that the standard is followed with every live fire training evolution. The basic system provided in the standard can be adapted to meet local needs. One state agency may decide that only permanent live fire training structures can be used to train their students, while another may only allow certified fire fighters to participate in live fire training using acquired structures. Currently, there are ongoing debates about the value of using gas-fired live fire training structures versus acquired structures. The debates stem from the question of whether or not the gas-fired structures provide actual fire conditions. This is one of the areas where the AHJ needs to make the determination based on what their local needs are. NFPA 1403 provides guidance on how to safely conduct evolutions using both types of structures, but it does not recommend one type of structure over the other. The AHJ should carefully review the overall objectives of the training as well as the experience levels of the participants, and make a decision from there, ensuring the safest environment is chosen.

The purpose of the standard is to ensure that the safest training possible is provided to the fire fighters. The training environment is more controllable than real fire emergency, and should be the safest place to fight a fire. This text will teach you how to ensure that the facilities you use are safe by following a thorough inspection process. Regardless of the type of live fire training you are conducting, the goal is to ensure that the exposure to hazards is limited and controlled.

There are certain types of live fire training evolutions that are not covered by NFPA 1403. These include live fire training evolutions involving the following:
- Ground cover or wildland fires
- Suppression of fires set for training fire cause and origin investigation

The Impact of NFPA 1403 on Live Fire Training

You, as the live fire training instructor, need to consider how NFPA 1403 impacts your training. NFPA standards are consensus standards that will be used to judge the actions of those conducting live fire training. The application of NFPA 1403 is handled in a variety of ways throughout the nation. Some states require that instructors conducting live fire training be trained or certified as live fire instructors. There are some jurisdictions where this training is required by state statutes. Other states rely on instructors to be knowledgeable of the standard and follow the guidelines provided. The process can be formalized through standard operating procedures (SOP) or standard operating guidelines (SOG), or may simply be a directive to instructors. Regardless of how the authority having jurisdiction handles the adoption of the standard, the instructor-in-charge must be thoroughly knowledgeable on the standard and assure that it is enforced on the training ground. It is the immediate duty of the instructor-in-charge to ensure full compliance with NFPA 1403 in all instances. The case studies presented in this chapter and those that follow show the need to follow the standard. The failure to follow the standard can result in legal actions, possible arrests of those in charge of the training, and needless fire fighter deaths and injuries.

A lot of thought should go into selecting instructors to conduct live fire training. This should be based on meeting the job performance requirements of NFPA 1041, as Fire Instructor I. NFPA 1403 also requires that all instructors be qualified by the AHJ to deliver live fire training, and the instructor-in-charge is responsible for full compliance with the standard. Some states have already adopted live fire training programs, and the ISFSI delivers live fire training to instructors throughout the nation. If training is going to involve an acquired structure, the instructors need to have experience fighting structural fires. Knowing how to read smoke, understanding fire behavior, and determining fuel quantities are all skills that are learned through experience and cannot be gained simply by reading a textbook. The training props used in live fire training can be as simple as untreated hay inside a container or as complicated as a control panel that works a variety of different devices. Again, knowledge of the training prop and its proper usage is necessary.

Following the standard does not guarantee that there will not be injuries during training. However, it will help minimize the hazards that are encountered. Every fire fighter is aware of the saying "everyone goes home," and safety should be the first priority for all participants. This applies to the training environment as well. At the end of the training, the instructor should be able to say that the participants learned something and that the live fire training was conducted in the safest possible manner.

Referenced Standards

NFPA 1403, *Standard on Live Fire Training Evolutions*, is vital when conducting any live fire training exercise. The instructor must be familiar with the information provided in the standard. No one is expected to be able to recite the standards, but the instructor should have access to them and be able to look up information as needed. If an answer can not be found in the standards, it will be up to the AHJ to decide how to handle the concern.

All of the knowledge needed for conducting live fire training evolutions cannot all be contained within NFPA 1403. Additional published standards should be consulted. Selected NFPA standards are covered briefly, highlighting areas that may impact live fire training.

NFPA 30, *Flammable and Combustible Liquids Code* addresses flammable and combustible liquids and is referenced in NFPA 1403 with regards to definitions of flammable and combustible liquids by their flashpoints. When dealing with live fire training, NFPA 1403 addresses not using flammable and combustible liquids in acquired structures or permanent live fire training structures. The authority having jurisdiction will need to make the decision if flammable or combustible liquids can be used for ignition purposes.

Even where state law allows less restrictive requirements, fire departments need to consider very carefully the possible impact of deviating from NFPA 1403. State legislation can protect a department from prosecution, but not from civil actions. Most importantly, while it will be argued vehemently by some that the use of combustible or flammable liquids is safe, the specter of improper use and the dangers of vapors and near-instantaneous fire spread should weigh very heavily in an informed decision. Regardless of the decision, there are other factors to consider such as air quality, amount of smoke produced, and environmental impact.

NFPA 58, *Liquefied Petroleum Gas Code* and NFPA 59, *Utility LP-Gas Plant Code* are also referenced in NFPA 1403. When conducting live fire training using liquid-propane (LP) props, the instructor is required to visually inspect and operate the LP prop prior to the training evolutions. The instructor must be trained on the LP prop by the manufacturers or their designee. However, to properly inspect the LP prop, knowledge of these two standards is also required. There are a variety of props nationwide that include those which are permanently piped, some that have flexible above-ground piping, and some that are actually LP tanks used to train for techniques on how to fight LP-tank fires. See Chapter 8, Nonstructural Training Props, for more Infomation on LP props.

Perhaps the most critical standard for live fire training, NFPA 1001, Standard for Firefighter Professional Qualifications, directly affects NFPA 1403. Prior to participating in live fire training, participants must be qualified in specific areas of training for Fire Fighter I, as outlined in NFPA 1001. It should be stressed that all participants must be trained in all job performance requirements for Fire Fighter 1. This chapter will briefly touch on these job performance requirements, however, it is up to the instructors to verify each participant's qualifications before putting them through live fire training.

NFPA 1142, Standard on Water Supplies for Suburban and Rural Fire Fighting impacts NFPA 1403. The instructor must know how to calculate the needed water supply to fight the fire based on type of construction, exposures, and water sources. The main goal is to assure that there is enough water to handle any emergency situation that may occur. Many live fire training evolutions have become full emergency responses due to fire spreading and changing conditions.

NFPA 1971, Standard on Protective Ensembles for Structural Fire Fighting and Proximity Fire Fighting is referenced in NFPA 1403. The live fire training instructor and safety officer are responsible to ensure that all participants' gear meets NFPA standards, and that they are in a safe, usable condition. This can easily be addressed by conducting a gear check prior to beginning the live fire training event, to check for signs of deterioration and fatigue. All pieces of personal protective equipment need to be inspected before being given to the students, including protective coats, trousers, hoods, footwear, helmets, and gloves. It is critical to check the outer lining as well as the inner lining, because a damaged inner lining may have been paired with a new outer lining mistakenly, which could leave a fire fighter exposed. After the gear check is completed, the students should dress completely and be checked again for any areas of skin showing.

The instructor needs to review NFPA 1975, Standard on Station/Work Uniforms for Fire and Emergency Service. This issue needs to be addressed prior to the live fire training evolution as some departments do not met this standard. Many problems can be solved by simply sending out a notice to the departments/participants about clothing requirements. This will be strongly influenced by the AHJ. The instructor should also review, if station or work uniforms are not worn, what clothing is to be worn under personal protective clothing. Clothing made of all-natural fibers should be considered.

The standard also requires that self-contained breathing apparatus (SCBA) be manufactured to meet NFPA 1981, Standard on Open-Circuit Self-Contained Breathing Apparatus (SCBA) for Emergency Services. Instructors need to be familiar with all components of the SCBA unit. Depending on the experience levels of the students, the instructor may want to have the group perform an inspection prior to the beginning of the training and before entering the building for live fire training. Instructors need to remind students to check tank air pressure against the gage reading, check gaskets, assure they have the proper amount of air in their cylinder, and make sure that all safety devices such as low pressure alarms, PASS devices, and bypass or purge valves are working properly. The instructor is responsible for the complete safety of the students including gear checks and SCBA functionality The extra time that it takes to assure that the SCBA is in good working condition and safe to use can prevent an emergency from happening.

NFPA 1403 requires that PASS devices be manufactured to meet NFPA 1982, Standard on Personal Alert Safety Systems (PASS). Again, it is the instructor's responsibility to assure that students have been trained on how to operate their PASS device. It is also the instructor's responsibility to assure that every student has a PASS device. If the unit is not equipped with an integral pass device on the SCBA, then an attachable device needs to be present. No student should ever enter live fire training without an active PASS device. Check to make sure the PASS is functioning before entry.

NFPA 1041, *Standard for Fire Service Professional Qualifications, f2012 edition,* is now a referenced publication. Instructors of live fire training should meet the qualifications of an Instructor I as a minimum. Safety and instructional techniques are especially important in live fire training, and the standard outlines the skills and knowledge that an instructor should possess.

NFPA 1407, *Standard for Training Fire Service Rapid Intervention Crews,* 2010 edition, is another important standard to be followed. Rapid Intervention Teams need to be assigned at live fire training events, and the members on this team need to understand the responsibilities and duties of that assignment and should be trained to meet this standard.

Using NFPA 1403

Strict adherence to NFPA 1403 is strongly advised, as it will lead to the safest live fire training environment possible. NFPA uses specific terminology throughout the course of their standard. Most of these terms will be specifically defined within their respective chapters in this book, however a brief overview of some of the more general terms is presented here for reference.

Should there be any confusion on what a **live fire** is, NFPA 1403 defines it as any unconfined open flame or device that can propagate fire to a building, structure, or other combustible materials. Those individuals participating in a live fire training evolution within the operations zone are defined as **participants**. By this definition, a participant could be a student, instructor, safety officer, visitor, or other person who is in the operations area.

Although there will be many instructors involved in the live fire training evolution, there are specific differences between an instructor and the instructor-in-charge. An instructor is an individual qualified by the AHJ to deliver fire fighter training who has the training and experience to supervise students during live fire training evolutions. Each AHJ will have to determine what qualifications instructors must obtain or possess to be involved in live fire training. The **instructor-in-charge** has more responsibilities than an instructor. They are responsible for ensuring that the evolution is conducted safely, ensuring the training structures are prepared to meet NFPA 1403, monitoring all fireground activities, ensuring there are enough properly trained live fire training instructors involved, and making sure that all training evolutions are conducted following the standard. This responsibility carries a lot of weight and requires someone willing to take on the additional responsibilities of live fire training.

The first question that you as an instructor must answer is, "Who is the authority having jurisdiction?" The answer to this question is not as straight forward as you would think. There are states who by law are the AHJ. Those states outline the requirements for instructor qualifications, equipment that may be needed, and guidelines on how to meet NFPA standards. The AHJ acts as the judge and jury regarding enforcement of standards and codes and all approvals.

NFPA 1403 uses the word *shall* to indicate a requirement that is mandatory, and the word *should* to represent a recommended procedure. The AHJ or instructor-in-charge can make the decision to follow the suggestion or not.

Instructors should have training on delivery techniques including demonstration. A **demonstration** is the act of showing a skill. Instructors should demonstrate a skill step by step in the correct format for the students, ensuring that they have practiced it so it will be shown the correct way.

Another important designated individual that must be present at live fire training is the **safety officer**. This individual is appointed by the AHJ as qualified to maintain a safe working environment at all live fire training evolutions. This position has increased responsibilities in live fire training evolutions. Overall, safety falls under the safety officer's jurisdiction, and he or she has the authority to shut down operations that create an unsafe environment.

This standard also addresses the need for emergency medical services (EMS) at the site of any live fire training. The definition of EMS is the provision of treatment, such as first aid, cardiopulmonary resuscitation (CPR), basic life support (BLS), advanced life support (ALS), and other prehospital procedures, including ambulance transportation, to patients.

Student Prerequisites

Live fire training is an opportunity for students to bring all of their previous training to bear in an environment that *approximates* or *simulates* an actual fire event. It is very important that the students be prepared for the context of the operations they will be expected to perform under live fire conditions. At this point, students should be proficient in raising and climbing ladders, advancing hose lines in open parking lots and directing hose streams to imagined ceilings, donning their SCBA, crawling with obscured vision, and other skills that are done singularly. Live fire exercises provide students an opportunity to draw on their knowledge base and put the skills they have learned in training into actual practice. Whether in a gas-fired or non-gas-fired live fire training structure, exterior prop, or acquired structure, this is a chance to experience the real thing. It is an exciting time for most students.

When working with less experienced students, there are several very important considerations in transitioning from singular skills to live fire training evolutions. Some examples are as follows:

- Conduct wet hose line drills in narrow spaces such as hallways, around corners, and with obstructions.
- Teach students how to advance and direct streams. Get them accustomed to working with a charged hose line while crawling, crouching, and duck-walking, as they will need to do in the exercises and in a real fire event.
- If possible, conduct drills in the acquired structure, or permanent prop, that will be used for training. Conduct obscuration, then smoke-only evolutions, before the live fire exercises.
- Conduct orientation drills **Figure 1-2**. Have students determine their whereabouts by landmarks that are visible from outside (windows, doors, etc.), and by interior flooring, furniture, looking for light from windows, etc.
- Conduct emergency evolutions. Conduct personnel accountability reports inside and outside of the building. Have students follow a hose line to the outside, and

Live Fire Tips

Every instructor involved in live fire training evolutions should have a clear understanding of what this type of training is and what it is not.

- It **is not** a test. There are more appropriate and meaningful ways to test a student's abilities without the danger of live fire.
- It **is not** an opportunity for instructors to engage in recreational training. Instructors must focus on the expected benefit for the student and refrain from the temptation to rebuild the Great Chicago fire in a 10' x 10' (3 x 3 m) room, in an attempt to achieve some spectacular visual or thermal effect.
- It **is not** baptism by fire. Students should not fear what instructors may do to them during the evolution. This indicates both a lack of professionalism and a lack of understanding regarding this type of training on the part of the instructor.
- It **is not** an opportunity to discover the limits of human endurance or tolerance of extreme temperature. Nor is it an opportunity to test the fire retardant tendencies of our personal protective equipment (PPE), or the integrity of critical structural building components.
- It **is** the end of training, not the beginning. No fire fighter should be placed in dangerous live fire conditions without receiving the necessary background training.

identify, by the coupling direction, which way they need to travel.

Florida has two requirements that should be considered for inclusion to any department or training center program:

1. Each fire fighter, regardless of tenure, shall be trained to constantly identify hazards and alternative escape routes during interior fire suppression operations, including training exercises.
2. Live fire training in any structure must include instruction of the student in planning for a secondary means of egress or escape in case of an unexpected fire condition change. Prior to live fire training evolutions, each fire fighter must identify two means of egress or escape from each area.

Figure 1-2 Orientation drills are an essential component of preparing students for live fire training.

Incident Report

Boulder, Colorado - 1982

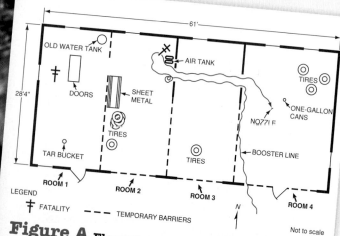

Figure A Floor plan of involved Boulder structure.

Figure B West wall of Boulder house showing door used for escape by lieutenant.

A search and rescue drill was being held in a large 28' × 61' (8.5 × 18.6 m) shed **(Figure A)**. The fire department had divided the shed into smaller rooms with temporary partitions made of combustible materials to provide better search scenarios. The shed had a wood frame with a ceiling made of combustible, low-density fiberboard tiles. This engine company was the third to participate that day.

Although this was set to be a smoke drill only, the smoke was produced using motor crankcase oil to burn tires on the floor in each of the rooms. Smoke bombs were also used. A road flare and small amounts of gasoline mixed with cleaning solvent were used to ignite each fire.

A water supply was not established and the nearest hydrant was 1100' (335.3 m) away. A reserve engine with a 500-gallon (1892.5-liter) booster tank was used to supply water. The primary engine had a 300-gallon (1135.5-liter) booster tank, with only a charged booster hose line used by the crew participating in the training exercise. The pump operator was one of the participating crew members, and both pumpers were left unattended.

Prior to the incident during the engine company's training, there were fires in the rooms for roughly two hours. Before the last crew entered the structure, another tire was added to each pile. The three-engine company members and a training officer were inside the structure when the preheated ceiling tiles ignited, trapping the four fire fighters. They became separated, and only the training officer and the company officer were able to make it out alive **(Figure B)**. The company officer received third degree burns on over 35 percent of his body. It is important to note that the training officer in charge had never previously been directly involved with this sort of training.

| **Post-Incident Analysis** | **Boulder, Colorado** |

NFPA 1403 Noncompliant
NOTE: This training fire occurred before the creation of NFPA 1403.

- Combustible ceiling tiles contributed to rapid fire spread (4.2.17)
- Rubber tires and flammable and combustible liquids used (4.3.3)(4.3.6)
- Multiple simultaneous fires on the floor in open containers (4.4.15)
- No interior hose line (4.4.6)
- No established water supply (4.2.23)
- No incident safety officer (4.4.1)
- Flashover and fire spread unexpected (4.3.9)
- No written emergency plan (4.4.10)
- Students not informed of emergency plans and procedures (4.4.10)
- No back-up hose lines (4.4.6.2)
- No EMS on scene (4.4.11)
- Instructor not trained or experienced with topic (4.5.1)
- Deviation from the plan (4.2.25)
- No preburn plan (4.2.25.2)
- No walk-through performed with students and instructors (4.2.25.4)

Incident Report

Hollandale, Minnesota - 1987

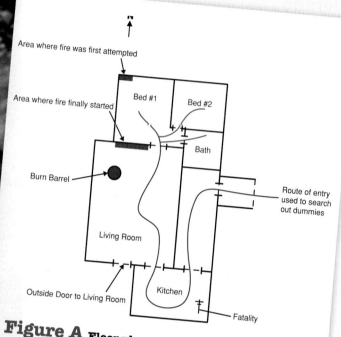

Figure A Floor plan of the Hollandale structure.

Figure B Photo of rear of house in the Hollandale fire. The kitchen is below the area with the flat roof. Note the minor amount of smoke and heat damage above door.

Search and rescue drills had been conducted for some time in a small two bedroom house without problems. Leaves burning in a 35-gallon (132.5-liter) drum were used to create smoke. The drum had a metal cover to reduce the danger of fire spread, and the drum was situated above the floor. Caution was exercised to prevent fire from spreading from the container. Two fire apparatus were on scene with 1400 gallons (5299 liters) of water on one of the fire units and 1000 gallons (3785 liters) in a portable tank. The participants carried a charged 1½" (38.1 mm) hose line.

After the interior search evolutions, the intent was to destroy the house by burning it down. Four crew members, wearing full protective clothing, equipped with breathing apparatus, and protected by a charged 1½" (38.1 mm) hose line, entered the structure using the east entry that had been used during the drill **(Figure A)**. They attempted to light a fire in the back bedroom but were unable to get it to ignite, most likely due to the dampness from the water. They moved to the living room and splashed #2 fuel oil on the wall and carpet. Two of the fire fighters moved to the kitchen while another used a propane torch to ignite the fire in the living room. The entire room flashed, and a third fire fighter crawled out, thinking his partner was with him. Three of the fire fighters crawled to the outside. The fourth made it to the kitchen and stopped, where he apparently removed his SCBA mask to call for help and then reentered the fire area and was killed. His Nomex hood, propane torch, and fuel can are later found in the kitchen. The three fire fighters that had exited the house reported the emergency and entry was immediately forced into the front entrance. The fire was quickly extinguished, and the victim was found dead.

It was uncovered in the investigation that the living room paneling was made of a light-weight combustible material, found in mobile homes at the time. The smoke barrel with live fire may have preheated the combustibles in the living room **(Figure B)**. The report noted that the NFPA 1403 standard should have been followed. Investigators believed that if the deceased had remained in the kitchen, he most likely would have survived.

The fire department was cited by the state OSHA for failure to "examine the building more closely, failure to brief the participants on the building's layout, failure to have a building evacuation plan, and failure to have a qualified instructor on the scene." This was the first fire fighter fatality after the release of NFPA 1403.

Post-Incident Analysis — Hollandale, Minnesota

NFPA 1403 Noncompliant

- Combustible wall paneling and carpeting contributed to rapid fire spread (4.2.17)
- Flammable and combustible liquids used (4.3.6)
- Flashover and fire spread unexpectedly (4.3.9)
- No direct egress from living room (4.2.24.5)
- No formal evacuation plan including evacuation signal (4.4.10)
- No incident safety officer (4.4.1)
- No interior hose line (4.4.6)
- No EMS on scene (4.4.11)
- No written emergency plan (4.4.10)
- Participants not informed of emergency plans and procedures (4.4.10)
- No preburn plan (4.2.25.2)
- No walk-through performed with students and instructors (4.2.25.4)

Incident Report

Milford, Michigan - 1987

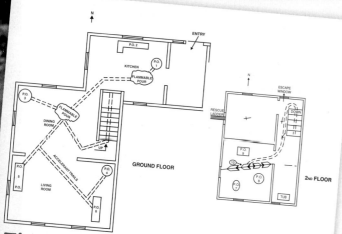

Figure A Floor plan of involved Milford structure.

Four days after the Hollandale fire, a training session for fire fighters to recognize physical evidence of arson fires trapped six fire fighters on the second floor of a 120-year-old house **(Figure A)**. The wood-framed house had low-density ceiling tiles with lightweight wood paneling on the first floor. There were numerous holes in the walls and ceilings.

Various arson scenarios were set up in each of the rooms on both floors using furniture, clothing, and other items. Both flammable and combustible fluids were used.

Multiple hose lines were charged and ready on the outside of the structure. Portable water tanks were set up and filled with tenders, prepared for a water shuttle operation.

Fire fighters toured the building to see the scenarios before ignition. Initial ignition efforts failed, and several windows on the second floor were broken to improve ventilation. A fire set on a couch in the southwest corner of the living room using flammable or combustible liquids was openly burning and had breached the exterior wall and entered the attic space.

One fire fighter was already inside trying to ignite the fires on the second floor. Four fire fighters were directed to enter the house, along with the assistant chief. Without a hose line, the fire fighters passed multiple burning fire sets that were producing little heat or smoke. Most likely, they did not see the fire in the living room or realize it had spread to the attic. They met with the assistant chief and another fire fighter on the second floor to ignite a fire in one of the bedrooms. At this time, the other upstairs bedroom was already burning. The assistant chief directed them to exit the house as fire conditions rapidly intensified. The escape route down the stairs to the first floor was cut off and, under very adverse conditions, the fire fighters were able to locate a window on the second floor for egress **(Figure B)**. Three of them were able to exit through the window and onto the first floor roof below. Reportedly, the SCBA face piece was melting off of the last fire fighter able to exit.

Outside, fire fighters observed the change in conditions and the rescue of the assistant chief and the two other fire fighters. They initiated suppression and rescue operations. Ladders were raised to the second floor windows in the now almost fully involved house.

Figure B Stairs to the second floor of the Milford house show significant fire involvement that blocked escape from second floor.

The first trapped fire fighter was located and removed in about 10 minutes. The others were all located on the second floor shortly thereafter. Per the NFPA report, "One of the fire fighters who was able to escape did not know that he would be part of an interior training until he was instructed to "suit up," and even then he was unsure of his specific assignment." Three fire fighters died in the fire, which was the first multiple death training incident in the United States since the release of NFPA 1403 in 1986.

Post-Incident Analysis Milford, Michigan

NFPA 1403 Noncompliant

- Combustible wall paneling and ceiling tiles contributed to rapid fire spread (4.2.17) (4.2.10.5)
- Flammable and combustible liquids used (4.3.6)
- Multiple simultaneous fires on two floors (4.4.15)
- No interior hose line (4.4.6)
- Numerous holes in walls and ceilings allowed fire spread (4.2.10.4)
- Interior stairs only exit other than windows (*note: a violation due to the limited normal means of egress which may have precluded the use of the upper floor without additional provisions put in place*) (4.2.12.1) (4.2.13)
- Flashover and fire spread unexpected (4.3.9)
- Fire chief claimed no knowledge of NFPA 1403 (4.5.4)

Safety Tips

Practice what you will want to do during live fire training as much as possible before the training occurs, so that students will be more comfortable when it comes time for the real thing.

The Student and the Live Fire Training Evolution

Every live fire training evolution should have an intended purpose other than simply throwing water on flames. The learning objectives and desired outcomes should be clear to all participants, regardless of experience levels. Under the watchful eye of the team of live fire instructors, students should be able to see the effectiveness of fire streams, when applied properly. They should be able to observe the results of the coordinated effort of the attack crew and ventilation crew, and see how ventilation impacts the fire attack effort. They should experience the benefit of proper radio communication, effective crew management, and situational awareness, while operating in and around the fire building. Students should also have the opportunity to observe the instructors as they manage the evolution through the use of the incident command system, personnel accountability system, resource status and management process, and situation status and management process. In short, the students should come away from the exercise with an appreciation for how their knowledge of firefighting methodology, and their skills in applying that knowledge led them to a desired outcome. It is particularly important to ensure that all students are adequately prepared for live fire training evolutions, having received prerequisite training in accordance with the standard.

The realism that can be achieved in the live fire environment is its greatest advantage Figure 1-3 . Conversely, that same realism, if not properly managed, can be its greatest disadvantage as the fire dynamic can suddenly spiral out of control. It has the potential to overwhelm the command and control systems and their available resources, with potentially disastrous results. Some of the inherent dangers of fighting fire in real structures are present during training burns. Failure to recognize these dangers has proven to be very costly. Accordingly, it is imperative that these training exercises be conducted with strict adherence to NFPA 1403, *Standard on Live Fire Training Evolutions*.

The standard was written to ensure that significant, actual firefighting experience could be gained in an environment that closely approximates the fire behavior encountered in the real world, but without all of the risks. Through proper planning and preparation, many of the unknowns that exist in the real world can be eliminated during training. For example, the contents of most occupancies today present the fire fighter with a wide array of fuels and other materials with varying burning characteristics, off-gas production, explosion potential, and various other hazards. Building features, layout, means of egress, and avenues of fire spread may not be known. Building contents are rarely known. The water supply capability may not be known or may be nonexistent. By following the standard, all of these variables are known. As a result, we are able to create a live fire training evolution that allows the student to observe actual fire development, apply appropriate fire streams, and observe the actual results of their tactical operations without taking unneeded risks. When this is done in a coordinated manner with appropriate types of ventilation, incident command, and personnel accountability systems, it can be some of the most meaningful training any fire fighter receives.

One of the most important aspects of conducting live fire training, particularly in acquired structures, is the preparation necessary to conduct the drill safely. Depending on a number of factors including the construction type, condition, and location of the building, the terminal objective of the drill, the number and skill level of the participants, and the available resources, this process could take months Figure 1-4 .

The instructor-in-charge of live fire training must remain committed to this process and must ensure that every concern identified in the standard is appropriately addressed and satisfied prior to beginning, and evaluated during, any live fire training evolution. It is vitally important to ensure that all students are

Figure 1-3 Live fire in acquired structures is considered by many to be the most realistic type of live fire training available.

Figure 1-4 Some acquired structures are going to take more work to prepare than others.

adequately prepared for live fire training evolutions, having received prerequisite training in accordance with NFPA 1001, *Standard for Fire Fighter Professional Qualifications*. Without this necessary training, the student is in no way prepared to go into a live burn.

Recommended Student Training

According to NFPA 1403, prior to being permitted to participate in live fire training evolutions, the student shall have received training to meet the job performance requirements for Fire Fighter I in NFPA 1001, *Standard for Fire Fighter Professional Qualifications*, related to the following subjects:

1. Safety
2. Fire behavior
3. Portable extinguishers
4. Personal protective equipment
5. Ladders
6. Fire hose, appliances, and streams
7. Overhaul
8. Water supply
9. Ventilation
10. Forcible entry
11. Building construction

It is important to recognize the significance of prerequisite training in each of the subject areas identified by the standard. Exposing participants to the perils of live fire training who have not been adequately prepared to anticipate and respond to developing conditions is not wise. Though the exercise may teach some valuable lessons, those lessons may come at a great cost **Figure 1-5**. When things go wrong during live fire training evolutions, they usually go very wrong, very quickly. One of our greatest safeguards against catastrophe is to ensure that everyone is well trained and well equipped for the undertaking. It should be restated that these are the *minimal* requirements for student participation in live fire training. Instructors should check with their AHJ to see if additional prerequisites are needed.

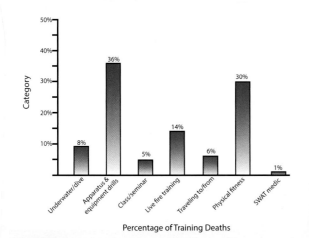

Figure 1-5 Live fire training accounted for 14 percent of the injuries sustained during training.

Safety

First and foremost, safety is an *attitude* that must become ingrained into the organizational culture. Everyone must be accountable for compliance with established safety rules, procedures, and policies. On the training ground, it starts with the instructor-in-charge, who is responsible for both compliance and enforcement. The instructor-in-charge will set the tone for the training evolution, and it should be expected that everyone will follow his or her lead. Our actions should always be consistent with our goal to conduct all training evolutions without injury. Therefore, safety violations must never be dismissed or overlooked, but should be corrected at once.

To ensure safe conduct during the training evolutions, students must come to understand and appreciate the measures that are in place for their personal safety and protection. In today's fire service, rules and policies related to the use of PPE and SCBA are generally understood and accepted by even the newest rookie. The intent of prerequisite training is to develop that same level of awareness, acceptance, and willingness to comply with all safety standards. The procedures are somewhat more challenging, but equally as important. They generally require repetitive drilling to develop muscle memory and proper technique. This training may focus on fundamental skills such as proper lifting, dragging, and carrying techniques. The training should be comprehensive enough to involve the proper use of all hand and power tools, especially where those tools are used for self-rescue and disentanglement. Whether the training involves a rule, policy, or procedure, it should be made clear that compliance is mandatory and not an option.

Safety is everyone's responsibility. All participants in the live fire exercise, including students, should be encouraged to speak up regarding something that appears to be an unsafe operation or practice. Though the instructor-in-charge has the authority and responsibility for the safety of the drill, all participants' concerns should be heard and evaluated.

Fire Behavior

One of the primary reasons to conduct live fire training is to observe the associated fire behavior in compartment (structural) fires, vehicle fires, and flammable liquid and gas facilities and installations. In order for this training to be effective, the student must have a thorough understanding of the physical science of fire, combustion, and the methods and techniques of extinguishment. It would serve little purpose to expose anyone to the perils of live fire without a sufficient base of knowledge to benefit from the experience.

In addition to an understanding of what fire is and how it develops, the student must have a thorough understanding of the four methods used to control and extinguish a fire. It is not enough for a student to have a superficial knowledge of fire behavior and fire extinguishment. The students' ability to regurgitate the methods of heat transfer (conduction, convection, and radiation) or the methods of extinguishment (cooling, smothering, starving, and chemical-free radical quenching) is not enough to save their lives in a live fire event.

Portable Extinguishers

In some cases portable fire extinguishers are the best, or perhaps the only, choice to control and extinguish a small fire or a fire

Near Miss REPORT

Report Number: 08-234
Date: 05/09/2008

Event Description: Our shift had put on a live fire training event for our new recruit class covering topics including fire behavior, ventilation, thermal imager orientation, and fire control. Near the conclusion of the afternoon round of burns, the captain of one of the participating engine companies wanted to provide a scenario for an on-duty firefighter who had been training regularly over the past several weeks in preparation for the engineers' exam. The scenario was laid out that a fire would be set on the first floor of the fixed burn facility simulating a room and contents residential fire, and the engineer in training would go through the necessary steps to charge an initial attack line for fire suppression. The training officer elected four of the recruits to act as the fire suppression crew tasked with deploying a 200' 1.75" cross-lay with a combination nozzle, making entry, and extinguishing the fire. A captain and firefighter were selected to oversee the interior operations and 2 firefighters were to oversee the exterior deployment.

The training division engine, which had been utilized all day for the training burns, was left in place with 2 charged hose lines already deployed, one for the interior crew who would be observing, and a back-up line. The hydrant was left hooked to the engine, but was shut off and manned awaiting the engineer's order to charge, as part of the scenario. The 500 gallon booster tank was full prior to the hydrant being closed. A third cross-lay was left bedded for the recruits to deploy as part of the scenario.

The scenario was initiated and the recruits attempted to deploy the cross-lay, but had some difficulty properly deploying it, resulting in entanglement and a partial deployment, leading the exterior firefighters to attempt to help the recruits straighten the line. Meanwhile, the engineer, for an undetermined reason, shut down the hose line to the interior observation crew, and before charging the line being deployed by the recruits, left the pump panel to turn on his own hydrant. At this time the fire conditions within the building were beginning to intensify. The interior observation crew, watching the fire from an adjacent room, realized that the suppression crew was not making a timely ingress. They attempted to apply some water to the fire to keep it in-check for their own protection, only to then realize their hose line was no longer charged. They attempted to make contact with the engineer over the radio, but he did not respond and was seemingly overwhelmed with the scenario at this point. With rapidly deteriorating conditions within the building and a dry hose line, the interior observation crew abandoned their position and exited the structure. Upon their exit they reported the conditions to the incident commander who immediately halted the exercise, ordered a certified engineer to take over the pump operations and sent the exterior observation crew, which was still attempting to assist the recruit hose team, in on the back-up line to extinguish the fire.

The fire was successfully extinguished by the back-up line without further incident. The interior observation crew had some minor heat damage to their turnout gear, which was immediately removed by awaiting firefighters. They were evaluated and treated on location for heat exhaustion.

Lessons Learned: This incident brings about several important issues including:
- Always underestimate the abilities of personnel who are not fully trained and certified
- There needs to be sufficient experienced personnel involved in scenarios and training to oversee the actions of those being tested
- When dealing with live fire evolutions, corrective actions must be taken immediately when potentially dangerous actions are taken, regardless of scenario parameters

Actions which could have been, and need to be taken to prevent this and similar incidents are:
- Always follow NFPA 1403 guidelines for any live fire event, including first-due scenarios
- Constantly man a back-up line
- Always have an experienced engineer overseeing pump operations any time personnel are involved with live fire
- Always have a rapid intervention or similar team in place
- Incident command must carefully observe all scene conditions, actions and radio traffic
- Have a safety officer in place at all times
- Ensure a constant and reliable water source for interior operations
- Utilize separate apparatus for safety lines and scenario lines

needing a special agent to extinguish. A student must be familiar with the use and operation of the various types of portable extinguishers.

Personal Protective Equipment
The fire fighter's personal protective equipment (PPE) is designed to offer limited protection from the heat of fire and gaseous products of combustion. Understanding the intended purpose, correct use, and operating limitations of PPE is essential to fire fighter safety and survival.

Live fire training instructors are responsible to ensure that all required PPE is in serviceable condition and properly worn.

Ladders
The student should be familiar with the different types of ground ladders used by the authority having jurisdiction. Proper use of ladders will help to ensure completion of assigned tasks and reduce the risk of personal injury.

Fire Hose, Appliances, and Streams
The student must possess the ability to determine and deploy an appropriate hose assembly and deliver an adequate fire stream for the fire attack.

Overhaul
The student must be familiar with the overhaul process, as there will generally be overhaul considerations at some point during the live fire training evolution. The student must be able to recognize hazards that may be present during overhaul and should also be able to spot indicators of possible structural collapse. Students must also be familiar with overhaul tools, like pike poles and thermal imaging cameras, and how they are used in overhaul techniques.

Water Supply
Establishing and maintaining a reliable water supply is a critical component of the live fire training evolution. The ability to secure a water supply from any potential source is a crucial skill, particularly if the water supply originally established for the training fails and another source needs to be quickly established.

Ventilation
Fires in enclosed spaces are directly affected by the amount of air available, the level at which the air is supplied, its volume and velocity, and the point on the time to temperature curve at which it occurs. Accordingly, ventilation can have a positive or negative impact on the fire attack effort, the search and rescue effort, the evacuation effort, or other interior and exterior operations. Students may endanger themselves or others by inadvertently or indiscriminately opening the structure in the wrong place or at the wrong time. The ventilation effort, therefore, must be carefully directed by the instructor-in-charge.

Ventilation is a critical component of any successful fire attack and warrants particular attention during the prerequisite training process.

Forcible Entry
It may become necessary to force entry to gain access to participants who have become trapped or pinned. It may also become necessary for participants to force their way out of an area due to fire spread or extension, structural collapse, hose line failure, or other unforeseen events. Participants must have an understanding of the construction and operation of typical doors, windows, and locking mechanisms, and the construction of typical walls, floors, and roofs in each of the five construction types. There are a variety of tools available to force or breach a structure. Participants must be versed in the application and use of all forcible entry tools on hand during the live fire training evolution.

In situations where crews must remove themselves from an environment where they have become lost, trapped, or pinned during the live fire training evolution, forcible entry techniques are necessary for survival.

Building Construction
Knowing the basic types of building construction is vital for fire fighters because building construction affects how fires grow and spread. Fire fighters must be able to understand the different types of building construction and how each type of building construction reacts when exposed to the effects of heat and fire. This will help them to anticipate the fire's behavior and respond accordingly. Also, an understanding of building construction will help determine when it is safe to enter a burning building and when it is necessary to evacuate.

The Last Session Before Live Fire Training Evolutions

Take the opportunity to review with the students and try to allay unnecessary concerns. Review hydration with the students. Students need to be reminded to start hydration a couple of days ahead of the live fire training, because dehydration is common among students showing up for live fire training.

Review the preburn plan and the order of events that will be followed. Remind the students of building "sides" protocol used locally and how to use it (e.g., 2nd window on side B). This is a good opportunity to explain that prior to the evolutions starting, all participants must conduct a walk-through of the structure to become familiar with the layout and to facilitate any necessary evacuation. Emergency procedures will be reviewed as well as the emergency egress routes. Let the students know to look for the marking of the primary exits and exit routes and make sure they know how to open windows. This may provide a good segue to discuss applicable real-life incidents.

Before the session draws to a close, students should be reminded to maintain a professional image during all facets of training. This is also the time to give out directions for parking, proper apparel, what to bring to the training, any limitations or directions on camera use, and finally student responsibilities after the last entry.

Live Fire Tips

Orientation Drill: Barrel Drill

The barrel drill is an orientation exercise that the Florida State Fire College has effectively used for almost 25 years **Figure 1-6**. It is conducted in the open air to reduce concerns for claustrophobia, and to represent wide open areas found in some occupancies. Students see the arrangement of the 55-gallon (208.2-liter) drums before starting. Students are paired up at one of the two starting points, and then proceed with their vision obscured, crawling in pairs to where they believe the next drum is in the course **Figure 1-7**. Instructors are positioned throughout the course to help the students find their way by calling out to them, banging on the drums, and other cues. The students go from barrel to barrel and then have to locate the barrels with the pallets on top, go underneath, and then proceed to the end line, still going barrel to barrel. The drill helps recruits improve their teamwork, communication skills, and orientation while their vision is obscured. According to "Rusty" Barrett, a Gainesville, Florida district chief and 20-year Florida State Fire College instructor "we are taking away their sight and making them use their other senses." They are told to use audible cues, such as noise of heavy machinery from a large mine behind the campus, traffic in the opposite direction, the feel of sun on one side of their body, tactile cues such as hard dirt around the area the barrels with the pallets are, or knowing where the chin up bars are, etc.

Figure 1-7 Recruit fire fighters advance from one barrel to another in the barrel orientation, constantly supervised by an instructor.

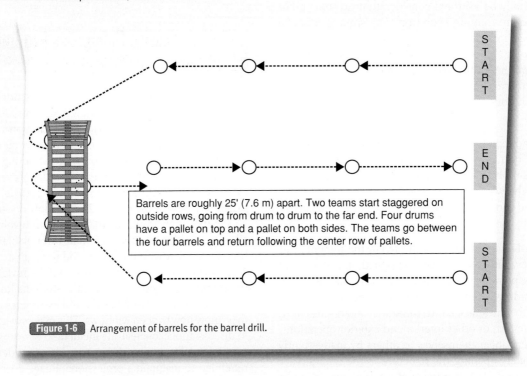

Barrels are roughly 25' (7.6 m) apart. Two teams start staggered on outside rows, going from drum to drum to the far end. Four drums have a pallet on top and a pallet on both sides. The teams go between the four barrels and return following the center row of pallets.

Figure 1-6 Arrangement of barrels for the barrel drill.

Wrap-Up

Chief Concepts

- From 1987 to 2005, while there was a decrease in structure fires in America, there was an increase in fire fighter injuries during training.
- The purpose of NFPA 1403 is to provide a process for conducting live fire training evolutions to ensure that they are conducted in safe facilities and that the exposure to health and safety hazards for fire fighters during live fire training is minimized.
- Strict adherence to NFPA 1403 is strongly advised, as it will lead to the safest live fire training environment possible.
- The instructor-in-charge is responsible for ensuring that the live fire training evolution is conducted safely, ensuring the training structures are prepared per NFPA 1403, monitoring all fireground activities, ensuring there are enough properly trained live fire training instructors involved, and making sure all training evolutions are conducted following the standard.
- Students must fulfill the prerequisite training requirements before being permitted to participate in live fire training.
- Safety is a primary concern in live fire training. Safety violations must never be tolerated and must be corrected immediately.
- The learning objectives and desired outcomes of the live fire training evolution must be clear to all participants. In order to prevent dehydration, students should begin hydrating a couple days prior to the live fire training evolution.

Hot Terms

<u>Authority having jurisdiction (AHJ)</u> An organization, office, or individual responsible for enforcing the requirements of a code or standard, or for approving equipment, materials, an installation, or a procedure.

<u>Demonstration</u> The act of showing a skill.

<u>Instructor-in-charge</u> An individual qualified as an instructor and designated by the authority having jurisdiction to be in charge of the live fire training evolution.

<u>Live fire</u> Any unconfined open flame or device that can propagate fire to a building, structure, or other combustible materials.

<u>Participant</u> Any student, instructor, safety officer, visitor, or other person who is involved in the live fire training evolution within the operations area.

<u>Safety officer</u> An individual appointed by the authority having jurisdiction as qualified to maintain a safe working environment at all live fire training evolutions.

Wrap-Up

References

Brown, T. 1987. Occupational Safety & Health Division Report B6968. Minnesota Dept of Labor and Industry.

Demers Associates, Inc. 1982. Two die in smoke training drill. *Fire Service Today*.

Klem, T. 1988. Fatal live fire training incident. *Fire Command*. May 1988:44–57.

Klem, T. J. 1988. Investigation report: Fatal live fire training incident—Milford, Michigan. Quincy, MA: NFPA, 1988.

Moynihan, B. 1982. Two fire fighters are killed at "routine" training session. *Fire Engineering*. August 1982:79–80.

Murtha, C. 2002. Learning from live burn tragedies. *The Voice*. October 2002:7–10

United States Fire Administration. 2003. *Trends and hazards in firefighting training: Special report TR100*. Emmitsburg, MD: USFA.

Live Fire Training Instructor in Action

You are a certified instructor with significant live fire training experience. You are called in two days before a recruit class burn in an acquired structure to fill in for an injured instructor. When you get to the drill site, the instructors gather for the preburn briefing. What information will you be looking for?

1. The instructor-in-charge tells you there will not be a walk-through of the structure because it will not be "realistic," and the recruits have already taken their certification written exams, so therefore, they do not have to follow the walk-through rule. Is he correct?
   A. The instructor-in-charge is wrong and you have to tell him.
   B. It is the instructor-in-charge's responsibility and only he or she can make that determination.
   C. The students' levels of experience dictate whether or not the walk-through is required.
   D. The walk-through is usually pretty useless in training.

2. The safety officer calls for a student walk-through and you join them to listen for:
   A. emergency procedures.
   B. secondary and emergency exits and procedure to exit.
   C. emergency evacuation signal.
   D. All of the above.

3. You have not worked with this particular department in a live fire training burn before, however you have responded to fires with their company and have noticed significant problems with their PPE and SCBA. Which standard should you consult to ensure compliance with this issue?
   A. NFPA 30
   B. NFPA 1001
   C. NFPA 1142
   D. NFPA 1971

4. What is live fire training?
   A. The first training that any fire fighter recruit must go through before becoming certified as a Fire Fighter I.
   B. The end all of training. No fire fighter should be placed in dangerous live fire conditions without receiving necessary background training.
   C. A way for students to test their capabilities in a high heat environment.
   D. The best way for instructors to show off what they know about fire behavior to recruit students. You simply cannot learn fire behavior in a classroom setting.

Critical Incident Planning

CHAPTER 2

NFPA 1403 Standard

NFPA 1403 contains no live fire training job performance requirements for this chapter.

Knowledge Objectives

After studying this chapter, you will be able to:
- Describe the purpose of a line-of-duty significant injury or death plan and its necessity.
- Describe the role of the National Institute for Occupational Safety and Health's Fire Fighter Fatality Investigation and Prevention Program (NIOSH FFFIPP) following a fire fighter fatality.
- Describe how to appropriately manage the news media following the injury or death of a participant during a live fire training evolution.
- Describe the repercussions that a line-of-duty significant injury or death can have on a fire department.

Skill Objectives

There are no skills objectives for this chapter.

You Are the Live Fire Training Instructor

After several years of working under the direction of other instructors, you have been given the assignment as instructor-in-charge of a live fire training evolution. This is the day you have been waiting for, your chance to showcase what you can do. Nearly all of the authority and responsibility now rests with you. As you prepare for this training, there are many questions that must be answered to satisfy the requirements of the standard. Will we have enough engines and tenders, and are we able to satisfy the water-supply and fire-flow requirements? Are all of the preburn documents completed, signed, and filed? Have we identified locations for command, staging, rehabilitation, medical, air supply, entry, evacuation, assembly, spectators, and the media?

Have we provided for emergency vehicle movement and access? Do we have adequate, written plans for setup, water supply, delivery, communications, operations, evacuation, medical monitoring, triage, treatment and transportation, contingencies, demobilization, and postevolution debriefing? Have we overlooked anything?

Certainly, all of these considerations are important, and this is by no means a complete list. However, there is another, often overlooked, section of NFPA 1403, *Standard on Live Fire Training Evolutions* that must be carefully considered.

1. Has the student prerequisite training component of the standard been met?
2. Are the participants, both students and instructors alike, *truly prepared* for this level of training evolution?
3. If something goes horribly wrong, what will you do?

Introduction

The fire service has conducted live fire training evolutions for many years. The repercussions of an injury or death in training, as with any other line-of-duty significant injury or death, can be long-lasting and can have detrimental effects on the individuals, their families, their co-workers, and the agency itself. Those outside of the fire service may be led to believe that the fire department is to blame for the casualty because it was not incurred during a response to protect somebody's life and property, but from something done by the fire department. The resulting investigation can last for months and can result in demotions, firings, or even jail time.

Such an investigation is one of the most difficult and important activities to conduct. This difficulty is compounded by the fact the investigation must usually be conducted under extremely stressful circumstances and often under pressure for the rapid release of information.

This chapter will discuss the necessary steps to be taken if such an incident occurs. It will also prepare the live fire training instructor on the effects to the fire department and what to expect from the media, investigators, and civil litigation.

Line-of-Duty Significant Injury or Death Plan

Every fire department needs to have a plan in place to deal with a serious injury or death of a fire fighter. The line-of-duty significant injury or death plan should be easily accessible and should be able to be implemented immediately. Such a plan should include the following:

- Notification of the family and liaison: The availability of practically instant communication from the scene with the use of texting, cell phones, Internet, and so on, has changed the long-time practice of a chief officer and chaplain personally delivering the news to a spouse or family. Now, the focus is on delivering the news immediately, before other communications can reach them. Although this does not directly affect the care of the fire fighter or the investigation, it is of extreme importance

to the family and the extended fire service family. There should be somebody assigned to the family throughout this time of need, to act as the family's single point of contact with the department and to handle any needs that may arise.
- Insuring care and hospital liaison: A fire department official, to represent the department at the hospital, is needed to ensure all agency and insurance information is immediately available to the caregivers. The liaison will also be able to provide information that the caregivers may not be familiar with, for example, the possibility of cyanide poisoning.
- Notification to the rest of the fire department: Word of the incident will spread quickly, and once the family is notified, the extended fire service family needs to be told as well. A good item for the plan is to determine a meeting site, as personnel will want to convene at the hospital.
- Securing the scene: As any crime scene would be secured, all equipment needs to be left as is until the scene is thoroughly photographed, documented, and investigated.
- Seizing evidence (at hospital and scene): Protective clothing, breathing apparatus, uniforms, and other apparel will most likely have to be secured for the investigation. The fire department representative can make sure those items are taken into custody from the scene or from medical services.
- Personnel statements: The involved personnel need to give official statements immediately, before talking to each other. This is not as much of a concern of covering up information as it is protecting against changing perspectives based on hearsay information that may or may not be correct.
- CISD: Critical incident stress debriefing (CISD) needs to be arranged for immediately, with follow-ups for involved personnel and families.
- Media: Getting pertinent information to the proper sources in a timely manner is a very important step in the plan. Model press releases should be developed and be ready to go with only minimal updates, if needed.
- Notifications: Notifications need to be provided as required by law, such as to Worker's Compensation, the Occupational Safety and Health Administration (OSHA) or state or provincial equivalent, municipal risk management, and so on.
- Emergency Operations Center: The plan may want to call for activation of the fire department's emergency operation center, to keep the flow of information better organized.

The line-of-duty significant injury or death plan needs to ensure proper care of the injured, and the family of the injured, protect and document the necessary evidence, and provide support for the involved personnel. Experiencing such a traumatic event will be a gut-wrenching situation, but not having a plan will make it all the worse.

The Investigation

The goals of the investigation are to identify deficiencies in policy, procedures, and other actions that contributed to the incident, and make corrective recommendations to prevent this type of incident from occurring again. In addition to the fire department, other entities will also investigate the incident. Other investigations that may occur include the following:
- Law enforcement: This can be local law enforcement, or a state or provincial law enforcement, to determine if there were any criminal offenses.
- Regulatory: Depending on your unit of government, this can be OSHA or your state or provincial equivalent. These agencies normally have the authority for "compulsion to testify," to take procession of evidence, and can levy fines. In some cases, the violations can be criminal.
- NIOSH: In the United States, the National Institute of Occupational Safety and Health (NIOSH) investigates incidents for the purposes of preventing similar incidents. NIOSH does not have law enforcement or regulatory authority, nor does it have the legal ability to compel individuals to testify.

The incident will most likely be investigated for criminal prosecution, if it is believed to have reached a threshold of criminal negligence or negligent homicide. One chief described the process of having multiple agencies investigating a fatal incident as "tearing the heart out" of the department. The possibility of criminal charges being issued adds to the internal discord of the department. As previously discussed, such fatality or injury incidents will normally require investigations from multiple agencies, which can cause suspicion among personnel because the interviews are conducted separately. One department described enduring multiple investigations as creating a sense of panic among the fire fighters. Many rumors of conspiracy theories evolved as a result, and those involved were concerned about whether any blame was being cast their way.

NIOSH Investigations

In 1998, Congress appropriated funds to the National Institute for Occupational Safety and Health (NIOSH) for a fire fighter safety initiative. As part of this initiative, NIOSH developed and implemented the Fire Fighter Fatality Investigation and Prevention Program (FFFIPP).

The overall goal of the NIOSH FFFIPP is to reduce the number of fire fighter fatalities. To accomplish this goal, NIOSH conducts investigations of line-of-duty fire fighter deaths to identify contributing factors and to generate recommendations for prevention. As part of this initiative, NIOSH conducts independent, on-site investigations of fire fighter line-of-duty deaths. The following are the goals of the FFFIPP:
- Better define the characteristics of line-of-duty deaths among fire fighters.

- Develop recommendations for preventing deaths and injuries.
- Disseminate prevention strategies to the fire service.

NIOSH investigations are voluntary, and names remain confidential, as the purpose is not to find fault or cast blame on fire departments or individual fire fighters. During a fatality investigation, investigators evaluate the incident through a review of records, such as police, medical, and victims' work and training records, as well as departmental procedures. Investigators also examine the incident site and equipment used, including personal protective equipment. Interviews are conducted with members of the fire department, department investigators, and family members (when appropriate). Together, these sources build a picture of the circumstances of the event.

Next, the investigators review best practices, NFPA standards, information from the United States Fire Administration (USFA), and the public health and fire service literature related to each case. They develop recommendations for prevention of future incidents and write a narrative report describing the event. References for each recommendation can be found in the NIOSH fatality investigation reports.

Near Miss REPORT

Report Number: 06-461
Date: 09/07/2006

Event Description: At the time of this event I was employed as the Training Captain for the City of [Name deleted]. I had been in the fire service for about 11 years however; I had only been a Training Officer for about 2 years.

The event occurred during a controlled training burn in a residential structure. The garage had been converted to a family room and the interior walls were covered with plywood, as was the ceiling. We had conducted multiple single room fires throughout the home with an emphasis on getting in quick for a direct attack. The final planned entry was for the family room area, and the intent was to allow the fire to spread to more than one room.

At the time the crews were directed to make entry, the fire was pushing out of the exterior door that went directly into the family room. Fire was coming out of the door down to at least the door knob level and the cement floor could be heard spalding. The fire grew rapidly as the crew made entry. Unfortunately, they entered in the same fashion as the previous evolutions, quickly and semi crouched without flowing water from the nozzle. The total time of entry was less than 6 seconds as the crew stumbled out. One firefighter suffered serious burns to his arms and shoulders and the SCBA mask was blackened by bubbles of plastic. I don't believe they ever discharged any water.

The lack of communicated expectations on my part and the lack of recognition on the part of the nozzleman were key factors that lead to this near catastrophe. I should have made it clear that this was a different type of fire problem. They should have observed the conditions were not suitable for immediate entry, and that some firefighting at the door would be necessary before making entry.

The other thing that I failed to recognize, although I was aware of the wall covering, was that the burning characteristics of plywood are much different from sheetrock. The added fuel caused an intense and extremely hot environment with rapid fire growth. In fact, it grew faster than expected which caused a later than desired entry although, I recognize that fact likely prevented multiple injuries or possibly the death of one or more Firefighters.

Lessons Learned: First and foremost was attention to detail and structural make up. The plywood walls were obvious, yet not accounted for. Firefighters in a training mode will likely do exactly what you tell them to do, so when the scenario changes you must communicate changes in expectation. The spawling of the concrete should have been recognized as a sign of intense heat, or at least a rapid heat increase.

External Reviews

Some departments have utilized external review panels. Although most are often fact-finding panels with an emphasis on avoiding reoccurrence, some have made recommendations to remove personnel. Opinions vary from impartiality to adding to the dynamics of the crisis, or fire fighters becoming products of victimization. External reviews can provide an unbiased look at the incident and help remove the potential for a perceived cover-up.

The News Media

There will be considerable pressure, internally and externally, to find a particular individual or action at fault for what happened. There could also be temptation to protect the agency or an individual's reputation by omitting or covering up details.

Dave Statter, a veteran TV and radio reporter and editor of a fire and EMS service blog, recommends:

If you know a fire fighter has been burned, release it. If you know for certain a flashover has occurred, say so. If you know that ten people were involved in the exercise, say so. There are plenty of things that are facts that will not change. There is no reason not to release them.

Donald Adams was the fire chief of Osceola County when a live fire exercise claimed the lives of two fire fighters. He explains the difficulty in coping with the sudden avalanche of interest from every type of news media available.

We had national and regional news crews, in addition to our local news crews, each wanting an interview. We attempted to focus on key points that needed to be covered. The media was attempting to portray Osceola County Fire in a negative light as a result of the deaths. This quickly shifted public opinion and perception of the organization into the perception of other organizations conducting similar types of training.

Because of the (law enforcement) investigation; we could not say anything about certain aspects of what happened, which quickly added to the perception that we were hiding something. This also added to distrust and even a greater negative perception. We learned that not being able to confirm or tell the media what happened only adds to the enormous negative impact of the story and attitudes creating a growing level of negative public emotion. No matter how positive of a message the Public Information Officer (PIO) delivers, the expectations already created by the media are nearly impossible for the organization to overcome for years to come.

Joe Farago, a former TV journalist who became a fire public information officer and also worked on a federal incident management team, warns that an event like a training fatality will result in the media hitting you like a tidal wave.

Think of the biggest incident you've ever been on, and that's what this will be like for reporters. Just like you, they'll bring in every available resource and won't back off until their job is done. This is a national story that people respond to on a visceral level, and news organizations will attack it from all sides. Multiple reporters from the same outlet will probably be assigned to get various perspectives. One might cover the basics of how and why it happened. Another will probably try to get emotional reactions from family and friends. Yet another will probably be assigned to do research on the history of these types of incidents and the identity of the fatality or fatalities. In the age of Facebook and MySpace, pictures and personal information are readily available for use on the evening newscasts and, more immediately, on the websites and blogs that most news organizations maintain. Competition in the news business is intense and there will be tremendous pressure on you for immediate information. If that information is not forthcoming, reporters won't wait for your official releases.

Joe Farago suggests that dealing with the media is the last thing on the minds of fire fighters in a training fatality. Emotions will be running high as reporters push for statements and reactions. The likelihood of saying something that may be regretted later is high.

As fire fighters, the way we survive in highly stressful environments is through planning and training. No matter the size of your department, have public information policies, procedures, guidelines, and a plan in place to deal with the media during a fatality. Train company officers in media relations and call them in for support. Know what you can and cannot say about the incident and avoid phrases like 'no comment' or 'can't comment' because of the reaction those words evoke in reporters, readers, and viewers. Deal directly and immediately with your internal audience of fellow fire fighters. You need to keep them informed and updated. Understand the size and scope of an event like this and establish a process for quickly approving limited initial statements and identifying agency personnel who are authorized to make those statements. Above all, be available, open, and truthful. You won't ever be able to control a story like this, but you can keep from making it worse.

The news media is going to be looking for information wherever they can get it. If the department is not forthcoming, members of other departments may be contacted for their thoughts, and those thoughts and opinions will be based

Live Fire Tips

Deal directly and immediately with your internal audience of fellow fire fighters. You need to keep them informed and updated.

Incident Report

Lairdsville, New York - 2001

Figure A A couch at base of stairs in Lairdsville duplex.

The training evolution in Lairdsville, New York involved a duplex that was to be razed. A search and rescue scenario with a rapid intervention crew (RIC) deployment was planned to simulate the entrapment of fire fighters on the second floor of one unit. For the purpose of the drill, the stairs to that unit would be considered impassable, so the RIC would have to enter from the other unit of the duplex and breach a second floor wall to reach the trapped fire fighters who were under light debris. Smoke from a burn barrel was used to obscure vision. The burn barrel was located in the back bedroom of the same unit as the trapped fire fighters, also on the second floor.

Neither of the two new fire fighters posing as victims had worn SCBA in fire conditions, and one (the deceased) was a brand new recruit fire fighter with no training. Both fire fighters wore full protective clothing and SCBA for this drill.

All of the crews assembled initially at the duplex. One engine was connected to a large water tanker on-site but no hose lines had been set up. A 1¾" (44.5 mm) hose line was supposed to be placed at the rear with another at the front doors. An engine and heavy rescue were located off-site and would simulate a response to the scenario.

Safety officers were assigned and placed throughout both sides of the duplex. One officer was assigned to light the fire and place the new fire fighters in the front bedroom on the second floor. He was also instructed to guide the RIC if necessary. Another officer was assigned to a room on the second floor of the duplex unit, which the RIC was to first enter through. He was assigned to guide the RIC if necessary, making sure they did not go through a wall opening that went to the outside. He was equipped with a 20-lb (9.1-kg) fire extinguisher. The 1st assistant chief was located on the first floor of the unit. The fire chief checked the scenario to make sure that accelerants were not used. He took command out front and the training was ready to begin.

The smoke barrel upstairs was ignited, but it was not creating the desired smoke conditions. Downstairs, the 1st assistant chief struck a road flare and ignited the foam mattress on a sleeper sofa that was located next to the open stairs **(Figure A)**. The fire spread very rapidly across the ceiling, producing heavy smoke and flames, and flames quickly extended up the stairs and out the front windows.

The fire spread was so rapid that the officer in the adjoining unit could not enter to reach the new fire fighters, and he had to exit using a ground ladder. The 1st assistant chief exited the duplex and went to the rear exterior but he could not find a hose line. He then went to the pumper and pulled a 200' (61 m) preconnected hose line to the rear of the structure.

The fire fighter on the same floor as the trapped fire fighters had left his position to investigate the sound of another road flare being lit. Upon seeing the fire, he went back and reached the two new fire fighters. The flames were now entering the second floor windows. He led the new recruits to the stairwell, which was now fully engulfed in flames from the first floor. He lost his gloves and immediately burned his hands. At this point, he lost contact with the two recruit fire fighters. He was able to reach the back bedroom with heavy smoke and tenable heat. He searched for the boarded up window and was able to break it open with his hands and jump from the second floor.

The 1st assistant chief advanced the hose line to the second floor, knocking down the fire. Arriving units were advised it was no longer a training and two RIC units were deployed to the second floor. One RIC forced entry through the involved unit's front door. One fire fighter was found and brought to the outside. The second RIC found the other fire fighter unresponsive. They removed him and transported him with advanced life care support to a local hospital, where he was later declared dead. The rescued fire fighter and the fire fighter who jumped from the window were both flown by EMS helicopters to a burn unit, due to the severity of their burn injuries.

During the ensuing investigation, the Lairdsville fire chief claimed he didn't know that a structure fire would be set as part of the training exercise, saying, "It was only supposed to be smoke." The 1st assistant chief was indicted and found guilty of criminally negligent homicide. He was sentenced to seventy-five days in jail and instructed to avoid contact with any fire department under a five-year term of probation. His attorney argued that the NFPA standard was not known by the department, and that it was the state's responsibility to distribute the guidelines to every volunteer fire fighter.

Post-Incident Analysis   Lairdsville, New York

NFPA 1403 Noncompliant

- Students not sufficiently trained to meet job performance requirements (JPRs) for Fire Fighter I in NFPA 1001 (4.1.1)
- Debris from the structure used as fuel material (4.3.2)
- Students acting as victims within the structure (4.4.14)
- Deviation from the preburn plan (4.2.25)
- Fire ignited without charge hose line present (4.4.19.2)
- Exterior hose lines not in place (4.4.6)
- Rapid Intervention Crew (RIC) not in place (NOTE: This is not a requirement of 1403, but is an OSHA requirement and is referenced in NFPA 1500, NFPA 1710, and NFPA 1720.)

on incomplete information. Comments made without due consideration, and out of duress, may end up hurting more than helping. When referring to the death of two of his fire fighters in a training exercise, Chief Donald Adams said that most of the unofficial communications came from fire fighters on the scene who notified other fire fighters, who then notified others.

> And with each new communiqué, the members of our organization were hit again and again. Even with early notification of the County PIO, the unofficial communications to the media and each new rumor were already impacting the normal reactions to the crisis. Without firsthand knowledge of what had happened, blame was directed to those participating in the training incident. I believe the unofficial communications brought on more grief and intense feelings of despair or even rage than would have if the unofficial communications could have been controlled.

Dave Statter also has a warning.

> The harder you work to suppress the timely release of factual information, no matter how damaging to your agency, the more difficulty you will find in getting out from under negative stories or restoring your reputation with the public and the reporters who cover your organization....while the facts are being withheld, all kinds of wild speculation is being presented to reporters. Bits and pieces of information come our way from a variety of sources. With the Internet, this has exponentially increased the amount of material that ultimately gets before the public through the news media or the "new media."

> The instant defense of the department and its people in a time of crisis is human nature. Some even think of it as a good thing and a positive leadership trait. I would ask, "Is it any better than the speculation that the same chiefs might claim is done by the news media?"

The Aftermath

Regulatory and Criminal Charges and Prosecution

The Lairdsville, New York case was the first one in which criminal prosecution with incarceration occurred. The laws of criminal negligence vary by state and province. In other cases, criminal investigations have occurred, sometimes going before the grand jury.

Retired fire fighter Walt Malo, now occupational safety and health program manager for the Florida State Fire Marshal, points out, "Defense, is the word. Most of the departments and institutions that I have dealt with after a fatality took to the defensive posture almost immediately. This is not to criticize, I fully understand; having been a municipal risk manager that is exactly what I would have instructed the chiefs and instructors to do, defend your position." While defense can provide some protection to the agency and some of the involved parties from prosecution or litigation, it never resolves the real issues.

Internal Strife

Any fatal or significant injury incident involving fire fighters can bring internal strife. Blame and bitter feelings, which are often fueled by incomplete, and sometimes false, information, can be directed towards the leaders and participants of the training. What is said externally by the media, comments by members of other departments or so-called experts, and even comments made to the media by department members or family members can further exacerbate harsh feelings inside the department. Comments made by involved individuals can make internal agency feelings toward individuals even worse.

Following one well-known fatal live fire training incident, the responders and participants were in a CISM debriefing hours after the incident. During the session, the personnel that had responded, treated, and transported the fire fighters that had died were accusatory towards the instructor-in-charge of the training evolution. Those accusing him did not participate in the training evolution and responded to the emergency after it happened, so they did not see the preparation, how the safety personnel were in place, or how the evolution had been run. While they were on the site briefly, they did not see it occur, nor did they understand what really happened. They just assumed that the instructors had done wrong and had not taken precautionary measures to ensure the fire fighters' safety in the evolution. The people who were present during the evolution were not the ones pointing fingers, but it fell on them to correct misstatements and accusations when they could. In this case, the CISM debrief caused harm and increased distrust. The debriefing must be controlled to not affix blame.

In several incidents, arguments and physical altercations have occurred. Following one fatal fire, there were several incidents of internal discord towards the training officer and safety officers inside the structure. The discord ranged from the chief officer level to the rank and file. There was a report of an altercation when fire fighters were sitting around a pub revisiting the events. According to the involved fire chief, the criminal justice system added to the internal discord as each investigation was announced with possible criminal charges attached. Even weeks after the event, investigations brought the nightmare of the incident back to life, reviving the same emotions that brought discord following the incident.

In another department, the discord took on an ethnic overtone, exacerbating strained relations. Depending on the circumstances, fire fighters' confidence in the department's administration and training staff can be shaken. In several cases, the involved training staff and the deceased or injured, belonged to the same bargaining unit, causing conflict there. According to Walt Malo, "After a catastrophic event, it is quite human to try to identify someone other than yourself as being at fault, or at the very least with the ability to prevent the occurrence." Referring to a serious non-training-related injury event, he relates how there was a great deal of finger-pointing that took place in the city's fire headquarters during the investigation by the state and NIOSH. His concern is that if one or more persons are slandered and the agency doesn't accept an institutional responsibility, then the issue doesn't get resolved. The same misjudgments and inappropriate tactics will continue, next time with possibly even worse results.

In the aftermath of the deaths of two fire fighters in Osceola County, Florida, the instructor-in-charge was portrayed negatively in the media by some of his co-workers, and strongly criticized in some of the fire service blogs. The instructor related how a retired chief officer and national fire service figure, who had lived through controversy and the loss of fire fighters under his command, had counseled him from his own experience. The retired chief told him not to go online or read the comments or blogs, and not to respond or try to correct them.

Certainly all discord cannot be controlled, and there has been plenty of shared blame in fatality and injury incidents. Correct information is most likely the strongest advocate, but most often the truth is not immediately clear, and opinions will be made through different on-scene perspectives and what transpired leading up to the event. Participants "lawyering up" sends a message that may damage the department's reputation, unless the lawyer can be a clear, noninflammatory spokesman that does not say anything to protect his client at the expense of the department. A clear and decisive move, such as a safety stand-down to review procedures or a strict directive to comply with the investigation from the top, can help the department's image during these trying times. Department members need to have confidence that the situation is being handled properly, with appropriate expediency in a nonbiased manner, that the injured are cared for, and that the dead are respected and their family supported.

News Media Scrutiny and Public Opinion

Most often, the public opinion will be based on what they learn from the news media. If the department is portrayed in a negative light, it can turn the public's opinion and undermine their confidence in the agency. In one case, a volunteer fire department affected by the influence of the media ceased to operate and was merged into an adjacent department.

What is said by the media can destroy many years of good work and confidence in the public's eye, as well as bring other decisions and actions into question. Unfortunately, many media outlets take the stance of, "If they screwed this up, what else went wrong, or what else have they done wrong?" This can further erode public opinion and confidence and will take an agency a long time to overcome. Almost anything the department does following one of these incidents will be looked at as suspect by the media.

Safety Tips

In the last three live fire training fatality incidents conducted by different fire departments, several similarities were found in the results:
- All three long-tenure fire chiefs resigned months after the events.
- The training chiefs (or heads of bureau) were replaced.
- Instructors were fired or retired soon thereafter.
- All three fire departments selected very well qualified and respected new fire chiefs from *outside* their departments.
- The fire chiefs believed not to have had control over their departments.
- All three fire departments lost public respect.

Employment

Such incidents have resulted in personnel being demoted or fired. In all of the recent cases, the long-term fire chiefs have left their positions after a fatality and many of the training officers faced termination, some were demoted, and several resigned. Remaining in the system becomes very difficult after a serious incident.

Civil Litigation

Where there has been inadequate planning, organization, or other untoward activities, there has been civil litigation against the fire department involved. According to an attorney who represented the widow of a fire fighter that died during a live fire training exercise, this is a "target-rich environment" because of fire departments that don't follow NFPA 1403.

Avoidance

Walt Malo says the following about avoidance:

Responding to a structure fire, you don't know exactly where the fire is, what contents are on fire, the condition of the building, and you don't get to assemble all of the personnel you want before you start suppression. You don't get the opportunity to conduct a walk-through of the building and learn the alternate escape routes, and review the emergency plans just for that building—before the fire starts. But if you look at the fatal training fires, NFPA 1403 was not followed, and important planning steps were skipped. While not 100 percent infallible, following the standard and the formal training available pretty much will allow you to avoid what this chapter covers.

Sam Spatzer is an attorney that has been involved in civil litigation concerning the failure to follow NFPA 1403. He suggests putting some of the basics of the standard into the recruit curriculum, so recruits know if they are not given a walk-through, or an emergency briefing, or there are not PPE inspections, that they need to ask. Spatzer went on to suggest that the "buck stops at the top," and, "if you want accountability, have the fire chief and training chief sign off on compliance to the NFPA 1403 standard and legal requirements. Take away the excuse of plausible deniability. The fire service has had the standards for some time now, but what good is it if it isn't followed? Enforcement must come internally from the top down, and not depend on the state fire marshal or other outside agencies for enforcement." He suggests that avoidance may come from even the newest recruit if they know what is supposed to be done, and if they have a way without fear of repercussion to let the leadership know of problems.

Wrap-Up

Chief Concepts

- Every fire department needs to have a line-of-duty significant injury or death plan in place. The plan should be easily accessible and should be able to be implemented immediately.
- The goals of the investigation of a fire fighter injury or death during training are to identify deficiencies in policy, procedures, and other contributing actions, and to make recommendations on how to prevent a similar future incident.
- The National Institute for Occupational Safety and Health (NIOSH) conducts investigations of line-of-duty and training fire fighter deaths to identify contributing factors and to generate recommendations for prevention.
- Managing the news media is vital. Stating and being forthcoming with the facts is essential in placing the fire department in a positive light.
- Communication is key in helping to mitigate the potential conflicts that may arise after an incident.

References

Foley, S. 2003. *Resources for fire department occupational safety and health,* Quincy, MA: National Fire Protection Association.

Ridenour, M., Noe, R., Proudfoot, S., Jackson, J., Hales, T., and Baldwin, T. 2008 *Leading recommendations for preventing fire fighter fatalities, 1998–2005.* NIOSH Publication 2009–100. Washington, DC: NIOSH.

Tarley, J., Mezzanotte, T., and Koedam, R. 2002. Firefighter fatality investigation report. Washington, DC: National Institute for Occupational Safety and Health.

Live Fire Training Instructor
in Action

Two fire fighters in a neighboring fire department have been seriously hurt during an acquired structure live fire training exercise. The rapid intervention team and instructors had to assist the crew and the instructors that were inside the house to escape after a flashover. The fire spread from the original house to a neighboring house and detached garage before additional emergency units arrived.

The involved fire department has requested the regional fire chief's association to send representatives to conduct an external investigation, and your chief wants you to be a member of the investigative team. You leave for the nearby town immediately.

1. A local television news reporter calls you about the incident. You should keep your comments:
   A. to yourself.
   B. brief and factual.
   C. slightly opinionated.
   D. detailed and lengthy.

2. The involved fire chief has alerted you that he is concerned that his personnel may say something to the media that will worsen the situation. His department does not have a policy that covers news media. He is also concerned that if he tells them not to talk to the press, it will appear as though a cover-up is forming. You advise him to remind personnel that:
   A. the first priority should be the injured personnel, and to allow time for the families to be advised of the extent of the injuries.
   B. the appropriate agencies have been notified and are already en route to investigate.
   C. they should set an assembly location out of public view where they can be kept informed.
   D. All of the above.

3. Based on the very limited information you have received at this point, should you have an expectation that law enforcement will have a presence for possible criminal charges?
   A. Yes, prior fires have resulted in criminal prosecution.
   B. Yes, but only to support the investigation.
   C. No, there should be no law enforcement presence.

4. What agencies or groups listed below do not have enforcement authority (will vary by jurisdiction)?
   A. OSHA (CANOSHA) or state or provincial equivalent
   B. NIOSH (United States only)
   C. NFPA
   D. The news media

Preparation and Training of Instructors

CHAPTER 3

NFPA 1403 Standard

4.6.9 Instructors shall take a personal accountability report (PAR) when entering and exiting the structure or prop during an actual attack evolution conducted in accordance with this standard. [pp 48–49]

Additional NFPA Standards

NFPA 1001 *Standard for Fire Fighter Professional Qualifications*
NFPA 1041 *Fire Service Instructor Professional Qualifications*

Knowledge Objectives

After studying this chapter, you will be able to:
- Describe the necessary prerequisite knowledge and skills for a fire fighter to become a live fire training instructor.
- Identify the requirements to be a live fire training instructor according to NFPA 1403.
- Describe the fire behavior knowledge necessary for a live fire training instructor, including fire dynamics and heat release rate.
- Describe the standard training evolutions for the live fire instructor candidate.

Skills Objectives

After studying this chapter, you will be able to:
- Demonstrate proficiency operating as a live fire training instructor working with interior crews with regards to crew safety and observing fire conditions.

You Are the Live Fire Training Instructor

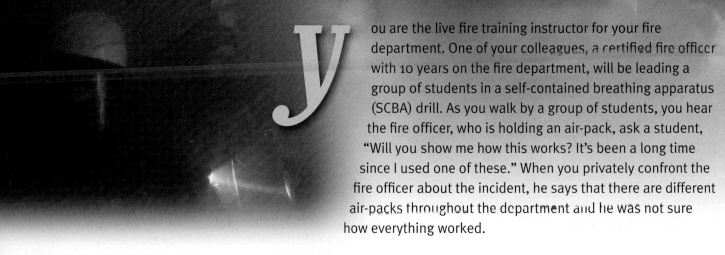

You are the live fire training instructor for your fire department. One of your colleagues, a certified fire officer with 10 years on the fire department, will be leading a group of students in a self-contained breathing apparatus (SCBA) drill. As you walk by a group of students, you hear the fire officer, who is holding an air-pack, ask a student, "Will you show me how this works? It's been a long time since I used one of these." When you privately confront the fire officer about the incident, he says that there are different air-packs throughout the department and he was not sure how everything worked.

1. How can you ensure that the instructors serving with you are truly certified?
2. How can you avoid compromising situations, like this one, in the future?
3. Should the fire officer participate in the training?

Introduction

According to NFPA 1403, *Standard on Live Fire Training Evolutions*, an **instructor** is an individual qualified by the authority having jurisdiction (AHJ) to deliver fire fighter training who has the training and experience to supervise students during live fire training evolutions. It is important to realize that ultimately, a fire chief can name anyone whom he or she wishes to be considered instructors for a live burn. Choosing less-than-competent individuals to fill these roles can lead to accidents and the accompanying liability. It is recommended that The AHJ should consult referenced NFPA 1041 standard, *Fire Service Instructor Professional Qualifications*, to determine if an instructor has the training and experience necessary to supervise students in a live fire training evolution. There is a difference between the learning environment and an emergency incident. There should be no surprises while performing a live fire training evolution.

The "You Are the Live Fire Training Instructor" relates how an inexperienced or untrained instructor can play a role in potential mishaps in a live fire training session. Untrained or otherwise unprepared instructors have contributed to injuries and even fatalities, as is seen in some of the incident reports in this text. NFPA 1403 is quite clear on the requirements to become a fire service instructor and that instructors must be qualified to deliver fire fighter training. NFPA also requires that live fire instructors be qualified by the AJH to deliver live fire training, and those instructors acting as the safety officer requires training on the application of the requirements of the NFPA 1403 standard. The safety officer needs to have training to understand fire behavior, human physiology, water supply, and other factors covered in this standard. A comprehensive training program needs to be considered for instructors acting as the instructor-in-charge and/or safety officer. A number of regional or local training centers, as well as several states, have training programs that regulate this process. In Florida, for example, a training program is legally required for all fire instructors participating in live fire training, and even requires state-issued fire instructor certification in addition to the training course.

Prerequisites to Becoming a Live Fire Training Instructor

NFPA 1403 now requires that the instructor-in-charge has received the training to meet the minimum job performance requirements for Fire Instructor I (NFPA 1041). However, training centers and fire departments can set their own requirements for their live fire training instructors, unless there are state or provincial requirements otherwise. In those states, the department can specify requirements above those required by the state. Florida requires live fire instructors to also be certified as a Fire Instructor I, which requires six years of fire service experience, completion of a 44-hour course, and a state certification test. Pennsylvania's requirements include eight years of fire service experience in addition to letters of recommendation.

Live fire training instructor candidates should have the following qualities:

- They have operated extensively in the environment they will teach in, and have an excellent understanding of all potential nuances and dangers the situation can present Figure 3-1.
- They have proven, practical teaching skills, not just in the classroom, but in turnout gear and in a dynamic learning environment.

Figure 3-1 Live fire training instructors need to be experienced and well trained in fighting fires.

- They possess extensive knowledge regarding NFPA 1403 and other related NFPA standards, Occupational Safety and Health Administration (OSHA) standards, fire behavior, building construction, smoke reading, and hostile fire events.
- They are physically fit enough to perform their duties including emergency actions.
- They are willing to lead a cultural change regarding the importance of participant safety, in both the training and emergency environments.

All of the above sound pretty straightforward, until you try and quantify these requirements and add detail to them. It should go without saying, but the live fire training instructor candidate needs to be an excellent fire fighter. At the least, a good start to achieving this goal is for the person to meet the requirements of NFPA 1001, *Standard for Fire Fighter Professional Qualifications*.

The candidate must have a solid foundation with the required knowledge, skills, and ability in the following topics:

- The history and orientation of the fire service
  - Fire fighter qualifications
  - Tactical priorities and fire department organization
  - Policies, procedures, and standard operating guidelines
  - Roles and responsibilities included in Fire Fighter I and II training
- Fire fighter safety
  - Causes of fire fighter injuries and deaths
  - Safety standards (NFPA 1500)
  - Safety and health programs
  - Safety during training
  - Fire fighter personal protective equipment
  - Protective clothing
  - Respiratory protection
  - Donning and doffing breathing apparatus
  - Inspection and maintenance of breathing apparatus
  - Use of breathing apparatus
- Fire service communications
  - Radio use
- Incident command system (ICS)
  - History and characteristics of the ICS
  - ICS organization
- Fire behavior
  - Chemistry of fire
  - Fire development and control
  - Smoke reading
  - Fuel loading
  - Heat saturation
- Building construction
  - Construction terminology, materials, and classifications
  - Hazards related to building construction
- Portable fire extinguishers
- Firefighting tools and equipment
  - Function and use
- Ropes and knots
- Response and size-up
  - Safety during emergency and nonemergency driving
  - Emergency operations
  - Scene safety
- Forcible entry
  - Breaching walls, floors, and ceilings
  - Forcing doors and windows
- Ground ladder functions and types
  - Inspection and maintenance of ladders
  - Handling, positioning, and safe use of ladders
- Search and rescue operations
  - Rescue and extrication
- Ventilation
  - Reasons for fire ground ventilation
  - Considerations
  - Tactical priorities
- Water supply
  - Rural and municipal systems
  - Fire hydrant types, location, and operation
- Fire hose operations, nozzles, and streams
  - Fire hydraulics
  - Care and maintenance
  - Couplings, appliances, and tools
  - Fire stream patterns and nozzles
- Fire fighter survival
  - Safe operation procedures
  - Self-extrication
  - Mayday and emergency procedures
- Salvage and overhaul
- Fire fighter rehabilitation
  - Function, causes, and need
- Fire suppression
  - Handling hose lines while advancing and operating
  - Basic fire tactics at the company level
- Preincident planning

Near Miss REPORT

Report Number: 07-904
Date: 05/09/2007

Event Description: During an NFPA 1403 compliant live burn exercise several events coincided to endanger a firefighter. The evolution was the fifth of the day, and from the point of view of fire behavior and drill sequence, was no different than any other. Accountability was being observed, a RIC was in place with a separate water supply, there was an assigned Safety Officer coordinating with the instructor and his assistant, who were lighting and maintaining the fire. There was constant staffing of the safety line to suppress the fire should the crew fail to do so appropriately. A crew was assigned to interior attack.

Before entering the structure they had time to decide to "take some heat" and to move into the burn room prior to knocking the fire down. This decision was made without the knowledge of the staff running the burn. The evolution was commenced as normal. As the crew entered the room, the rookie firefighter on the nozzle at the front of the crew felt that she was taking too much heat and wanted to exit. She had training that said, "Do not put too much water on the fire as it will cause steam burns" so she did not suppress the fire. She tried to back out, but the firefighters behind her were pushing into the room and did not allow her egress. Because she did not feel it was appropriate to suppress the fire and couldn't back out normally, she abandoned the nozzle, pivoted on her foot and departed the burn room over the top of the rest of her crew. Because she didn't stay low in the room for that instant, she was exposed to high heat during that time. She was not burned, but several items of her gear including her turnout jacket were damaged beyond repair. When this occurred, the instructor who was watching from a different doorway immediately extinguished the fire.

Lessons Learned: Several causes for the incident were identified and procedures were put in place to prevent them in the future.

The crew was "made up" of several folks from different departments with varying experience. Upon examining those experience levels after the incident, it was determined that there was not enough overall experience to leave this crew to their own judgment. Specifically, an experienced officer was needed on each crew in an evolution of this type. We changed our procedure to include a series of questions about experience from each crew and have tailored the amount of instructor involvement to the answers to these questions. The questions include fire experience and live burn experience, which members are rookie firefighters, and which crews have less experienced officers, whether acting or appointed.

In essence, by making a decision prior to entering about moving into the room prior to knocking the fire down, the crew was freelancing and prevented them from making an appropriate decision. Our safety briefing now defines such behavior as freelancing and emphasizes avoiding it.

The last change we made is to limit the instructor's movement during the evolution. We determined that allowing the instructor to use "judgment" to leave the entrance to the fire room was a causative factor and we now require a designated instructor at the door to the fire room with the fire attack crew during every evolution.

I believe that live burns are essential training and that they can be executed safely. This non-injury incident taught our department several lessons.

- Emergency medical care
  - Infection control
  - Emotional stress
  - CPR
  - Bleeding and shock
  - Burns
- Hazardous materials
  - Properties and effects
  - Recognizing and identifying
  - Proper protective equipment and decontamination

Not only does the AHJ need to make sure that the candidates have the skills, ability, and knowledge to succeed, but they must also have the right attitude. This is one of the reasons why some states require written letters of recommendation.

Fire fighter training is all about the students, not the instructors. It is not the time or place for the instructors to show off their skills. It is time for the instructors to show the students the correct method for hose placement, advancement, and stream application and to discuss what is occurring around them, while in a hostile environment.

It is imperative that the candidate understands that accountability must be continually maintained for all personnel operating on the scene. In order for the training process to run safely, the candidates must understand the fundamentals while making safety their ultimate priority. If they do not operate in such a manner during emergency responses, how do you expect them to act during a live fire training evolution? Those individuals have no business being live fire training instructors. Hopefully, if this is the case, it is discovered before the training has started.

Training for the Live Fire Training Instructor

Productive training has to include practical applications delivered in a method that allows the candidate to demonstrate increasing proficiency with less and less hands-on oversight, until the candidate is able to teach or perform the skills on his or her own proficiently and consistently under evaluation.

Fire Behavior and Structural Fire Dynamics

When looking at the fire dynamics in a structure, the candidate must understand fire behavior and all of the elements that go along with it. Live fire training instructors have to be able to effectively teach fire behavior and recognize and disclose adverse and changing conditions that could endanger participants. Reading smoke or fire behavior, understanding room and contents versus deep seated structure fires, and recognizing normal fire spread and extreme or hostile fire behavior are all skills that live fire training instructors need to be able to observe and correctly identify during suppression operations. It is very important for live fire instructors to understand terminology and fire behavior. Under the definitions section of the standard, there are many terms relating to fire behavior. It is very important for the instructor to be knowledgeable on how fuel loads will affect fire behavior. It is also important to understand the breakdown of fuels and understand consequences if too much fuel is used which could result in flameover, flashover, or backdraft if not properly ventilated. Furthermore, the importance of fire behavior is strongly realized in that both the instructor-in-charge and the safety officer have the responsibility to assess rooms for fire growth potential. If fire growth or fire behavior presents a hazard to participants the instructor-in-charge or safety officer can terminate the incident. The instructor-in-charge and safety officer are also responsible for documenting the fuel load.

Candidates need to understand that most fuels in structural, room, and contents fires start out as solid materials. As these materials are heated they go through a process called **pyrolysis**, which is the decomposition or off-gassing of a material that, once it reaches its ignition temperature, will burn openly. During this process, smoke, or the incomplete combustion of materials, is produced. This smoke contains many toxic gases, such as carbon monoxide, hydrogen chloride, hydrogen cyanide, phosgene, and chlorine. The most abundant of these gases is carbon monoxide, with a flammable range of 12.4 percent to 74 percent, and an ignition temperature of 1128°F (609°C).

Live Fire Tips

Remember from fire fighter I and II training that 1 gallon (3.785 liters) of water will absorb over 9000 BTU. This number is found by using the following formula:

212°F − 60°F = 152 BTU per pound of water
152 BTU × 8.35 (weight of 1 gallon of water)
= 1269.20 BTU

Due to the law of latent heat of vaporization, water will absorb an additional 970.3 BTU per gallon.

970.3 BTU/gallon × 8.35 (weight of 1 gallon of water in pounds) = 8102.01

Add 1269.20 + 8102.01 = 9371.21 BTU per gallon. One gallon of water will absorb a little over 9000 BTU.

Table 3-1 **Heat Release Rate of Common Materials**

Material	BTU per min
1 lb. (0.45 kg) of wood	7000–8000
cotton mattress	7969–19,922
Styrofoam	18,400
gasoline (2 sf)	18,400–22,768
gasoline (gallon)	115,000–125,000
dry Christmas tree	28,460–36,998
propane per pound	21,857
propane per gallon	91,800
polyurethane mattress	46,105–149,699
polyurethane sofa	177,590

Data from NFPA 921, 2004 edition.

Heat Release Rate

When looking at materials in today's homes, most are not made of wood, but of many different types of fuel materials, each having a different heat of combustion and **heat release rate**. Heat release rate is the amount of heat energy released over time in a fire. The scientific community uses Celsius, joules, calories, and kilowatts to measure heat release rate. Remember that 1 BTU (British Thermal Unit) per second is roughly equal to 1 kilowatt (1.055), and that 1 BTU is equal to 1055 joules or 252 calories. BTU is basically the amount, not the intensity of heat. In the fire service, BTUs are the standard unit of measure.

Table 3-1 lists some common materials and the approximate heat release rate BTU per minute. Looking at the table, it is clear why NFPA 1403 says to only use class A materials with known burning characteristics for the live fire training evolution and why other materials must be removed. A pound of wood will generate between 7000 to 8000 BTU per pound. On the other hand, the same volume of polyurethane will generate 16,000 to 20,000 BTU per pound. A pound of polyurethane will give off up to three times the heat of a pound of wood. Polyurethane's rate of burning is faster than that of wood. Further, the smoke from the polyurethane or other plastics contains greater toxins and generally more unburned fuels, increasing the dangers considerably. Knowing the heat release rates of different materials are extremely important for the safety of the participants.

Smoke

Usually, the higher the BTU, the more smoke that is produced in a fire. It is critical for candidates to understand fire behavior and to learn how to read smoke. Remember, smoke is fuel and fuels behave in relatively predictable ways. Smoke volume, smoke velocity, smoke density, and smoke color are key factors in reading smoke.

Smoke volume and color help determine how much fuel and what type is burning. Smoke velocity will help determine the pressure that is being built up, whether it is a lazy smoke (laminar), or it is agitated (turbulent) smoke **Figure 3-2**. Smoke density can help determine how much fuel is in the smoke. The denser the smoke, the more fuel there is to burn. Smoke color, from white to grey to black, is also an indicator of fire intensity, with black smoke being the hottest of the three. Lastly, the rate of change of smoke in a structure may help determine if a flashover is about to occur **Figure 3-3**. This can be observed from inside or outside, however, it is more easily read from outside the structure, where the instructors can see the whole picture.

Inside the structure, the smoke may become darker, more aggressive, and much hotter, and it may bank to the floor. Fingers of flame or rollover may not be seen, due to the density of the smoke and the height of the ceiling. It is the candidate's duty to recognize these signs as potentially dangerous conditions, and cool the upper atmosphere, or get out.

On the outside, the smoke may go from lazy to aggressive, become darker, and **vent point ignition** could occur **Figure 3-4**. Vent point ignition occurs when the smoke is at its ignition temperature, but is lacking oxygen and is too rich to burn while in the structure. The smoke receives oxygen and falls within its flammable range as it exits an opening and bursts into flame. This is a sign of impending flashover. Remember that carbon monoxide is the most prevalent gas in the smoke, with an ignition temperature of 1128°F. How rapid is the rate of change of these factors and what may be influencing them? These could be signs of flashover and must be carefully but quickly analyzed by the candidate. If a flashover occurs, it only takes seconds to injure or kill.

Figure 3-2 Agitated, turbulent smoke is indicative of a imminent flashover.

Safety Tips

A thermal imaging camera (TIC) can help the candidate to see through the smoke to get a better understanding of the temperatures at different levels in the room. A TIC can also be used by a candidate to monitor students, their locations, and their reactions to the environment in which they are in. To become proficient with their use, a candidate must be trained on the use of thermal imagining cameras to the point where using one becomes second nature.

Figure 3-3 Flashover.

Figure 3-4 Vent-point ignition occurs when smoke and unburned gases escape to the outside and then ignite.

Also, remember that the more times you burn within a structure, the more the contents and the structure are absorbing heat. This will reduce the amount of radiant heat it can absorb and lessen the time required for the material to off gas, and more importantly, lessen the time for flashover to occur.

Student Psychology: Fire Fighter Style

The candidate must have a solid understanding of student psychology and know how to modify his or her own training techniques to maintain student motivation. Whether a live fire training instructor is leading rookies or certified fire fighters into a live fire, he or she must consider both the same. One may have no live fire experience while the other may have fought many fires in his or her career. In both cases they are students in a learning environment and instructors will be just as responsible for both. Just because someone has been on the job for a long time, it cannot be assumed that they have the experience to go with it. Remember the saying, "Do they have 21 years of experience or do they have 20 one-year experiences?" There is a huge difference and the live fire training instructor candidate must understand this difference.

A brand new fire fighter is more likely to listen and do exactly as they are told, whereas a seasoned fire fighter "knows all they need to know." Do not be afraid to "un-train" a trained fire fighter of his or her poor practices. Often the experienced individual will be much more difficult to accept new or changing practices or concepts, whereas the newer member is often eager to absorb knowledge.

To a degree, an instructor's teaching style and demeanor will depend on the students. The instructor must reassure the students while keeping a close watch on their behavior and body language. As an example, a student who is continually readjusting their SCBA mask may not be comfortable wearing it. Therefore, the instructor should be close to that student when entering the burn building. A candidate must learn the subtle signals and clues of body language in order to discern which students need to be closely monitored.

Training Evolutions

In order to become live fire training instructors, candidates must demonstrate a proficiency at performing standard live fire training evolutions. To maintain a high-level of safety, interior training evolutions designed to take place inside a live fire training structure or acquired structure are first practiced in an exterior prop or drill tower. A demonstration of the training evolution is first performed and then the candidates practice repeatedly with less and less hands-on direction. When the candidates are ready to move inside of a live fire training structure or acquired structure, interior training evolutions are first practiced in a clear atmosphere and progress into a limited and then a no-visibility atmosphere. The following are standard training evolutions that candidates will practice and master:

- Advance with mock student crews to learn positioning in relation to the crew.
- Candidates will practice rotating student crews in the nozzle position to master maintaining crew accountability and protection.
- Candidates will practice withdrawing student crews while maintaining accountability and monitoring fire conditions.
- Candidates will also practice positioning and mobility in order to be able to assist the student crews in case of problems.
  - Candidates should be able to reach and touch students for communication, reassurance, or to provide direction, and ultimately, remove students if needed **Figure 3-5**.
  - Candidates should be able to reach the nozzle to adjust the pattern, open or close it, or take control **Figure 3-6**.

Live Fire Tips

Do they have 21 years of experience? Or do they have 20 one-year experiences?

Figure 3-5 The candidate needs to be able to reach and touch the students for communication or reassurance, or to provide assistance and direction.

Figure 3-6 The candidate should be able to reach the nozzle to adjust the pattern, open or close it, or take control of it.

Incident Report

Pennsylvania State Fire Academy - 2006

Figure A The Pennsylvania State Fire Academy's non-gas-fired live fire training structure.

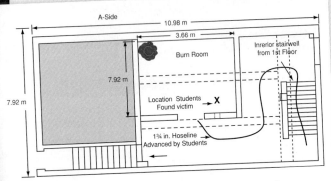

Figure B The floor plan of the basement in the non-gas-fired live fire training structure at the Pennsylvania State Fire Academy.

The Pennsylvania State Fire Academy was conducting a Suppression Instructor Development Program (ZFID), an instructor development program for live fire training. The course covers instructional components including academy polices, NFPA 1403, building fires in the permanent live fire training facility, the instructor's role in student safety, emergency and rescue operations, and more. Instructors taking the course are already Instructor I and Fire fighter II certified, and have completed the incident safety officer course. Candidates act as students, ignition officers, incident safety officers, incident commanders, and instructors, each with an academy instructor evaluating.

The 2½-story residential non-gas-fired structure was built in 1993 and is used for structural firefighting training **(Figure A)**. The building has a basement, ground floor, second floor, and an attic space. Burn rooms are located on the second floor and in the basement, where this incident occurred **(Figure B)**. The burn room in the basement has only one entrance, which can be accessed by either the interior or exterior stairwell. Only wooden pallets and excelsior are used for fuel.

As part of the ZFID program there would be six scripted (planned) evolutions in the burn building. Certain problems or emergencies were planned, and while the candidates had a basic understanding of the types of events, they did not know what event would occur during each evolution. The candidates were evaluated on their ability to handle the events.

The fire fighter who lost his life in this incident had already served as the evaluator of an instructor candidate in previous evolutions that day. On this last evolution, he was assigned to be the ignition officer for the burn room in the basement. He had over thirty years in the fire service and had been an adjunct instructor at the facility for seven years. He was wearing full protective clothing including coat, pants, helmet, hood, gloves, boots, and self-contained breathing apparatus (SCBA) with integrated personal alert safety system (PASS), and he had a portable radio.

The fatal event occurred during the last evolution. The scenario was to simulate a fire that had not been knocked down enough from the previous evolution. An instructor waiting to monitor the last crew was in the burn building. He reported good visibility with some smoke. While waiting, he heard the basement door open and saw the victim come out of the burn room, and lay on the floor pulling and moving his bunker gear about. The instructor asked the victim twice if he was ok. The victim answered, "It is hot as hell down there!" The instructor asked the victim if he was alright and if he wanted to go outside. Although he responded that he was okay, he was told to go outside. He declined, got up, and said that he was fine and would see the other instructor in the basement when the crew arrived. The victim went back down the basement steps. (*Note: It is estimated that 1½ minutes elapsed from the time the victim came up the stairwell until he went back down to the basement*).

It is believed that the victim returned to the burn room and was carrying pallets to add to the fire. In NIOSH's report and following National Institute of Standards and Technology's (NIST) experiments, they hypothesized that excessive heat in the burn room caused a catastrophic failure of the lens of the victim's SCBA face piece. The face piece failure was a result of the heat conditions within the burn room and not a manufacturer's problem. The victim dropped the pallets and fell forward. The victim then struggled on the ground and crawled towards the exit.

As the next crew approached the burn room while advancing a hose line, they reportedly heard the victim moaning. They found him struggling on the floor in the right corner of the burn room. One of the candidates declared a Mayday by radio and the candidate serving as the incident commander deployed the RIC. There was initial confusion whether this was part of the scenario or an actual emergency. The crew that found him immediately removed him outside and emergency care was initiated. He was transported initially to a local hospital and then transferred by EMS helicopter to a regional trauma/burn unit, where he passed away two days later.

This was the first fatality of an instructor in a permanent live fire training prop, and only the second in such a facility. The fire fighters did not report the PASS device activating, as the victim was still moving. In this case, the PASS device did sound once removed from the victim and due to heat damage it could not be shut off, continuing to sound its alarm two days later.

Post-Incident Analysis — Pennsylvania State Fire Academy

NFPA 1403 Noncompliant

- No noncompliance issues with NFPA 1403

NFPA 1403 Compliant

- Structure in good condition (6.2.2)
- Instructors had experience in the fire service and with live fire training (6.5.1)
- RIC crew ready
- An ambulance staffed by two EMTs on scene (6.4.11)

Other Contributing Factors

- Instructor working alone
- Repeated entries by instructor
- No monitoring of interior temperatures
- No method to rapidly ventilate building
- Basement with no "at grade" exit

- While monitoring and encouraging students, candidates must be prepared to take action to guard students from harm Figure 3-7.
- Candidates will need to practice remaining on the side of the hose away from the wall, as to not get pinned against it Figure 3-8.
- Candidates will review practical evolutions in evacuations and missing students.
- Candidates will practice using thermal imaging cameras (TIC) for monitoring students' location and performance. Candidates will also learn to read heat signatures and fire spread indicators.
- Under live conditions, candidates must understand how to shield themselves from radiant heat and how to protect themselves during interior training Figure 3-9. Candidates must be able to demonstrate the proper methods of training students to use features to protect themselves from hostile fires. These methods include using doors, corners, and furniture, and staying low on staircases.

Figure 3-9 Candidates need to know how to protect themselves from radiant heat when necessary.

- Candidates will practice reviewing heat saturation inside personal protective equipment (PPE). This type of heat saturation most often occurs in the upper areas of the body where the PPE is pressed against the body, such as the upper arms, shoulders, and upper back. Moving about to reinstitute the air pockets helps.

Accountability, Safety, and Practice for When Things Go Wrong

The candidate must be able to maintain the integrity of the crew and have accountability for them throughout the entire evolution, from arriving at the burn site, initial setup, burning, overhaul, demobilization, and leaving the burn site. All participants must understand that crew accountability is in place for one reason: safety. It must not be compromised for any reason.

A live fire training instructor must keep track of his or her students, just like a company officer keeps track of his or her crew. To prepare candidates a **personnel accountability report (PAR)** is conducted outside of the structure ensuring all participants are present, properly geared up, and ready to go. A head count must be done prior to entry, once inside, and then also upon exiting. Any time a student cannot be accounted for, it must be reported immediately. If this happens, all participants must exit the structure as the evolution has now gone from a live fire training evolution to a Mayday situation, or one where a fire fighter is missing, lost, or in need of immediate assistance. In any case, the live fire training evolution needs to be stopped and the unaccounted-for fire fighter must be found.

In addition to conducting practice PAR drills, candidates may practice the following scenarios without fire:
- The crew is having difficulty advancing the hose during line advancement. Candidates must determine if the hose stuck on a corner and if somebody needs to be positioned at the corner to feed the hose.
- The fire stream is insufficient due to the wrong nozzle selection on variable gallon nozzles. Candidates need to recognize the problem and correct it.

Figure 3-7 It may be necessary for the candidate to take action to keep the student out of harm's way.

Figure 3-8 The candidate needs to remain on the side of the hose away from the wall as to not get pinned against it.

Figure 3-10 Once a fire has gotten into the attic or other concealed spaces, it is no longer a training session.

Safety Tips

1. Vent point ignition is being communicated from outside: This is an indicator that flashover is possible and the instructor must be aware of this and cool the atmosphere.
2. Heavy black smoke with intense heat is suddenly banking down: This could be an indicator of an impending flashover.
3. The fire has gotten into the cockloft, attic, or some other concealed space: This is not a live burn evolution anymore.
4. Sudden changes in heat, flames, or visibility could also be an indicator of flashover.

- Kinks in the hose result in poor quality of fire stream. Candidates need to recognize the problem and correct it.
- A fire fighter or instructor gets pinned against the wall. Candidates need to determine what actions need to be taken.
- Contact is lost with one of the students. Candidates need to locate the student rapidly and safely.
- Students become bunched up and cannot move. Candidates need to separate them.
- A PASS device goes off. Candidates need to check for a problem, and if there is a problem, make a Mayday call.
- A fire ensues above and behind an attack line crew. Candidate need to advise the attack crew and get it under control.
- The room and contents fire has grown out of control. Candidates need to determine what actions need to be taken.
- The fire has spread into the cockloft, attic, or some other concealed space **Figure 3-10**. Candidates need to determine what actions need to be taken.
- An emergency requires quick action. Candidates must demonstrate an understanding of the difference between emergency traffic, like a power line down on the Charlie side, and a Mayday call, like a fire fighter lost, fallen, trapped, or low on air.

Florida's Live Fire Training Laws: Instructor Requirements

Following two fire training incidents, one in an acquired structure which claimed the lives of a fire fighter and a lieutenant and one in a permanent maritime prop where a fire fighter recruit was killed, Florida now requires compliance with NFPA 1403 by law. Florida has also set requirements beyond NFPA 1403 and for instructor training and certification. At the time of the three deaths, NFPA 1403 was already adopted in the Florida Administrative Code by the Department of Environmental Protection, but it was not enforced. The new requirements encompassed recruit and in-service training and included training in permanent live fire training structures, acquired structures, and exterior props.

Florida's program now has two primary levels: the Live Fire Training Instructor (LFTI) and the Live Fire Adjunct Trainer (LFAT). The LFAT program is sixteen hours and is required for all primary positions except the instructor-in-charge and the safety officer. Anyone who will be leading crews inside a structure or in the immediate proximity of the exterior are required to complete the program. The course includes crew management, accountability, safety and emergency operations, training methods, training objectives, position assignments and responsibilities, basic compliance to NFPA 1403, and other state requirements. The course is roughly 50 percent practical, with actual drills for working with interior crews in hallways, instructor positioning with regard to students, and performing the practical evolutions expected during live fire evolutions.

The instructor-in-charge and the safety officer must be certified Live Fire Training Instructors. This forty-hour course includes practical evolutions and a comprehensive review of NFPA 1403 and Florida's requirements. It is designed for the leadership roles in organizing and conducting training. The program must be run through one of the state's certified training centers, and personnel attending must already be state-certified as a Fire Instructor I and Fire Fighter II. The program includes permanent live fire training structures and acquired structures, and exterior props with considerable time spent on the safe operation of props and training evolutions. It includes how to build and monitor class A fires, training- and suppression-related physiology, and crew accountability and safety. An important component of this class is that individuals must develop preburn plans and conduct live fire training evolutions as the instructor-in-charge or safety officer of a functional assignment.

Live Fire Training Instructor II and Live Fire Master Instructor are the two other tiers that exist in Florida. The Live Fire Training Instructor II can instruct the Live Fire Training Instructor and Live Fire Adjunct Trainer courses to certified instructors under the auspices of a certified training center.

To gain certification, the candidate must show competency and responsibility to teach all positions and must also complete a position task book under the direction of a Live Fire Master Trainer. The Live Fire Master Instructor is a Live Fire Training Instructor II appointed by the certified fire training center director, to oversee the training of instructors and to oversee live fire training operations for a center. This individual also serves as the contact point between the state and the certified instructors.

Pennsylvania Suppression Instructor Development Program (ZFID)

Pennsylvania also has an excellent program that utilizes formal training with mentoring and observed evaluation. The program is limited to training people to operate in their permanent prop at their facility, but utilizes a good format and process.

Pennsylvania's ZFID program started in 2000 for the purpose of evaluating instructor candidates to teach in the structural live fire training prop at the Pennsylvania State Fire Academy (PSFA) only.

Prerequisites for candidates to attend the program include at least eight years of fire service experience, three letters of recommendation, 150 hours of "structural track" training, completion of the National Fire Academy's incident safety officer class, and certification as a Fire Fighter II and Fire Instructor I.

This 40-hour program reviews specific PSFA policies, including the intent of NFPA 1403. The program features presentations by instructor candidates, a review of the PSFA facilities, and a walk-through of the burn building.

The candidates act as students, ignition officers, safety officers, incident commanders or instructors-in-charge, and live fire training instructors. As candidates perform their roles in each position, they are shadowed and evaluated by an adjunct instructor. Candidates are also shown the safest methods of stacking pallets, spreading excelsior, and starting the fire.

The first four evolutions are "wet" (charged) hose lines only, with no live fire. For the next twelve evolutions, live fire is utilized. Each evolution is slow and methodical so that each candidate can acquaint themselves with the evolution process. The fire sets are smaller and fires are less intense than those in their structural firefighting classes, since candidates have previously proven their skills as interior fire fighters. Fires occur in the basement and division 2 of the "residential" live fire training structure.

ISFSI Live Fire Trainer Program

The International Society of Fire Service Instructors (ISFSI) has developed a live fire training instructor program and credentialing system that allows local fire departments and training centers to learn how to be compliant to the NFPA 1403 standard and how to prepare candidates in a structured, documented way **Figure 3-11**.

One program is offered for both permanent live fire training structures, and another for acquired structures. Both classes have considerable practical aspects and require the completion

Figure 3-11 The International Society of Fire Service Instructors.

of a position task book by an approved evaluator that has the candidates demonstrate their skills and verifies their abilities. Upon successful completion of the training program and the position task book, the candidate gains credentials from ISFSI.

The National Wildfire Coordinating Group

Whichever system a department or training center chooses to utilize, it needs to be documented and thorough. One of the arguably best organized systems is the use of task books by the forest fire service. This system of training may be used by your program. Task books are developed with the knowledge to be exhibited, or performance objectives, and the skills to be demonstrated. Trainers work with the candidates in classroom, practical situations, or combinations of methods to secure the knowledge in an organized and documented manner, with less coaching as the candidate gains proficiency until the candidate takes the lead position and is monitored for success.

Through the National Wildfire Coordinating Group (NWCG), a performance-based qualification system is in place that provides standards for minimum training and qualifications for a responder from different agencies and areas for national mobilization. This allows for the different responders to operate cohesively together on incidents and sets a level of expected proficiency that has been demonstrated.

Position task books (PTB) have been developed for each credentialed position. The PTBs contain all of the critical tasks that are required to perform the assignment and provide a means to document a candidate's ability to perform each task. During training, the trainer or coach instructs and mentors the candidate via the classroom, practical evolutions, and other methods. Much of the standard structural fire service training utilizes similar guides, based on NFPA standards, where the instructors sign off on the successful demonstration of practical skills by a fire fighter trainee in an evaluative setting. Similarly, the NWCG qualifications require the successful completion of a training program, including the associated practical skills and testing, but it does not guarantee an individual will be qualified or credentialed to perform in a specific position.

Once the training has been completed to the trainer's satisfaction, and the PTB has been completed, to become credentialed the candidate must perform "tasks pertaining to tactical decision making and safety require position performance on a wildland fire. Remaining tasks may be evaluated through other means such as a simulation, or emergency or nonemergency incidents/events." This means that there is an additional step and actual application of the knowledge and skills on real incidents, or sometimes in simulated exercises, is required. Even the quality of experience is carefully evaluated.

Strict criteria are in place for trainer and coach qualifications. He or she must be fully credentialed in the role they are training and have completed the necessary instruction and supervised teaching. The roles of the trainer and evaluator are different, and must remain separate for the integrity of the process.

NFPA 1403, *Standard on Live Fire Training Evolutions, 2012 Edition,* Chapter 3 Definitions, New Terms:

Backdraft A deflagration resulting from the sudden introduction of air into a confined space containing oxygen deficient products of incomplete combustion.

Combustible Capable of burning, generally in air under normal conditions of ambient temperature and pressure, unless otherwise specified. Combustion can occur in cases where an oxidizer other than oxygen in air is present (e.g., chlorine, fluorine, or chemicals containing oxygen in their structure).

Conduction Heat transfer to another body or within a body by direct contact.

Convection Heat transfer by circulation within a medium such as a gas or a liquid.

Deflagration Propagation of a combustion zone at a velocity that is less than the speed of sound in the unreacted medium.

Flameover (rollover) The condition in which unburned fuel (pyrolysate) from the originating fire has accumulated in the ceiling layer to a sufficient concentration (i.e., at or above the lower flammable limit) that it ignites and burns. Flameover can occur without ignition of or prior to the ignition of other fuels separate from the origin.

Flashover A transition phase in the development of a compartment fire in which surfaces exposed to thermal radiation reach ignition temperature more or less simultaneously and fire spreads rapidly throughout the space, resulting in full room involvement or total involvement of the compartment or enclosed space.

Fuel load The total quantity of combustible contents of a building, space, or fire area, including interior finish and trim, expressed in heat units or the equivalent weight in wood.

Pyrolysate Product of decomposition through heat; a product of a chemical change caused by heating.

Radiation Heat transfer by way of electromagnetic energy.

Ventilation-controlled fire A fire in which the heat release rate or growth is controlled by the amount of air available to the fire.

Wrap-Up

Chief Concepts

- It is critical to assign competent and qualified instructors to supervise students in live fire training evolutions.
- It is up to the authority having jurisdiction to determine the qualifications of live fire training instructors, ensuring they have more than just an understanding of basic fire behavior. If there is any doubt to the qualifications and competence of an instructor, they should not be used for live fire evolutions.
- NFPA 1001, *Standard for Fire Fighter Professional Qualifications* is the minimum qualification that an instructor must possess. The instructors should also meet the standards set forth in NFPA 1041, *Fire Service Instructor Professional Qualifications*, especially in high hazard training.
- Candidates need formal training to become live fire trainers, preferably mentored and supervised until competency and effectiveness is gained.

Hot Terms

Instructor An individual qualified by the authority having jurisdiction (AHJ) to deliver fire fighter training, who has the training and experience to supervise students during live fire training evolutions.

Heat release rate The amount of heat energy released by a material over time in a fire.

Personnel accountability report (PAR) A verification by person in charge of each crew or team that all of their assigned personnel are accounted for.

Pyrolysis The process of decomposition of a material into other molecules when it is heated.

Vent point ignition Smoke is at or above its ignition temperature and is lacking oxygen. The smoke will ignite as it exits the opening and falls within the flammable range.

References

Berardinelli, S. 2008. *Fire fighter fatality investigation report F2005-31*. Washington, DC: National Institute for Occupational Safety and Health.

Dunkle, T. 2001. *The day the impossible happened*. Philadelphia, PA: Pennsylvania Office of the State Fire Commissioner.

Hankins, J. 1997. *Hazards associated with the storage of flexible polyurethane foam in warehouse situation*. Factory Mutual Research.

National Wildfire Coordinating Group. 2009. *Wildland fire qualification system guide—PMS 301*. http://www.nwcg.gov/pms/docs/pms310-1.pdf.

Live Fire Training Instructor in Action

You have been newly assigned as the live fire training instructor for an upcoming training involving your department and a neighboring department. How can you use the knowledge you have acquired in your training classes to help you answer the following questions about live fire training?

1. What standard is followed during a live fire evolution?
   A. NFPA 1001
   B. NFPA 1021
   C. NFPA 1041
   D. NFPA 1043

2. What standard should the authority having jurisdiction (AHJ) follow when determining if an instructor is qualified to supervise students in live fire training?
   A. NFPA 1001
   B. NFPA 1021
   C. NFPA 1041
   D. NFPA 1043

3. What is one of the most important components that the instructor must have a complete understanding of when working in a live fire environment?
   A. Building construction
   B. Fire behavior
   C. Ventilation
   D. Water supply

4. Smoke volume can help determine:
   A. the pressure that is being built up.
   B. how much fuel is in the smoke.
   C. how much fuel is burning.
   D. how hot the smoke is.

5. While inside an acquired structure, the smoke becomes darker and more aggressive, and high heat is banking to the floor. These are signs of:
   A. improper ventilation.
   B. backdraft.
   C. flashover.
   D. rollover.

6. Which of the following reasons describes why crew accountability is performed?
   A. To make sure everybody shares in the work
   B. For the safety of all participants
   C. So the accountability board matches
   D. None of the above.

7. Which of the following would be a reason to use a Mayday call?
   A. Fire fighter lost
   B. Fire fighter fell or trapped
   C. Fire fighter low on air
   D. All of the above.

Fire Fighter Physiology

CHAPTER 4

NFPA 1403 Standard

4.6.5 The instructor-in-charge shall provide for rest and rehabilitation of participants operating at the scene, including any necessary medical evaluation and treatment, food and fluid replenishment, and relief from climatic conditions. (See Annex D.) [pp 64–65, 70–71, Appendix D]

Additional NFPA Standards

NFPA 1582 *Standard on Comprehensive Occupational Medical Programs for Fire Departments.*

NFPA 1583 *Standard on Health-Related Fitness Programs for Fire Department Members*

NFPA 1584 *Standard on the Rehabilitation Process for Members During Emergency Operations and Training Exercises*

Knowledge Objectives

After studying this chapter, you will be able to:
- Describe the cardiovascular and thermal responses to firefighting.
- Describe how firefighting activity and turnout gear affects cardiovascular and thermal strain.
- Describe how fitness levels affect cardiovascular and thermal strain.
- Describe how to prevent injuries, such as heat illness, during firefighting activity and training.
- Describe the warning signs for heat illnesses that may occur in firefighting activity and training.
- Describe how high aerobic fitness is necessary to safely and effectively perform firefighting activity.
- Describe the risk factors for cardiovascular disease.
- Describe the importance of modifiable risk factors for cardiovascular disease, and describe ways to decrease those factors.
- Describe the goals of on-site rehabilitation.

Skills Objectives

There are no skills objectives for this chapter.

You Are the Live Fire Training Instructor

You have just been given an assignment to be the instructor-in-charge of a live fire training evolution. Three local fire departments will be present to take part in the exercises. The training will be conducted at a regional training facility and is scheduled for early July. There are many aspects that need to be planned for before this training can be a success. As the training date is approaching, you ask yourself the following questions:

1. What are the primary physiological threats that I must consider in order to ensure the safety of the instructors and students?
2. How will a well-run rehabilitation sector help me address the safety of my students?

Introduction

Firefighting is an inherently dangerous and physically demanding occupation. Every day, fire fighters are faced with potentially life-threatening challenges, including burn injury, asphyxiation, collapse, and entrapment. Less appreciated, however, are the physiological consequences that threaten fire fighters. The combination of strenuous work, heavy and encapsulating personal protective equipment (PPE), hot and hostile fire conditions, and high adrenaline levels leads to significant levels of cardiovascular and thermal strain during firefighting.

Live fire training is necessary to prepare fire fighters for the dangerous and challenging environment in which they are expected to perform. Live fire training places fire fighters in high heat environments with live fire conditions. These intense training sessions can create high levels of cardiovascular and thermal strain, and thus increase the risk for heat-related injuries and sudden cardiac events. Live fire training instructors, in particular, are often exposed to severe heat conditions for prolonged periods of time. The exposure to such severe conditions creates a challenging and potentially dangerous situation for both instructors and students. It is important that the risks to fire trainees and fire instructors are managed effectively, so that they do not outweigh the benefits. The focus of this chapter is to help you understand the following:

1. How the body responds to firefighting activity
2. The dangers associated with these responses
3. Effective strategies to minimize the risk of heat-related or cardiac-related events

Cardiovascular and Thermal Strain of Firefighting

As a result of the combination of heavy work, heavy and encapsulating PPE, and hot and hostile environmental conditions, firefighting creates significant physiological strain, affecting nearly every system of the body. The greatest risks to the fire fighter come from the ensuing cardiovascular and thermal strain. The NFPA 1403, *Standard on Live Fire Training Evolutions*, mandates that live fire training evolutions be done in such a way as to minimize the exposure to health and safety hazards for the fire fighters involved. In terms of human physiology, this means recognizing the thermal strains and cardiovascular strains associated with firefighting and pursuing measures to minimize those risks.

Research studies have found that strenuous firefighting activities lead to near maximal heart rates that remain high for extended periods of time during fire suppression activities. As early as the mid-1970s, there were studies reporting the heart rates of on-duty fire fighters. The researchers documented one individual with a heart rate over 188 beats per minute for a 15-minute period compared to the average adult's heart that beats between 60 and 100 times per minute. These findings have been confirmed and extended by recent studies focusing on the physiological responses to firefighting.

Researchers at the University of Illinois have conducted a series of studies to document the effects of firefighting activities on the cardiovascular system. The research studies were conducted in a live fire training structure in moderate temperate conditions (≈ 120°F–150°F [49°C–65°C]). Fire fighters completed three repeated trials (≈ 7 minutes) of simulated firefighting activity lasting a total of 21 minutes with a 10-minute rest between the second and third trial. The study resulted in average heart rates of approximately 190 beats per minute, which is equal to the age-predicted maximal heart rates for these fire fighters **Figure 4-1**. **Stroke volume**, which is the amount of blood pumped with each beat of the heart, decreased by about one-third by the end of the third trial. This finding is especially troubling because the amount of blood being pumped out of the heart is decreased at the very time it is most needed. During firefighting, blood needs to be delivered to many areas of the body, such as the following:

- The working muscle to support contraction
- The skin to cool the body
- The heart to support increased work associated with elevated heart rate
- Other vital organs including the brain

Core temperature, or the temperature of the central part of the body, rapidly increases during firefighting activity. In one study, fire fighters who completed three repeated bouts of firefighting activity, each bout lasting approximately 6 minutes with a 10-minute break between the second and third trial, experienced an average increase in body temperature of 2.5°F (1.7°C). A British study that measured the core temperature of fire instructors reported an average increase in core temperature of 1.8°F (1°C) over the 40 minutes of live fire training evolutions. Mean core temperature increased to 101.3°F (38.5°C), and 8 of the 26 instructors had core temperatures over 102°F (38.8°C) after just 40 minutes of data collection. In a companion study, researchers investigated the ability of live fire training instructors to perform a simulated rescue after 40 minutes of a live fire training evolution. Ten minutes after the live fire training evolution, fire instructors were required to drag a 187 lb (84.8 kg) dummy 98 ft (29.9 m). In six out of seven trials they were able to do so. The authors concluded that *most* of the fire instructors were able to perform a rescue task after live fire training evolutions, but they were approaching their physical limit.

In addition to elevated body temperature, firefighting also causes profuse sweating. The body sweats in an attempt to cool itself through a process called evaporative cooling. Unfortunately for fire fighters, PPE creates a warm, moist, and stagnant air layer next to the skin, which severely limits the evaporation of sweat. Without the evaporation of the sweat, the body becomes unable to utilize evaporative cooling as a method for temperature control. Profuse sweating can also decrease plasma volume, placing additional strain on the cardiovascular system and further impairing thermoregulation. Researchers found a 15 percent reduction in plasma volume following approximately 18 minutes of strenuous simulated firefighting activity. Sweat loss of 2.8 (1.3 kg) pounds per hour have been reported during exercise in a hot environment while wearing PPE. Sweat loss is a major concern during live fire training because individuals may be engaged in training over a relatively long time period. Furthermore, training on consecutive days may present an additional challenge if individuals do not fully rehydrate in between training days.

Live Fire Tips

Firefighting training causes significant cardiovascular and thermal disruption, including the following:
- Increased core temperature
- Profuse sweat loss
- Near maximal heart rate
- Decreased stroke volume
- Decreased plasma volume

These physiological changes can lead to life-threatening pathological conditions including heatstroke and sudden cardiac events in extreme cases.

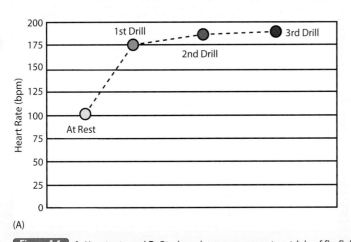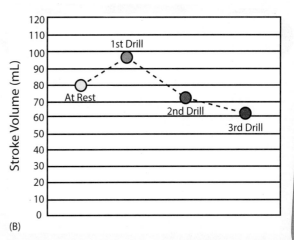

Figure 4-1 **A.** Heart rate and **B.** Stroke volume responses to 3 trials of firefighting drills. Each drill consisted of ≈ 7 minutes of firefighting activity.

Factors Affecting Cardiovascular and Thermal Strain

The magnitude of the thermal and cardiovascular strain experienced by a fire fighter depends on several interrelated environmental and personal factors **Figure 4-2**.

Environmental Conditions

Environmental conditions are a major contributing factor to the physiologic stresses of firefighting. In most cases, this means that the heat of the fire contributes to the total heat stress of the individual. Performing any task in a hot, oppressive environment creates greater physiological strain than performing the same task in moderate conditions. The environment in which fire fighters and live fire training instructors conduct live fire training varies greatly from a moderate level of heat to conditions so severe that they can only be tolerated for a brief period of time before damage to PPE can occur and physical injury can result.

An instructor must consider the potential for burn- and heat-related injuries when working in different conditions. Heat exposure can be in the form of the ambient environment or direct exposure to flames or another heat source. Heat exposure poses different challenges to the human body depending on the absolute temperature, or heat flux, and the duration of exposure. High ambient conditions combined with direct radiant heating will increase the effects on the human body. The effects of heat and the length of time the human body can sustain such conditions will depend on the intensity of the heat source and the duration of the exposure.

Provided below is a chart that identifies routine, hazardous, extreme, and critical exposures that fire fighters may encounter **Figure 4-3**. Individuals should not be exposed to critical conditions in training settings. It is important to keep in mind that even exposure to "routine" conditions presents a considerable challenge to the human cardiovascular and thermoregulatory systems.

Work Performed

On the fireground and in training scenarios, there are a number of different tasks performed, including throwing ladders, climbing stairs with heavy loads, performing a search, advancing a line, forcing open a door, and overhauling a room. The individual tasks, and the intensity at which they are performed, have a strong influence on physiological responses to firefighting. Clearly, the work being performed has a major influence on the physiological response of the body—the more strenuous the work and the longer the duration, the greater the cardiovascular response. Physical work creates metabolic heat, which leads to an increase in body temperature, and thus adds additional cardiovascular strain.

Personal Protective Equipment

PPE is necessary to protect fire fighters from burn and inhalation injuries. Because of its weight and restrictive properties, PPE adds to a fire fighter's work. PPE also interferes with heat loss, because of encapsulation. Thus, PPE adds to the cardiovascular and thermal strain associated with firefighting. In one laboratory study, fire fighters were asked to walk in full PPE with a self-contained breathing apparatus (SCBA). The fire fighters reached heart rates around 178 beats per minute, which is, on average, 55 beats per minute higher than walking in a station uniform alone.

Wearing PPE is absolutely essential to safety in training evolutions, however, it does add to the physiological stresses that the live fire training fighter encounters. Thus, a live fire training instructor should be aware of the amount of time that trainees are in full gear. During debriefing periods, breaks, and rehabilitation, gear should be removed to allow for recovery.

To combat the dangers of heat illness, it is important that instructors and students remove their PPE and cool down. The

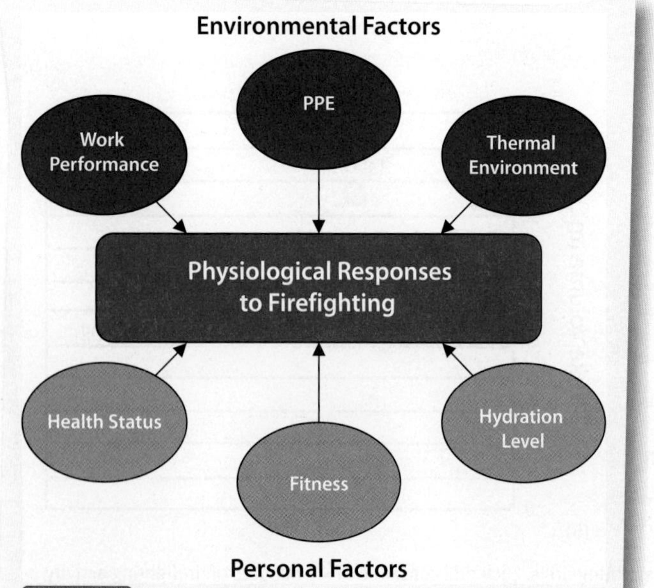

Figure 4-2 Factors affecting physiological responses to firefighting.

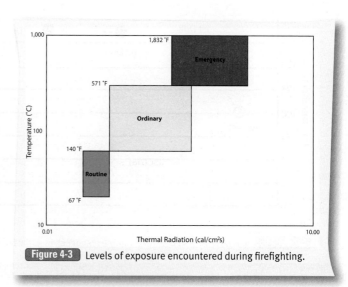

Figure 4-3 Levels of exposure encountered during firefighting.

Safety Tips

PPE is essential to protect fire instructors and students from burn injuries and smoke inhalation and should be worn in compliance with manufacturer's recommendations and local policies during live fire training. However, the live fire training instructor must also understand how the weight and insulative properties of PPE add to heat stress and cardiovascular strain. PPE should be doffed, when appropriate, to allow body temperature to decrease.

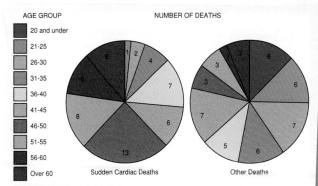

Figure 4-4 Training fatalities by age and cause of death (1996–2005). Data from *"U.S. Firefighter Deaths Related to Training, 1996–2005"* by Rita F. Fahy, National Fire Protection Association, 2006.

air outside is much cooler than the air inside of a fire fighter's gear, which will cause sweat to evaporate as gear is removed, giving the fire fighter a feeling of being cool. However, body temperature does not decrease as quickly as the mind perceives it to. In fact, body temperature often continues to rise, even after firefighting activity has ended. Therefore, to decrease body temperature, it is necessary to doff the bunker coat and pants.

Individual Characteristics

A fire fighter's age, gender, and body size all affect the physiological response to firefighting activity. In general, the risk of heart attack while performing firefighting work increases as the age of the fire fighter increases. The number of fatalities due to sudden cardiac events versus other causes during training over a 10-year period is shown in **Figure 4-4**. Note that as age increases, so does the percentage of fatalities due to sudden cardiac events. What may be surprising is the number of cardiac events that occur in young fire fighters. Instructors must not become complacent when working with young fire fighters.

Excess body fat creates additional cardiovascular strain on a fire fighter by adding to the metabolic work that must be done to move his or her body mass. Excess body fat also increases the thermal strain by providing insulation, impeding the range of motion and mobility, and interfering with heat dissipation.

Medical Conditions

Firefighting activities can be extremely strenuous. Fire fighters should be in good health to be able to operate safely on the fireground or in a training situation, because of the combination of heavy work, severe heat, and mental stress. A live fire training instructor who is not medically cleared for firefighting presents a risk not only to himself, but also to the students under his or her command. A fire fighter with a preexisting cardiovascular disease is more susceptible to cardiac events on the fireground because of the additional cardiovascular strain associated with firefighting. In fact, a retrospective analysis of sudden cardiac events over a 10-year period found that 75 percent of fire fighters who suffered fatal cardiac events had preexisting cardiovascular conditions. High blood pressure, high cholesterol, and obesity are factors that greatly increase the risk of cardiovascular mortality and should be taken very seriously. Diabetes and pre-diabetes (impaired glucose tolerance) are especially dangerous because they are associated with several cardiovascular risk factors.

NFPA 1582, *Standard on Comprehensive Occupational Medical Programs for Fire Departments*, identifies the 13 essential tasks of firefighting and provides guidelines for the medical clearance of fire fighters. In order to ensure that fire instructors can safely engage in live fire training and meet the responsibilities they have to their students, they should have a medical examination that meets the NFPA 1582 standard.

Fitness Level

Firefighting is physically demanding work and a high level of fitness is necessary to safely and successfully perform the job duties. A fit fire fighter can perform the same amount of work with less cardiovascular and thermal strain than a less fit fire fighter. A fit fire fighter also has a greater energy reserve needed to perform more work. Fire fighters need to possess a high degree of muscular fitness (muscle strength and endurance) and cardiovascular fitness (aerobic fitness). Cardiovascular fitness is especially important because it does the following:

- Increases the efficiency of the heart
- Improves work capacity
- Increases plasma volume
- Improves thermal tolerance
- Decreases tendency of blood to clot
- Enhances the ability of blood vessels to dilate to allow more blood to be supplied to muscles

The components of a health-related fitness program are outlined in NFPA 1583, *Standard on Health-Related Fitness Programs for Fire Department Members*. As an instructor, you have an obligation to pay careful attention to your own fitness level because of the following:

- Your fitness level directly affects your risk of heat-related illness or sudden cardiac events during training.
- Your fitness level affects your ability to perform strenuous firefighting activity, including the possibility of needing to rescue a participant.
- You are a role model for your students and your approach to fitness will affect your students' attitude toward fitness and their respect for you.

Hydration Status

Firefighting leads to a large amount of sweat loss due to the heavy work, hot conditions, and the impermeable nature of the PPE. During strenuous work in hot conditions, or in protective clothing, humans can lose more than two quarts of sweat per hour. This sweat loss contributes to a decrease in plasma volume, adds additional strain on the cardiovascular system, and decreases the ability to work in the heat, also known as **thermal tolerance**. Instructors need to pay careful attention to their own hydration status as well as ensuring their students consume adequate fluid during training drills.

Live fire training instructors need to be well hydrated before they begin training drills and need to be vigilant about consuming water or sports drinks throughout the training evolutions. Likewise, instructors must be diligent in providing opportunities for students to properly rehydrate during the training drills. Thirst is an inadequate mechanism of ensuring that adequate fluids are consumed to avoid dehydration. Therefore, the instructor should have structured breaks during which rehydration is emphasized.

The human body is comprised of approximately 60 percent water, and proper hydration is necessary for biological function. Water is constantly lost from the body by urinating, sweating, and breathing, so it is critical to replenish the supply. A sedentary person requires approximately 30 ml (1 oz.) of water per kilogram (1 kg = 2.2 lbs) of body weight per day. Thus, a man who weights around 154 lbs (70 kg) needs to consume about 2100 mL (2 qts) of water to remain hydrated when he is not engaged in strenuous work or sweating profusely. Exercise or work under hot conditions, such as those routinely encountered during firefighting, can dramatically increase the need for fluid. Humans routinely lose between 1–2 liters (1 L = 1.06 quarts) of sweat per hour when working in hot and humid environments. Given that training often occurs over a period of several hours, this high rate at which sweat is lost can lead to severe dehydration. Fire fighters should be mindful to consume adequate fluids to replace what is lost during normal fluid turnover plus what is lost with sweating. Furthermore, fluid ingestion should be done primarily with water or a low-calorie sports beverage. Caffeinated beverages such as coffee, tea, and soda can act as a diuretic and increase water loss, exacerbating rather than helping dehydration. An additional concern is that many drinks, including soda and sweetened tea, include a lot of sugar, which is often associated with excess body weight, another factor that increases heat stress.

Given the detrimental effects of dehydration, and evidence that many fire fighters are dehydrated even before beginning work, it is a good idea to monitor your hydration status. The easiest way to do this is the use of a simple urine chart **Figure 4-5**. These charts can be placed in rest rooms to serve as a reminder to instructors and students that urine should be a light color and should not have a strong odor. Urine color is affected by many factors, including medication and diet, so it is wise to have individuals follow a hydration program that ensures that they are well hydrated. From there, one should be able to determine the proper color of their urine. Students and instructors should be reminded regularly to consume fluid that ensures light urine or urine that is consistent with their urine color in a well hydrated state.

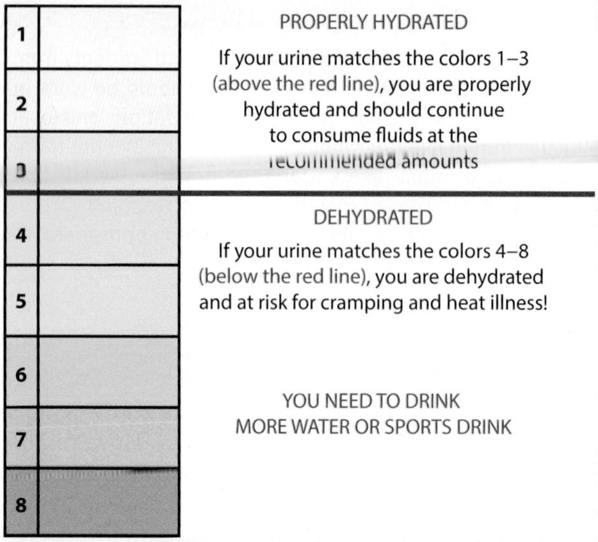

Figure 4-5 Urine color and hydration.

PROPERLY HYDRATED
If your urine matches the colors 1–3 (above the red line), you are properly hydrated and should continue to consume fluids at the recommended amounts

DEHYDRATED
If your urine matches the colors 4–8 (below the red line), you are dehydrated and at risk for cramping and heat illness!

YOU NEED TO DRINK MORE WATER OR SPORTS DRINK

Safety Tips

Dehydration is dangerous, it increases the risk of heat illness and sudden cardiac events, and unfortunately it is very common among students undergoing live fire training. Several major reports in the fire service have documented that students commonly show up for training dehydrated. It is important that efforts be made to get students and instructors to begin training in a well-hydrated state. To prevent excessive dehydration, instructors must provide frequent water breaks and should have policies that mandate fluid ingestion, including water and sports drinks. Instructors and participants should be encouraged to drink more fluids the day before attending live fire training to prehydrate.

Heat Emergencies

Heat illnesses can be fatal, and there are far too many instances of fire fighters suffering fatal heatstroke during training. Instructors have an obligation to be alert to the dangers of heat illness and to help decrease the number of fatal heat illnesses that occur in training. The United States Fire Administrations report, *Emergency Incident Rehabilitation*, released in 2008, documents seven case studies in which fire fighters have died due to heat illness following work or training. These tragic case studies should be read by every fire instructor because of the lessons that can be learned from them, the first of which is that fatal heatstroke can occur during training. Another important and related lesson is that heatstroke can affect young and old fire fighters alike. In fact, young recruits are especially vulnerable because of their extreme motivation to succeed in the training environment. Under these conditions, many young fire fighters push themselves far beyond what is physiologically safe.

Working in protective clothing with high ambient temperatures presents a serious challenge to the thermoregulatory and cardiovascular systems. If the cardiovascular system cannot meet the simultaneous demands of supplying adequate blood to the muscles and maintaining thermal balance, heat illness may ensue. Heat illness covers a spectrum of disorders from heat rash to life-threatening heatstroke. Heat-related illnesses occur because the body is unable to maintain thermal balance.

Thermal Balance

Normally, the body regulates its internal body temperature within a narrow range, despite wide variations in environmental temperatures. The process by which the body regulates body temperature is called **thermoregulation**. The human body typically regulates its temperature around 98.6°F (37°C). It is important to maintain a temperature within approximately 1.5°F (0.7°C) of this temperature because changes in body temperature dramatically affect biological function. Greater temperature changes can alter chemical reactions and, ultimately, directly damage body tissue.

Body temperature results from a balance between heat gain and heat loss. Heat can be gained from the environment, when the ambient temperature is higher than the body temperature, and from metabolic heat produced by the body as a result of muscular work. Heat can be exchanged from the body through four processes: radiation, conduction, convection, and evaporation. The extent of heat gain or loss through these processes depends on environmental conditions such as ambient temperature, relative humidity, and wind speed.

The effectiveness of heat exchange between an individual and the environment is affected by five factors:

1. Thermal gradient
2. Relative humidity
3. Air movement
4. Degree of direct sunlight
5. Clothing worn

The greater the difference between two temperatures, known as the **thermal gradient**, the greater the heat loss is from the warmer of the two. Typically, the body is warmer than the environment, so heat moves down its thermal gradient to the environment. More heat is lost in cooler environments because the thermal gradient is greater. In firefighting situations, heat may be gained by the body.

The body's use of evaporative cooling techniques are decreased in high humidity conditions because the air already contains an abundance of water vapor. Evaporative cooling is largely determined by the relative humidity; on humid days evaporative cooling is limited. Although a fire fighter may sweat profusely when humidity is high, the body doesn't properly cool down because the sweat does not evaporate as effectively.

Air movement increases convective heat loss from the skin to the environment. Thus, on windy days more heat is lost from the body. Conversely, when a fire fighter's skin is covered by heavy PPE, heat loss is minimized.

Direct sunlight can add considerably to the radiant heat load of an individual, just as shade or cloud cover can often provide significant relief from heat. Measures should be taken to avoid having students spend prolonged periods of time in PPE in direct sunlight on hot days, as it can cause dangerous increases in body temperature.

When the body is in thermal balance, the amount of heat lost equals the amount of heat produced, and body temperature remains constant. However, when heat produced (and absorbed) exceeds heat loss, an increase in body temperature occurs. This increase in body temperature places considerable strain on the cardiovascular system, hastens fatigue, promotes sweating and loss of plasma volume, and may lead to serious heat illnesses.

Heat Illness

Heat illness includes a spectrum of disorders, resulting specifically from the combined stresses of exertion and high heat situations. Heat illness affects many systems of the body and can cause elevated core body temperature and impaired thermoregulation. It varies greatly in severity from heat rash, sunburn, and heat cramps, to heat exhaustion and heatstroke. One of the greatest challenges in dealing with heat illnesses is distinguishing among disorders because they frequently overlap and can evolve into different forms over time. A summary of the causes, signs, symptoms, and treatments for heat illnesses is provided in **Table 4-1**. This section describes the most common and serious heat illnesses encountered during live fire training, namely heat cramps, heat exhaustion, and heatstroke. Symptoms of heat illness may be nonspecific, particularly in the early stages. As heat illnesses progress, so does the severity of signs and symptoms.

Heat Cramps

Heat cramps are an acute disorder consisting of brief, recurrent, and excruciating pain in the voluntary muscles of the legs, arms, or abdomen. Typically, the muscles have recently been engaged in intense physical work and are fatigued. Heat cramps may result from a fluid/electrolyte imbalance. Individuals who sweat profusely or who lose large quantities of sodium (salt) may be more susceptible than others.

Table 4-1 **Heat Illness Classifications**

Classification	Cause	Signs and Symptoms	Treatment	Prevention
Heat Rash (also called prickly heat or miliaria)	Hot, humid environment; clogged sweat glands	Red, bumpy rash with severe itching	Change into dry clothes and avoid hot environments. Rinse skin with cool water.	Wash regularly to keep skin clean and dry.
Sunburn	Too much exposure to the sun	Red, painful, or blistering and peeling skin	If the skin blisters, seek medical aid. Use skin lotions (avoid topical anesthetics) and work in the shade.	Work in the shade; cover skin with clothing; apply skin lotions with a skin protection factor of at least 15. Fair people have greater risk.
Heat Cramps	Heavy sweating depletes the body of salt, which cannot be replaced just by drinking water.	Painful cramps in arms, legs, or stomach that occur suddenly at work or following work. Heat cramps are serious because they can be a warning sign of other more dangerous heat-induced illnesses.	Move to a cool area; loosen clothing and drink commercial fluid replacements (sports drinks). If the cramps are sever or don't go away, seek medical aid.	Reduce activity level and/or heat exposure. Drink fluids regularly. Workers should check each other to help spot the symptoms that often precede heatstroke.
Heat Exhaustion	Fluid loss, inadequate salt, and the cooling system begins to break down	Heavy sweating; elevated body temperature; weak pulse; normal or low blood pressure; person is tired and weak or faint, has nausea and vomiting, is very thirsty, or is panting or breathing rapidly; vision can be blurred.	**GET MEDICAL AID.** This condition can lead to heatstroke, which can kill. Remove gear, move the person to a cool shaded area; loosen or remove excess clothing; provide sports drink. Use active cooling (forearm immersion, misting fans, or cold towels) to lower core body temperature.	Reduce activity level and/or heat exposure. Drink fluids (water and sports drink) regularly to compensate for sweat loss. Monitor participants frequently to assess symptoms.
Heatstroke	Elevation of body temperature due to a breakdown of thermoregulatory mechanisms. Caused by depletion of salt and water reserves. Heatstroke can develop suddenly or can follow from heat exhaustion.	Body temperature is very high and any of the following: the person is weak, confused, upset, or acting strangely; has hot, dry, red skin; a fast pulse; headache or dizziness. In later stages, a person can pass out and have convulsions.	**IMMEDIATELEY TRANSPORT TO A MEDICAL FACILITY.** This is a life-threatening emergency. If transport is delayed, immediately immerse body in cold water.	Reduce activity level and/or heat exposure. Drink fluids (water and sports drink) regularly to compensate for sweat loss. Monitor participants frequently to assess symptoms.

Modified from NFPA 1584.

Heat Exhaustion

Heat exhaustion is characterized by elevated body temperature and decreased pumping capacity of the heart (cardiac output). Heat exhaustion is caused by severe fluid loss and the inability of the cardiovascular system to properly compensate for the demands of blood flow to both muscles and the skin. A fire fighter with heat exhaustion will likely have some combination of the following signs and symptoms:

- Rapid and weak pulse
- Fatigue and weakness
- Profuse sweating
- Confusion or disorientation
- Dizziness or fainting

If untreated, heat exhaustion can lead to moderate to severe multiple-organ damage. If heat exhaustion is not aggressively treated, it can lead to heatstroke. A fire fighter who is suspected of suffering from heat exhaustion should discontinue activity and doff his or her gear. The fire fighter should receive a medical evaluation and be cooled and hydrated. Depending on the medical evaluation and the severity of the symptoms, the fire fighter may need to be transported to the hospital for further medical care.

Heatstroke

Heatstroke is a life-threatening emergency that requires immediate medical care and rapid cooling. Heatstroke involves multiple-organ damage and is often characterized by central nervous system dysfunction, including disorientation, seizure, and cardiovascular collapse. Signs that may indicate heatstroke include the following:

- Elevated skin and core temperature
- Rapid heart rate
- Vomiting
- Diarrhea
- Hallucinations
- Seizures
- Coma

Heatstroke involves a complete failure of the thermoregulatory mechanisms. If heatstroke is suspected, medical treatment must be summoned immediately and the individual must be cooled as quickly as possible using water, ice, or a fan.

■ Risk Factors for Heat Illness

A risk factor is something that increases a person's chances of developing an infection or disease. Many factors affect a fire fighter's risk for heat illness. Many of these factors, namely fitness level, excess body fat, hydration level, and medical conditions, were discussed earlier and are also summarized in **Table 4-2**. However, given their importance in increasing the risk of serious heat illness, they warrant review. A fire fighter who is unfit, overweight, or has preexisting medical conditions is at a much greater risk for suffering from heat-related illnesses. Furthermore, there are a large number of medical conditions, both serious and seemingly minor, that can affect the body's fluid balance and thermoregulatory capacity. Two factors that are often overlooked are gastrointestinal problems and skin irritations. Gastrointestinal problems include anything that causes loose bowel movements and loss of fluids and electrolytes. Skin irritations can interfere with the skin's role in sweat loss or heat dissipation. There are dozens of prescription and over-the-counter medications, ergogenic aids, supplements, and illegal drugs that can increase the risk of dehydration or heat-related injuries. Commonly used medications that can increase the risk of heat illness include, antihistamines, antidepressants, ephedrine, diuretics, and beta-blockers.

Individuals who have previously suffered heat illness appear to be more likely to suffer subsequent bouts. It is unclear if the same factors that originally caused the heat illness are responsible for the subsequent bouts, or if a damaged thermoregulatory system caused by heat illness makes repeat bouts more likely. Regardless of the mechanisms, it is prudent to ask students if they have suffered from heat illness in the past. If they have, additional caution is warranted.

Surprisingly, data from the military shows convincingly that one of the greatest predictors of heat illness is the weather conditions on the day *prior* to training. While a live fire training instructor will likely be aware that he or she must consider environmental conditions on the day of training, it is also imperative to consider the previous day's conditions and activities. This information is particularly relevant to students undergoing several subsequent days of training, but is also applicable to students who are training after a duty day or after strenuous work on an off-duty day. Alcohol use, both chronic and recent, increases the risk for heat illness. Efforts should be made to ensure that students attending special training sessions limit alcohol consumption on the day preceding training.

Table 4-2 Risk Factors for Heat Illness

Risk Factor	Description
Poor physical fitness	Poor fitness decreases thermal tolerance and increases risk for heat illness.
Excessive body weight	Excess body fat decreases heat dissipation, increases the amount of metabolic work that must be done to move the body, and increases the risk for heat illness.
Dehydration and salt depletion	Dehydration increases the risk of heat illness.
Chronic disease	Disease such as diabetes mellitus, cardiovascular disease, and congestive heart failure increase risk for heat illness.
Minor illness	Fire fighters who were already suffering from a minor illness, inflammation, or fever have an increased chance of heat injury due to a previously compromised autoimmune system; subjects with some form of gastroenteritis are particularly at risk because they may already be dehydrated and have salt and mineral imbalances within their bodies.
Skin problems	Skin irritations such as rashes, prickly heat, sunburn, burns, psoriasis, eczema, and poison ivy increases susceptibility to heat illness.
Medications, both prescription and nonprescription	Many medications affect the body's hydration level, ability to process fluids, and other bodily functions relative to dealing with the heat. Diuretics reduce fluids in the body.
Age	Individuals over 40 years of age, even those in relatively good physical condition, have an increased risk for heat illness. However, young people are susceptible too, especially when highly motivated.
High level of motivation	People who are highly motivated and committed to performing given tasks at all costs may overlook the signs of heat illness and increase their chance of overextending themselves. Fire fighters engaged in highly charged emergency scene operations or high-stakes training are clearly at increased risk of heat illness.
Prior heat exposure	The body needs more time to fully recover from exposure to heat and stress multiple times per day. Repeated exposure to heat day after day can also increase the risk of heat emergencies.
Prior heat injury	A heat illness may result in expedited heat illness and increases the likelihood of subsequent heat illness.
Recent alcohol use	Recent alcohol use will increase the likelihood of dehydration and heat illness and can impair the person's judgment.
Genetics	People who have genetic mutations, such as cystic fibrosis and malignant hyperthermia, have increased risk of heat illness.
Lack of heat acclimatization	Unaccustomed activity in the heat increases risk of heat illness.

Repeated exposure to work in the heat leads to acclimatization. Lack of acclimatization, or unaccustomed work in the heat, increases the risk for heat illnesses. A live fire training instructor should be mindful of the extent to which students are used to working in the heat. Extra caution should be taken when training is undertaken with individuals who have not been working in the heat. This often occurs during the first hot days of summer.

Prevention of Heat Illness

Although instructors must be able to recognize and respond to heat illness, it is far preferable to prevent heat-related illness by using sound judgment and observing basic recommendations.

NFPA 1403 states that the training session shall be curtailed, postponed, or canceled, as necessary, to reduce the risk of injury or illness caused by extreme weather conditions. Thus, one of the primary responsibilities of the instructor-in-charge is to determine if training should continue or be modified in some way to prevent undue risk of heat-related injuries. The annex to NFPA 1403 provides useful information to be considered when making these determinations. A HUMIDEX chart, used by the Canadian government, is very useful because it combines air temperature and humidity into one number that reflects the perceived temperature **Table 4-3**. This chart relies on measurements that are readily available (temperature and humidity) and provides guidelines to the level of caution associated with various perceived temperatures. This chart is very similar to the Heat Stress Index used by the US government. An example of Heat Stress Preventive Guidelines based on the HUMIDEX is provided in **Figure 4-6**.

It is critically important that fire instructors be mindful of the need for periodic breaks. Fire fighters need to have time to cool down, rehydrate, and allow their bodies to recover. In some instances students are asked to assist with evolutions in some way when they rotate out of specific training tasks. This may limit the fire fighter's ability to adequately recover.

It is absolutely essential that aggressive rehydration programs be implemented during training drills. Clean, cold water and sports drinks should be readily available and drinking should be encouraged.

Live Fire Tips

In order to prevent heat illness, fire instructors should do the following:
- Be alert to conditions that increase an individual's risk for heat illness.
- Consider curtailing, postponing, or canceling training when severe weather increases the risk of heat-related injuries.
- Provide frequent breaks for those involved.
- Encourage rehydration among all participants.
- Establish effective incident-scene rehabilitation.
- Monitor students for signs and symptoms of heat illness.

Table 4-3 HUMIDEX Chart

Air Temp °C \ % RH	20	25	30	35	40	45	50	55	60	65	70	75	80	85	90	95	100
43	47	49	51	54	56												
42	46	48	50	52	54	56											
41	44	46	48	50	52	54	56										
40	43	45	47	49	51	54	56	58									
39	41	43	45	47	49	51	52	54	56	58	59						
38	40	42	43	45	47	49	50	52	54	56	57	59					
37	38	40	42	43	45	47	49	50	52	54	55	57	58				
36	37	38	40	42	43	45	47	48	50	51	53	55	56	58	59		
35	35	37	39	40	42	43	45	46	48	49	51	53	54	56	57	58	
34	34	35	37	39	40	42	43	45	46	47	49	50	52	53	55	56	58
33	33	34	36	37	38	40	41	43	44	46	47	48	50	51	52	54	55
32		33	34	35	37	38	40	41	42	44	45	46	48	49	50	51	53
31		31	33	34	35	37	38	39	40	42	43	44	46	47	48	49	50
30		30	31	32	34	35	36	37	39	40	41	42	43	45	46	47	48
29		29	30	31	32	33	34	35	36	37	38	39	40	41	42	43	46
28			28	30	31	32	33	34	35	36	37	38	39	40	41	42	44
27			27	28	29	30	31	32	33	34	35	36	37	38	39	40	41
26			26	27	28	29	30	31	32	33	34	35	36	36	37	38	39
25			25	26	27	28	29	30	31	32	33	34	35	35	36	37	
24			24	25	26	27	28	28	29	30	31	32	33	33	34	35	
23			23	24	24	25	26	27	28	28	29	30	31	31	32	33	
22			22	22	23	24	25	25	26	27	27	28	29	29	30	31	
21					21	22	22	23	24	24	25	26	27	27	28	29	
	20	25	30	35	40	45	50	55	60	65	70	75	80	85	90	95	100

Legend:
- 54+ Extreme Danger
- 46–53 Danger
- 40–45 Extreme Caution
- 30–39 Caution

NOTE: The supervisor in charge of the facility or workplace is responsible for implementing these heat stress prevention guidelines. He or she shall determine the level of PPE required.

ALERT LEVEL 1 HUMIDEX 30–39

Caution: Fatigue and faintness are possible with physical activity or prolonged exposure. The most likely at risk this level are those performing heavy work for extended periods of time.

(1) Encourage all staff to increase water intake, be observant of signs and symptoms of heat stress (both in themselves and co-workers), and implement precautionary measures to prevent heat-related disorders.
(2) Addtional rest breaks should be introduced to reduce heavy exertion and allow for cooling.

ALERT LEVEL 2 HUMIDEX 40–45

Extreme Caution: Heat cramps, heat exhaustion, or sunstroke are possible with physical activity or prolonged exposure. An increased number of workers are at risk at this level, including those performing moderate physical exertions.

(1) Postpone optional activities, or reschedule them to cooler times of the day when possible.
(2) Introduce additional rest breaks for workers performing moderate work.
(3) Further reduce heavy work.
(4) Consider cessation of nonessential operations involving heavy physical activity.
(5) Minimize using bunker suits whenever possible.

Note: All training activities are considered nonessential except recruit training. The following safety precautions shall be implemented when conducting training within this Humidex range.

(1) Limit recruit live fire burns to occur between 0700–1200 hours only.
(2) Provide increased rest breaks for all work loads.
(3) Limit heavy work to less than 15 minutes per hour.
(4) Initiate rehabilitation at the beginning of the incident.
(5) Use active cooling where possible (forearm immersion, misting fan, and/or air conditioning).

ALERT LEVEL 3 HUMIDEX 46–53

Danger: Heat cramps, heat exhaustion, or sunstroke are likely. Heat stroke is possible with physical activity or prolonged exposure. Even those performing light work might require addtional rest breaks.

(1) Significantly reduce both heavy moderate work.
(2) Minimize using bunker suits whenever possible.
(3) Consider cesstion of nonessential operations involving moderate physical activity in this environment.
(4) Cease all nonessential operations involving heavy physical activity.

Note: All outdoor training activities are considered nonessential and shall be rescheduled or cancelled.

ALERT LEVEL 4 HUMIDEX 54 or greater—EMERGENCY HEAT ALERT

Extreme Danger: Heat stroke or sunstroke imminent, danger of DEATH. This is an extremely dangerous humid level, where all individuals are at risk of heat-related disorders, regardless of the workload.

(1) Minimize using bunker suits whenever possible.
(2) Discontinue all nonessential services performed in this environment.
(3) For essential operations, do the following:
    (a) Provide increased rest breaks for all workloads.
    (b) Limit heavy work to less than 15 minutes per hour.
    (c) Initiate rehabilitation at the beginning of the incident.
    (d) Use active cooling (forearm immersion, misting fan, and air conditioning).
    (e) Call for additional crews to facilitate rehabilitation.

Note: All outdoor training shall be rescheduled or cancelled.

Figure 4-6 Heat stress prevention guidelines.

Incident scene rehabilitation is a mechanism to ensure that fire fighters have scheduled rest breaks that provide an opportunity to cool down, rehydrate, and recover. Incident scene rehabilitation should also provide for medical monitoring of fire fighters. The goals of incident scene rehabilitation and some recommendations for effectiveness are provided later in this chapter.

Incident Report

Poinciana, Florida - 2002

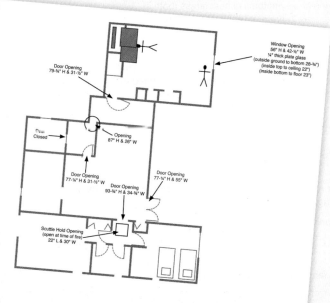

Figure A Poinciana floor plan.

Figure B Two consecutive offset turns and this narrow hallway made it difficult for fire fighters to maneuver.

Just 10 months after the incident in Lairdsville, and less than three weeks after the sentencing of the Lairdsville chief officer, a live fire training session trapped two fire fighters in Poinciana, Florida, near Kissimmee.

The 1600 ft^2 (148.6 m^2) cement block house had three bedrooms, one of which was converted from a one-car garage with the large door removed, the wall blocked up, and a window installed **(Figure A)**. The room had block walls, one door, and a fixed ¼" (6.4-mm) thick, commercial-grade glass window. The exit from the room was through two consecutive offset turns, through a 26" (66 cm) opening, then through two small rooms, and out another door into the dining room **(Figure B)**. A breezeway between the old garage and the main house had also been enclosed. Unlike many acquired structures, this house was in good condition and was part of an entire neighborhood being razed for a new campus. Several area fire departments were involved in a series of training events, and the structures available included other houses, as well as a motel.

All of the instructors were experienced in the fire service and had previously worked together on training fires. Safety crews were not briefed at the beginning of this exercise because of their past involvement. The training officer walked all of the participants through the structure and explained the safety aspects and goals of the training evolution, which were to conduct a search and rescue with an actual fire burning and find a mannequin in bunker gear hidden somewhere in the house and remove it. The first search team consisted of an experienced lieutenant and a trained recruit (state certified Fire Fighter II), who were to search without a hose line. Two suppression teams with 1¾" (44 mm) hose lines were in position in the house with four interior safety officers broken into two teams.

The fire was started in the converted bedroom near the only doorway to the fire room **(Figure C)**. Two piles, practically vertical with pallets, wood scraps, and hay, were almost adjacent to each other, one inside and one outside the open closet. After the fire started, and with the instructor-in-charge's agreement, a foam mattress from one of the other bedrooms was added to the pile.

The search and rescue (SAR) team entered the structure at the front (east) door with a suppression team following. The suppression team stopped in the small room located between the dining room and the bedroom, where the training fire was located. The search team continued into the burning bedroom, encountering deteriorating conditions with high heat and no visibility due to heavy smoke.

On the exterior, a second suppression team waited at the front doorway, with two more fire fighters assigned as the rapid intervention crew (RIC), and a third, uncharged 1¾" (44 mm) hose line also was available outside. One fire fighter was stationed on the exterior waiting for orders to ventilate.

With near zero visibility and increasing heat conditions, two of the interior safety officers monitoring the activities of the fire room area later stated that they heard the lieutenant of the search team ask his partner if he had searched the entire room. They heard the answer "yes." Shortly thereafter, one of the safety officers yelled into the fire room asking if the search team was out. Although someone answered "yes," no one knows who replied. The interior safety officer assumed that he had missed the search team's exit from the fire room, as there were several fire fighters present by the dining room. He began to search the rest of the structure in an attempt to find them.

Figure C Interior of fire room in Poinciana.

The instructor-in-charge ordered the front window of the fire room to be broken. After the window was broken out, heavy black smoke followed, which ignited very quickly with flames forcibly venting from the window. The suppression crew closest to the fire room applied water in short bursts, but increasing heat and steam forced the safety officers and the suppression crew to back out. Both safety officers were forced to exit the structure, after receiving burn injuries.

The second suppression team was ordered to replace the first team and engage the fire. The second team and an interior safety officer entered the fire room and extinguished the fire.

During this time the instructor-in-charge called the missing fire fighters several times and received no answer.

While the second suppression team was overhauling the fire area, they found a body in fire fighter bunker gear facedown on the floor. Both of the suppression team fire fighters initially thought it was the rescue mannequin, not realizing it was the lieutenant of the SAR team.

After no radio response from the SAR team, the instructor-in-charge ordered a personnel accountability report (PAR) and ordered the RIC to enter the structure and find the SAR team.

The suppression team that found the lieutenant, dragged him to the front window of the fire room, and removed him to the outside. The missing fire fighter was also found inside that window, and he too was removed to the outside. The entire event, from the time the SAR crew entered to when the first of the two fire fighters were located, was under 14 minutes. The two fire fighters died despite working in teams with experienced instructors, two interior staffed hose lines, two interior safety teams, an RIC with its own hose line, a participant walk-though, and other safeguards in place.

Incident Report

Poinciana, Florida - 2002

The National Institute for Occupational Safety and Health (NIOSH) investigated the fire. Two separate investigations were conducted by the state fire marshal, one for criminal violations, and an administrative investigation of state training codes. No criminal charges were filed, however the findings included the following:

- "All of the participants stated…they did not have any concerns regarding the conditions of the fire inside the structure and it appeared to them as normal fire behavior."
- Although NFPA 1403 was already required in state code, it was under the environmental laws and so the law enforcement department of the state fire marshal's office did not have the authority to enforce it.
- The fire was started in a room with too much fuel for the size of the room, including a foam mattress that was added after ignition. Flashover was precipitated by too high a fuel load and inadequate ventilation.
- National Institute of Standards and Technology (NIST) determined that the mattress was contributory to considerable smoke production; however in testing, the room flashed over in roughly the same time without the mattress present.

Post-Incident Analysis — Poinciana, Florida

NFPA 1403 Noncompliant

- No written preburn plan prepared (4.2.25.2)
- Accountability was not maintained at the point of entry to the fire room (4.5.6)
- Safety crews did not have specific assignments or emergency procedures (4.2.25)
- Fire started near the only doorway in and out of the fire room (4.4.16)
- Excessive fuel loading (4.3.4)
- No communications plan in place (4.4.9)

NFPA 1403 Compliant

- Experienced instructors had worked together on training fires before (4.5.1)
- Primary and secondary suppression crews inside with hose lines (4.4.6.2)
- A walk-through of the structure was conducted (4.2.25.4)
- Crews were alerted of the possibility of a victim (4.2.25.1)
- Structure in good condition (4.2.1)

Other Issues

- Radio communications less than optimal
- Egress limited by offset turns, a 26" (66 cm) opening, and multiple rooms and turns

Cardiac Emergencies

Sudden cardiac events are the leading cause of line-of-duty deaths among fire fighters. Each year approximately 45 to 50 percent of fire fighter deaths are caused by sudden cardiac events. In addition to the fatalities, approximately 800 to 1000 fire fighters suffer nonfatal heart attacks while on-duty.

Heart attacks occur when a blood clot blocks a coronary artery and deprives the heart muscle of oxygen. Heart attacks usually occur in individuals with underlying cardiovascular diseases. Sudden cardiac death results from a sudden loss of function of the heart, usually because of a lethal arrhythmia. A heart attack can lead to a sudden cardiac death, and often in the fire service these terms are used interchangeably when citing fatality statistics. This chapter uses sudden cardiac events as an inclusive term that includes all cardiac events that occur during or shortly after firefighting.

Firefighting appears to act as a trigger for sudden cardiac events in vulnerable individuals. A study by a Harvard research team found that fire fighters spend only a small percentage of their time engaged in firefighting activity, but a large percentage of their fatalities occur within a short time after firefighting activity. Based on a national sample of fire fighters, it was estimated that fire fighters spend approximately 1 percent of their time engaged in firefighting activity. Based on the time spent in firefighting and the percentage of deaths that occur shortly thereafter, the researchers calculated that a fire fighter is approximately 100 times more likely to suffer a sudden cardiac event after firefighting than during station activity.

Sudden cardiac events also occur during fire fighter training. In fire fighters above the age of 35 years, approximately 68 percent of fire fighter fatalities during training were the result of sudden cardiac events. A live fire training instructor must understand the factors that increase the risk of sudden cardiac events, and do everything in their control to lessen these risks.

Risk Factors for Developing Cardiovascular Disease

Risk factors can be modifiable or nonmodifiable. A modifiable risk factor is one that can be minimized, for example through diet, exercise, or modified personal habits. Several nonmodifiable and modifiable risk factors for cardiovascular disease are presented in **Table 4-4**. Age is an example of a nonmodifiable risk factor. For instance, males are more likely to suffer cardiovascular disease at a younger age than females; therefore, being over 45 years old is considered a risk factor for males and being over 55 years old is a risk factor for females. Another example of a nonmodifiable risk factor is heredity or family history.

Modifiable risk factors are important because altering them can directly influence a person's likelihood of developing infection or disease, and, in this case, cardiovascular disease. The more risk factors an individual has, the greater the likelihood that he or she will suffer from cardiovascular disease. The good news is that with information and support, and encouragement by coworkers and family, most fire fighters can reduce their risk for cardiovascular disease by following reasonable guidelines for healthy living.

Decreasing Risk Factors for Cardiovascular Disease

Cardiovascular disease is a major threat to the health and safety of fire fighters. In order to stay healthy, and address the risk factors for developing cardiovascular disease, a fire fighter should adopt healthy lifestyle habits. In short, to reduce the risk of suffering a heart attack or stroke, it is imperative that fire fighters attempt the following:

- Do not smoke, or quit smoking.
- Follow a regimen of moderate aerobic exercise.
- Eat a balanced diet, avoiding excess saturated fats and simple sugars.
- Maintain a normal body weight.

The risk factors that are influenced by each of the above recommendations are highlighted in **Table 4-5**. Notice the benefit of physical activity in eliminating or favorably impacting five of the six modifiable risk factors.

Prevention of Cardiac Emergencies

Live fire training often involves performing strenuous muscular work while wearing heavy personal protective clothing under hot and hostile conditions. This level of exertion can trigger a sudden cardiac event in individuals with underlying cardiovascular disorders. In order to minimize the risk of cardiac emergencies, fire instructors should do the following:

- Be aware of the risk factors for cardiovascular disease.
- Work to ensure that students have medical clearance to engage in structural firefighting.

Table 4-4 Risk Factors for Developing Cardiovascular Disease

Nonmodifiable Risk Factors	Major Modifiable Risk Factors
Age	Cigarette smoking
Heredity	Hypertension
Race	Cholesterol-lipid fractions
Gender	Obesity
	Diabetes mellitus
	Physical inactivity

Data from the American Health Association

Table 4-5 Recommendations for Decreasing CV Risk Factors

Recommendations	Risk Factor Influenced
Exercise moderately	Decreased blood pressure
	Improved lipid (chol) profile
	Decreased body fat
	Diabetes (improved glucose tolerance)
	Eliminates physical inactivity
Eat a balanced diet	Improved lipid (chol) profile
	Decreased body weight
	Diabetes (improved glucose tolerance)
	May decrease blood pressure
Quit smoking	Smoking

- Be aware of signs and symptoms of a heart attack.
- Establish incident scene rehabilitation that provides medical monitoring of personnel.

Heart attacks can occur suddenly and be associated with intense symptoms, or they may have a gradual onset. A fire fighter complaining of severe chest pain and radiating pain in the arm is showing signs of a possible heart attack. In less obvious cases, a fire fighter may convey vague information such as needing to rest or not feeling well. It is very important that information of this sort be taken seriously, and that a fire fighter who makes such complaints receives medical attention. In the case of a heart attack, the sooner treatment is initiated, the greater the chance of survival. Perhaps the most important initiative that can be taken to prevent cardiac emergencies is to have an effective incident scene rehabilitation area staffed with dedicated, trained emergency medical personnel responsible for monitoring the medical conditions of participants.

Incident Scene Rehabilitation

<u>Incident scene rehabilitation</u> for live fire training is defined as an intervention designed to lessen the physical, physiological, and emotional stresses of firefighting with a goal of improving performance and decreasing the likelihood of on-scene injury or death. Effective incident scene rehabilitation can mitigate the effects of some of the detrimental physiological problems of firefighting by providing rest, rehydration, and cooling of fire fighters. Incident scene rehabilitation can also help identify medical problems early, and may prevent potentially serious consequences by providing appropriate medical monitoring.

Goals and Purpose of Incident Scene Rehabilitation

NFPA 1584, *Standard on the Rehabilitation Process for Members During Emergency Operations and Training Exercises*, is the standard for on-scene rehabilitation during training and during emergency situations. The standard indicates that rehabilitation operations are required whenever training exercises pose a safety or health risk to members involved in the training. The primary goal of rehabilitation is to ensure that the physical and mental condition of members operating at the scene of an emergency or a training exercise do not deteriorate to a point that affects the safety of others or that jeopardizes the integrity of the operation. In order to achieve this, the NFPA standard identifies nine guidelines that need to be addressed in establishing incident scene rehabilitation:

1. Relief from climatic conditions
2. Rest and recovery
3. Active and/or passive cooling or warming as needed
4. Rehydration (fluid replacement)
5. Calorie and electrolyte replacement
6. Medical monitoring
7. Emergency medical services (EMS) treatment in accordance with local protocol
8. Member accountability
9. Release

These guidelines play a vital role in improving fire fighter performance and decreasing injuries and death during live fire training, especially those related to heat illnesses and cardiac emergencies. Perhaps most important in addressing heat illnesses and cardiac emergencies are the provisions for cooling, rehydration, electrolyte replacement, and medical monitoring.

Cooling

Students and instructors who have been involved in live fire training or have been working in personal protective clothing need to implement cooling techniques to lower their elevated body temperatures. Personal protective clothing should be removed and passive or active cooling techniques should be implemented in conjunction with fluid consumption. The decision on whether to use passive or active cooling techniques depends largely on the magnitude of the individual's heat stress. Active cooling techniques, such as forearm immersion, misting fans, and towels soaked in cool water, provide for greater cooling and a more rapid decline in body temperature. For these reasons, active cooling techniques are preferred in conditions in which individuals are exposed to high heat. Given that some individuals may become overheated in almost any live fire training environment, it is advisable to have these techniques readily available.

After exiting a live fire training structure, fire fighters will often claim that doffing their helmet, hood, gloves, and bunker coat is enough to make them feel cooler. Doffing protective gear does bring the skin in closer contact with cooler air than the training environment, and it provides a perception of the body "cooling down." However, as long as much of the body is still encapsulated in the protective gear, there is limited cooling of the internal body and core temperature is likely to remain elevated. The goal of cooling should be to decrease core body temperature, not just to make the fire fighter feel cooler.

In cases of suspected heat exhaustion or heatstroke, it is critical that aggressive cooling be initiated. Heatstroke is a life-threatening emergency, and if it is suspected, the individual's gear should be removed and the fire fighter should be transported to the hospital immediately. Measures to cool down the fire fighter such as ice packs and cold towels should be taken en route.

Rehydration and Electrolyte Replacement

NFPA 1584 indicates that individuals undergoing scheduled events, such as training, should maintain proper hydration and should prehydrate with an additional 16 oz. (500 ml) of fluid within two hours prior to the event. During rehabilitation, fire fighters should consume enough fluid to ensure that they are not thirsty. Fluid intake, beyond what satisfies thirst, may be beneficial given that thirst is an inadequate method for determining the body's hydration status.

Calorie and electrolyte replacement is also required for training that lasts more than three hours or where fire fighters are working for more than one hour. Commonly available sports drinks are a convenient way to provide carbohydrate and electrolyte replacement during training.

Again, it is worth emphasizing the importance of beginning firefighting activities in a well-hydrated state. Rehydration is necessary after strenuous work that causes profuse sweating. However, the fire fighter is put at risk when he or she is not adequately hydrated at the onset of an activity. It is nearly impossible to catch up with the body's fluid needs if starting at a deficit.

Medical Monitoring

Medical monitoring is a process of monitoring fire fighters who are at risk of adverse health events, such as heat illness or cardiac-related events. Medical monitoring includes a combination of obtaining vital signs, assessing individuals, and applying clinical judgment. Although it is important to monitor vital signs, this alone is insufficient. Vital signs that should be measured by trained emergency medical services (EMS) personnel include heart rate, respiratory rate, blood pressure, and body temperature.

Heart rates can easily reach maximal levels during live fire training. However, heart rate values should return to normal levels during rehabilitation. After 20 minutes of rehabilitation, a heart rate over 100 beats per minute is abnormal and may warrant further medical evaluation. Likewise, respiration rates should return to normal values, about 12–20 breaths per minute following 20 minutes of rehabilitation.

Blood pressure varies considerably among individuals, and the blood pressure response to training can vary widely as well. Systolic blood pressure measurements above 160 mm Hg, or diastolic blood pressure measurements above 100 mm Hg, after 20 minutes of rehabilitation are abnormal and should be continuously monitored or evaluated medically. Hypotension, or low blood pressure, can also be a problem following firefighting activity. The combination of elevated body temperature and sweat loss can cause a low blood pressure. This becomes problematic if pressure is not high enough to adequately profuse body tissues, especially the heart and the brain.

Body temperature is another important variable because elevated body temperature causes several detrimental responses. Body temperature is very difficult to accurately measure in rehabilitation settings. Oral temperatures are often artificially low because of heavy breathing, and the ingestion of cold fluid decreases the temperature of the oral cavity. Handheld tympanic (ear) thermometers are convenient but they often underestimate deep body temperature and they are greatly affected by environmental temperatures. Thus, tympanic temperatures may be low even when an individual has a high body temperature. If temperature is assessed in the field, the value must be interpreted based on the entire clinical picture. Given the difficulties in obtaining accurate measurements, it is certainly possible that a fire fighter with a low measured temperature is still suffering from heat stress. In other words, thermometers deployed in the field may detect individuals with high temperatures that need further examination, but they may also fail to detect individuals who have high temperatures. Thus it is critical that individuals providing rehabilitation consider the possibility of heat stress even in the absence of high measured temperatures.

It is important to monitor vital signs because they provide clues to how a fire fighter is responding to rehabilitation. However, the information gained from monitoring vital signs should not replace the judgment of EMS personnel. A fire fighter should be evaluated by EMS personnel if he or she is not responding in appropriate ways; is acting confused, weak, or overly fatigued; does not appear normal in color or alertness; or is excessively sweating. A fire fighter who is experiencing chest pain, shortness of breath, dizziness, or nausea should be immediately transported to a medical facility for evaluation. Abnormal vital signs may be an indication to transport a fire fighter to a medical facility, but a normal set of vital signs in conjunction with an abnormal clinical presentation should not be reason enough *not to* seek further evaluation or to transport to a medical facility.

Wrap-Up

Chief Concepts

- Firefighting is dangerous and physically demanding, therefore there are many strains placed on the fire fighter's cardiovascular and thermoregulatory systems.
- Personal protective equipment, high stress, and physical strain all cause stress on the body's regulatory systems.
- Instructors need to consider the amount of heat exposure received by themselves and their students.
- Several interrelated factors affect the physiological responses to firefighting, including environmental conditions, PPE, work performed, individual characteristics, fitness levels, medical conditions, and hydration status.
- Instructors should be aware of the amount of time that participants are in full turnout gear, and should allow time for them to remove their gear and recover.
- Personal fitness (aerobic and muscular) is important to safely and successfully complete firefighting activity.
- Instructors need to ensure the hydration of their students and themselves. Thirst is not an adequate way of measuring a body's need for fluid.
- Watch out for signs and symptoms of heat illnesses and deal with issues immediately before they get worse.
- Cardiac events are the leading cause death in fire fighters in the line of duty. Some risk factors cannot be changed, but modifiable risk factors are particularly important because they can be altered to improve your health and help avoid cardiovascular disease.
- The best way to avoid cardiovascular disease is to not smoke, follow a regimen of moderate aerobic exercise, and eat a balanced diet.
- Incident scene rehabilitation must be established at live fire training. Rehabilitation provides medical monitoring of personnel, time for fire fighters to recover, cool down, and rehydrate.

Hot Terms

<u>Core temperature</u> The body's internal temperature.

<u>Incident scene rehabilitation</u> A function on the fireground that cares for the well being of the fire fighters. It includes physical assessment, revitalization, medical evaluation and treatment, and regular monitoring of vital signs.

<u>Stroke volume</u> The amount of blood pumped with each contraction of the heart.

<u>Thermal gradient</u> The rate of temperature change with distance.

<u>Thermal tolerance</u> The body's ability to cope with high heat conditions.

<u>Thermoregulation</u> The process by which the body regulates body temperature.

Wrap-Up

References

Barnard, R. J., and H. W. Duncan. 1975. Heart rate and ECG responses of fire fighters. *Journal of Occupational Medicine* 17:247–250.

Eglin, C. M., S. Coles, and M. J. Tipton. 2004. Physiological responses of fire-fighter instructors during training exercises. *Ergonomics* 47:483–494.

Eglin, C. M., and M. J. Tipton. 2005. Can fire fighter instructors perform a simulated rescue after a live fire training exercise? *European Journal of Applied Physiology* 95:327–334.

Fahy, R. 2005. U.S. fire ighter fatalities due to sudden cardiac death 1995–2004. *NFPA Journal* July/August:44–47.

Kales, S. N., E. S. Soteriades, C. A. Christophi, and D. C. Christiani. 2007. Emergency duties and deaths from heart disease among fire fighters in the United States. *New England Journal of Medicine* 356(12):1207–1215.

Lazarus, G., C. Smeby, Jr., and D. Casey. 2002. *Incident investigation of two firefighters deaths during a training fire at Poinciana, Florida*. BFST Safety Investigative Report 02–01.

Madrzykowski, D. 2007. *Fatal training fires: fire analysis for the Fire Service*. National Institute of Standards and Technology, October 2007.

National Fire Protection Agency. 2006. U.S. fire fighter deaths related to training, 1996–2005. http://www.nfap.org.

Selkirk, G. A., and T. M. McLellan. (2004). Physical work limits for Toronto fire fighters in warm environments. *Journal of Occupational and Environmental Hygiene* 1(4):199–212.

Smith, D. L., T. S. Manning, and S. J. Petruzzello. 2001a. Effect of strenuous live-fire drills on cardiovascular and psychological responses of recruit fire fighters. *Ergonomics* 44(3):244–254.

Smith, D. L., S. J. Petruzzello, M. A. Chludzinski, J. J. Reed, and J. A. Woods. 2005. Selected hormonal and immunological responses to strenuous live-fire firefighting drills. *Ergonomics*, 48(1):55–65.

Smith, D. L., S. J. Petruzzello, M. A. Chludzinski, J. J. Reed, and J. A. Woods (2001b). Effect of strenuous live-fire firefighting drills on hematological, blood chemistry and psychological measures. *Journal of Thermal Biology* 26:375–379.

Smith, D. L., S. J. Petruzzello, J. M. Kramer, S. E. Warner, B. G. Bones, and J. E. Misner. 1995. Selected physiological and psychobiological responses to physical activity in different configurations of firefighting gear. *Ergonomics* 38(10):2065–2077.

Live Fire Training Instructor in Action

As a live fire training instructor you are responsible for ensuring the safety of students. Live fire training presents multiple risks that must be managed. Relative to the physiological responses to firefighting you must be acutely aware of the normal responses to firefighting and the potential for life-threatening events related to heat illness and cardiac emergencies.

1. Which of the following factors lead to high cardiovascular and thermal strain during firefighting activity?
   A. The strenuous and dangerous nature of the work
   B. Heavy and encapsulating PPE
   C. Psychological stress
   D. All of the above

2. How does PPE (personal protective equipment) affect a fire fighter?
   A. It only positively affects a fire fighter, by protecting him or her from direct contact with open heat sources.
   B. It interferes with evaporative cooling.
   C. It greatly increases heart rates.
   D. Both B and C are correct.

3. Which of the following is a life-threatening emergency requiring immediate medical attention?
   A. Heat rash
   B. Heat cramps
   C. Heat exhaustion
   D. Heatstroke

4. What is the leading cause of fire fighter fatality during training?
   A. Heat cramps
   B. Heat exhaustion
   C. Heatstroke
   D. Sudden cardiac events

5. What effect does moderate exercise have on modifiable risk factors?
   A. It only decreases body fat.
   B. It increases body fat.
   C. It has a positive impact on several modifiable risk factors.
   D. It has a negative impact on several modifiable risk factors.

6. Which of the following is a goal of incident scene rehabilitation?
   A. Provide medical monitoring and treatment
   B. Establish standards for food and fluid replacement
   C. Provide cooling
   D. All of the above

Planning for Live Fire Training

CHAPTER 5

NFPA 1403 Standard

4.4.1 A safety officer shall be appointed for all live fire training evolutions. [pp 89–91]

4.4.3 The safety officer shall have the authority, regardless of rank, to intervene and control any aspect of the operations when, in his or her judgment, a potential or actual danger, potential for accident, or unsafe condition exists. [pp 89–91]

4.4.4 The responsibilities of the safety officer shall include, but not be limited to, the following: [pp 89–91]

(1) Prevention of unsafe acts

(2) Elimination of unsafe conditions

4.4.5 The safety officer shall provide for the safety of all persons on the scene, including students, instructors, visitors, and spectators. [pp 89–91]

4.4.6 The safety officer shall not be assigned other duties that interfere with safety responsibilities. [pp 89–91]

4.4.7 The safety officer shall be knowledgeable in the operation and location of safety features available for the live fire training structure or prop, such as emergency shutoff switches, gas shutoff valves, and evacuation alarms. [pp 89–91]

4.4.8* Additional safety personnel, as deemed necessary by the safety officer, shall be located to react to any unsafe or threatening situation or condition. [p 91]

4.6.3 It shall be the responsibility of the instructor-in-charge to coordinate overall fireground activities to ensure correct levels of safety. [p 90]

4.6.4 The instructor-in-charge shall assign the following personnel: [pp 88–89]

(1) One instructor to each functional crew, each of which shall not exceed five students

(2) One instructor to each backup line

(3) One additional instructor for each additional functional assignment

4.6.5.1* Instructors shall be rotated through duty assignments. [p 94]

4.6.7 Additional instructors shall be designated when factors such as extreme temperatures or large groups are present, and classes of long duration are planned. [p 84]

4.6.8 Prior to the ignition of any fire, instructors shall ensure that all protective clothing and equipment specified in this chapter are being worn according to manufacturer's instructions. [pp 91, 94]

4.6.10 Instructors shall monitor and supervise all assigned students during the live fire training evolution. [p 84]

4.6.11 Awareness of weather conditions, wind velocity, and wind direction shall be maintained, including a final check for possible changes in weather conditions immediately before actual ignition. [p 87]

4.7.1 A fire control team shall consist of a minimum of two personnel. [pp 90, 94]

4.7.1.1 One person who is not a student or safety officer shall be designated as the "ignition officer" to ignite, maintain, and control the materials being burned. [pp 90, 94]

4.7.1.1.1 The ignition officer shall be a member of the fire control team. [pp 90, 94]

4.7.1.2* One member of the fire control team shall be in the area to observe the ignition officer ignite and maintain the fire, and to recognize, report, and respond to any adverse conditions. [pp 90, 94]

4.7.2 The decision to ignite the training fire shall be made by the instructor-in-charge in coordination with the safety officer. [pp 90, 94]

4.7.3 The fire shall be ignited by the ignition officer. [pp 90, 94]

4.7.4 The fire control team shall wear full personal protective clothing, including SCBA, when performing this control function. [pp 90, 94]

4.7.5 A charged hose line shall be available when the fire control team is igniting or tending to any fire. [pp 90, 94]

4.7.6 Fires shall not be ignited without an instructor visually confirming that the flame area is clear of personnel being trained. [pp 90, 94]

4.8.1 All students, instructors, safety personnel, and other personnel shall wear all protective clothing and equipment specified in this chapter according to manufacturer's instructions whenever they are involved in any evolution or fire suppression operation during the live fire training evolution. [pp 90–91, 94]

4.8.2* All participants shall be inspected by the safety officer prior to entry into a live fire training evolution to ensure that the protective clothing and SCBA are being worn correctly and are in serviceable condition. [pp 90–91, 94]

4.8.3 Protective coats, trousers, hoods, footwear, helmets, and gloves shall have been manufactured to meet the requirements of NFPA 1971, *Standard on Protective Ensembles for Structural Fire Fighting and Proximity Fire Fighting*. [pp 90–91, 94]

4.8.4 SCBA shall have been manufactured to meet the requirements of NFPA 1981, *Standard on Open-Circuit Self-Contained Breathing Apparatus (SCBA) for Emergency Services*. [pp 90–91, 94]

4.8.7* All students, instructors, safety personnel, and other personnel participating in any evolution or operation of fire suppression during the live fire training evolution shall breathe from an SCBA air supply whenever they operate under one or more of the following conditions: [pp 90–91]

(1) In an atmosphere that is oxygen deficient or contaminated by products of combustion, or both

(2) In an atmosphere that is suspected of being oxygen deficient or contaminated by products of combustion, or both

(3) In any atmosphere that can become oxygen deficient, contaminated, or both

(4) Below ground level

4.9.1 A method of fireground communications shall be established to enable coordination among the incident commander, the interior and exterior sectors, the safety officer, and external requests for assistance. [p 88]

4.9.2* A building evacuation plan shall be established, including an evacuation signal to be demonstrated to all participants in an interior live fire training evolution. [pp 86–87]

4.10.1 Basic life support (BLS) emergency medical services shall be available on site to handle injuries. [p 88]

4.10.1.1 For acquired structures, BLS emergency medical services with transport capabilities shall be available on site to handle injuries. [p 85]

4.10.2 A parking area for an ambulance or an emergency medical services vehicle shall be designated and located where it will facilitate a prompt response in the event of personal injury to participants in the evolution. [p 85]

4.10.3 Written reports shall be completed and submitted on all injuries and on all medical aid rendered. [p 95]

4.12.10.1 An exercise stopped as a result of an assessed hazard according to 4.12.10 shall continue only when actions have been taken to reduce the hazard. [p 87]

4.13.1 Areas for the staging, operating, and parking of fire apparatus that are used in the live fire training evolution shall be designated. [p 85]

4.13.2 An area for parking fire apparatus and vehicles that are not a part of the evolution shall be designated so as not to interfere with fireground operations. [p 85]

4.13.3 If any of the apparatus described in 4.13.2 is in service to respond to an emergency, it shall be located in an area that will facilitate a prompt response. [p 85]

4.13.4 Where required or necessary, parking areas for police vehicles or for the press shall be designated. [p 85]

4.13.5 Ingress and egress routes shall be designated, identified, and monitored during the training evolutions to ensure their availability in the event of an emergency. [p 85]

4.15.1 A preburn plan shall be prepared and shall be utilized during the preburn briefing sessions. [pp 83–88]

4.15.1.1 All features of the training areas shall be indicated on the preburn plan. [p 85]

4.15.2 Prior to conducting actual live fire training evolutions, a preburn briefing session shall be conducted by the instructor-in-charge with the safety officer for all participants. [p 87]

4.15.3 All facets of each evolution to be conducted shall be discussed. [p 87]

4.15.4 Assignments shall be made for all crews participating in the training session. [p 87]

4.15.5 The location of the manikin shall not be required to be disclosed, provided that the possibility of victims is discussed in the preburn briefing. [p 95]

4.15.6 Prior to conducting any live fire training, all participants shall have a knowledge of and familiarity with the prop or props being used for the evolution. [p 87]

4.15.7 Prior to conducting any live fire training, all participants shall be required to conduct a walk-through of the acquired structure, burn building, or prop in order to have a knowledge of and familiarity with the layout of the acquired structure, building, or prop and to facilitate any necessary evacuation. [p 87]

4.15.8.1 The persons in charge of the properties described in 4.15.8 shall be informed of the date and time of the evolution. [p 85]

5.2.1 In preparation for live fire training, an inspection of the structure shall be made to determine that the floors, walls, stairs, and other structural components are capable of withstanding the weight of contents, participants, and accumulated water. [p 83]

5.5 Rapid Intervention Crew (RIC). A RIC trained in accordance with NFPA 1407, *Standard for Training Fire Service Rapid Intervention Crews*, shall be provided during a live fire training evolution. [p 86]

9.1.1* The following records and reports shall be maintained on all live fire training evolutions in accordance with the requirements of this standard: [p 95]

(1) An accounting of the activities conducted

(2) A listing of instructors present and their assignments

(3) A listing of all other participants

(4) Documentation of unusual conditions encountered

(5) Any injuries incurred and treatment rendered

(6) Any changes or deterioration of the structure

(7) Documentation of the condition of the premises and adjacent area at the conclusion of the training exercise

Chapter 5 Planning for Live Fire Training

Additional NFPA Standards

NFPA 1001 *Standard for Fire Fighter Professional Qualifications*
NFPA 1971 *Standard on Protective Ensembles for Structural Fire Fighting and Proximity Fire Fighting*
NFPA 1975 *Standard on Station/Work Uniforms for Fire and Emergency Services*
NFPA 1981 *Standard on Open-Circuit Self-Contained Breathing Apparatus (SCBA) for Emergency Services*
NFPA 1982 *Standard on Personal Alert Safety Systems (PASS)*

Knowledge Objectives

After studying this chapter, you will be able to:

- Identify prospective structures for live fire training.
- Define the aspects of the preburn plan for live fire training.
- Describe the planning steps for live fire training to ensure compliance with NFPA 1403.
- Describe the planning and legal requirements of NFPA 1403 for live fire training.
- Identify participants and their roles in live fire training.
- Identify water supply needs for live fire training.
- Identify parking and areas of operation for live fire training.
- Describe the general order of operations for live fire training.
- Describe the roles and duties of the safety officer during live fire training.
- Identify safety hazards in live fire training.
- Describe the steps needed to mitigate safety hazards during the preparation and operation phases of live fire training.
- Identify the need for staff and participant rotation during live fire training.

Skills Objectives

After studying this chapter, you will be able to:

- Prepare a written preburn plan for an acquired structure to use for live fire training exercises for compliance with NFPA 1403.
- Prepare a written preburn plan for a permanent live fire training structure to use for live fire training for compliance with NFPA 1403.
- Develop a building plan for live fire training in any type of structure.
- Develop a site plan for live fire training.
- Develop emergency plans for live fire training.
- Develop a communication plan for live fire training.

You Are the Live Fire Training Instructor

You are the live fire training officer in your fire department. The department owns a Class A prop. You are instructed to conduct an in-service training for current fire fighters. Several of the department's chiefs want to run the training like a "regular fire," letting fire fighters pull up, advance lines, search for victims, and finally put out the fire.

1. What immediate problems do you see with this method?
2. While still remaining complaint with NFPA 1403, what is the closest to the chiefs' directives that you can get?

Introduction

With so many pieces needing to come together in order for a live fire training evolution to be valuable and safe, planning and organization can ensure that no piece is overlooked. Although some instructors may wish to skip over the planning and get right to the live action, be advised not to take this vital part of the process lightly.

NFPA 1403, *Standard on Live Fire Training Evolutions* uses the term **preburn plan** as an overall plan on how the live fire training evolution will be conducted. The preburn plan should include approximately 30 items. This number may vary depending on what type of structure will be used. This chapter will cover planning issues common to live fire training evolutions.

Initial Evaluation of the Site

The different types of structures used in live fire training are covered in detail in Chapter 6, Acquired Structures, Chapter 7, Gas-Fired and Non-Gas-Fired Structures, and Chapter 8, Nonstructural Training Props. Regardless of the type of structure being used, before any planning is done, the first question asked should be, "Will the acquired structure or permanent live fire training structure allow for what the fire department wants to accomplish with this training?" It is important to specifically define the goals of the training evolution(s) to ensure that the structure will meet those needs.

Along with the department's training needs, access to the training site will also play a determining role in the initial evaluation. A logical initial consideration should be the location of the structure relative to the location of the fire department. There may be a wonderful, state-of-the-art live fire training structure three hours away from the department, but is it feasible to get personnel and apparatus there while ensuring proper coverage back home? Especially with volunteer departments, distance can quickly rule out training traditionally scheduled for evenings. In addition, career departments can incur overtime costs, either for covering on-duty training participants, or when sending off-duty fire fighters to training.

Adjacent properties and infrastructure are of vital concern when training with live fire. The obvious concerns with adjacent structures are instances where fire can spread or where smoke can adversely affect such structures as schools, patient care facilities, child care facilities, commercial businesses, or residential dwellings. Infrastructures, such as roadways, airports, and railroads, are also concerns for live fire training. Some examples could include fire schools located on community college properties that cannot accommodate smoke during the week when classes are in session, or a fire academy located on county fair property being used for fairs or events. These are the types of planning issues that must be considered before setting a date for the training evolutions.

Once the location of the live fire training evolutions is finalized, it is then time to develop the preburn plan.

Developing the Preburn Plan

The first preburn plan developed by a department will always be the most difficult. A standard operating procedure (SOP) needs to be put in place to ensure the plan is a policy that must be followed. Once the SOP and a preburn plan are developed, they will serve as guides for future exercises. It is strongly recommended to use clear language that all participants can understand when developing the preburn plan. This is a critical piece of the planning stage because everyone needs to understand every element of the preburn plan.

Injuries and fatalities have occurred in the past when there was a deviation from the preburn plan and when leadership was not fully aware of the plan. It is crucial that once a preburn plan is in place, that it be followed strictly to ensure the safety of all involved. Items that should be a part of any preburn plan are as follows:

- Learning objectives
- Participants
- Water supply needs
- Apparatus needs
- Building plan
- Site plan
- Parking and areas of operation
- Emergency plans
- Weather
- List of training evolutions
- Order of operations
- Emergency medical plan
- Communications plan
- Staffing and organization
- Safety officer
- Staff and participant rotation
- Personal protective equipment (PPE) use
- Agency notification checklist
- Demobilization plan

Learning Objectives

Those involved in the planning need to determine what they wish to accomplish in order to develop the specific learning objectives of the training evolution(s). This can be more difficult than it sounds. The learning objectives need to be defined so that the individual evolutions can be specifically designed to meet those objectives. The instructors should be fully briefed and prepared to meet the learning objectives.

The participants and their experience must be analyzed to match their skill sets to the evolutions. Obviously, personnel fresh out of Fire Fighter I training have less skills and experience than seasoned personnel. But even with fully trained personnel, live fire training is rarely the time for learning new skills, except under the most controlled circumstances. The average engine company will have considerable differences in the experience and training levels of their members, and this differential needs to be considered.

Participants

All personnel participating in live fire training evolutions must be properly trained before the evolution. Prerequisite training for students is covered in NFPA 1001, *Standard for Fire Fighter Professional Qualifications* and is also briefly covered in Chapter 1 of this book. Proper methods of instructor training are covered in Chapter 3.

The participating student-to-instructor ratio, according to NFPA 1403, shall not be greater than 5 to 1. This will ensure that there are enough knowledgeable eyes watching over the students' actions. If there are large groups present, the classes should be planned for a long period of time. When there are situations involving extreme temperatures, additional instructors may be needed. It is the duty of all instructors involved to monitor and supervise all students throughout the course of the live fire training evolutions. Without proper monitoring and supervision, the results can be catastrophic.

Live Fire Tips

Since the training may include fire fighters from various skill levels, the objectives will most likely be different for each level of experience. The recruits in training have very limited fire experience, so their objectives will be quite different from those of veteran fire fighters. The experience levels of the instructors must also be considered when planning the objectives.

Water-Supply Needs

Water-supply requirements for acquired structures and permanent live fire training structures will be discussed in detail in Chapter 6, Acquired Structures, Chapter 7, Gas Fired and Non-Gas-Fired Structures, and Chapter 8, Nonstructural Training Props. However, an initial determination should be made if enough water is available on-site. Even some permanent live fire training structures may not have sufficient water supply, and static sources such as portable drafting tanks, drafting from a retention pond or other source, or tender shuttles may be necessary. If there is not enough water available on-site, then other means must be secured.

Apparatus Needs

Fire apparatus includes engines (pumpers), aerial devices, and possibly water tenders. An engine will be needed as the primary supply for attack lines, with a second engine being used for the back-up line(s). Depending on the site arrangement, it may be necessary to have one engine at the hydrant to pump the supply line(s) and another at a static water source. This will be dependent on hydrant locations, available fire flow, and the need for alternate water sources.

Building Plan

A building plan is a necessary piece of the planning stage. It will serve as a blueprint to be used by instructors and students, before and during the live fire training evolutions. The diagram of the building plan needs to include the dimensions of the building and the dimensions of all of the rooms, windows, and doors. This diagram does not need to be drawn to scale. All features of the training areas, including primary and emergency ingress and egress routes, and areas that are off limits to participants should be clearly labeled in the diagram. Each room that will have burn evolutions in them should be numbered, on the diagram and in the structure itself, in the sequence in which they will be burned. Fuel loading should be indicated and then confirmed and finalized on a hard copy of the building plan the day of the evolutions. The building plan should include the exact locations of each burn set, and how much and what type of material will be used, for example, "four pallets horizontally stacked with a bale of excelsior."

Follow local protocol when marking the sides of the actual building for clarity and for the purpose of reinforcing the procedure to the participants. In general, the accepted protocol is to label the front of the building as side A, and moving clockwise around the building, label sides B, C, and D in order **Figure 5-1**.

Figure 5-1 Physically mark the building's sides, doors, and windows.

Figure 5-2 Signs can be a great help on the day of the live fire evolutions.

There are local and regional variants, so it is best to check first before assigning designators. Each door and window also needs to be numbered and marked on the plans *and* the actual building. Number each window and door sequentially on *each* side, so that each is uniquely identifiable. The front door is typically labeled as Door A1 and the first window on the left side is typically Window B1. Again, local protocol should always be followed. Be sure to also mark north on the plan.

Site Plan

A site plan must be developed that shows how the building is oriented on the property and all other aspects of the layout, with specific measurements and north labeled. Whenever possible, the building plan, site plan, and any other plans should all be on the same axis, drawn in the same direction in relation to the paper. The structure, with the sides identified, command post, rapid intervention crew (RIC), rehabilitation/medical area, operations area, staging area, placement of apparatus, primary and secondary water sources, hot zones, self-contained breathing apparatus (SCBA) refill, emergency personnel accountability report (PAR) meeting place, parking area, exposures or out buildings, driveway, hazards in terrain, underground and overhead, significant changes in elevation, fencing, and gates all need to be clearly identified on the site plan. Any hazardous areas such as the locations of overhead power lines, septic tanks, and natural or man-made barriers should be identified as well. On the day of the evolutions, bring extra copies that only include the permanent features, in case the weather causes changes and functional assignments have to be relocated.

Parking and Areas of Operations

Areas for the staging, operating, and parking of all involved fire apparatus need to be planned for and noted on the site plan.

Signs, traffic cones, caution tape, and other markings can be a great help on the day of the evolutions **Figure 5-2**. Critical areas that need to be clear of civilian vehicles should be marked "Reserved" or barricaded the day before the evolutions.

EMS with transportation capabilities must be present on the scene of acquired structure live fire training to handle emergencies of participants. Parking for emergency vehicles *not* involved in the evolutions also needs to be set aside so as not to interfere with fireground operations. Ambulances and emergency medical services need to be conveniently parked where they can quickly access participants in the event of an injury. Ambulances and EMS vehicles also need to have clear, easy exit access, so these areas should also be marked with "No parking" signs.

The vehicles of participants and spectators can quickly congest the entrance and exit routes to the training site which may be needed for emergency response vehicle access. This congestion can quickly spread to the surrounding streets. Therefore, the entrance and exit routes must be designated and monitored during the training evolutions to ensure their availability in the event of an emergency.

The site plan needs to be made available to the various involved communication (dispatch) centers that may be needed for dispatch assistance in case of problems. For live fire training in acquired structures, emergency units should be alerted to the date and time so that they are not dispatched under the impression that it is training as opposed to an actual emergency. In many areas, police, fire, and EMS are provided by different agencies and could be in different centers. Sometimes, there are even multiple providers, such as local and county law enforcement. Regardless of how they are structured, all of the aforementioned need to be informed of the evolutions. This should be done prior to the day of the evolutions, so that routes can be planned for units that may be dispatched to the training site. Proper routing is just as important for emergency vehicles needing to leave the training site. Participating personnel may not be familiar with the neighborhood. The width of the roads, tree canopy, traffic congestion, and other factors are all important. The entrance and exit routes for emergency and regular use and the various parking areas should all be provided to participants ahead of time.

Emergency Plans

Unlike an emergency response, a live fire training session allows for all necessary personnel to meet before the fire is ignited. Hose

Near Miss REPORT

Report Number: 7-851
Date: 04/08/2007

Event Description: Our department was involved in live fire training in an acquired structure; the building had been prepped strictly following NFPA 1403. The structure is a warehouse covering an entire city block and was mostly non-combustible construction with an attached cinder block and concrete office. Due to the size of the building (approximately 150,000 sq. ft.) and the construction (the area for live burns was the separate office occupancy attached to the warehouse area) electric utilities were left on in the warehouse but isolated from the office. We completed over 100 live fire evolutions in this occupancy. On the last day of this event, we had completed 3 training evolutions, each involving 4 companies. The training staff set 5 pallets and ½ a bale of straw in the office in a room that had not had a fire in it on any previous day. The straw was ignited and the evolution was begun, the initial engine company made entry and knocked down the fire approximately 5 minutes after initial ignition. As the training officer, I was observing from the stairway directly above the entry point and watched 2 firefighters on the nozzle on their knees extinguishing the remainder of the fire when a sudden bright flash occurred at the ceiling above their heads with arcing throughout the room for approximately 5 seconds. Emergency traffic was called and the evolution was immediately stopped by the training officer. The building was evacuated and no one was injured. Investigation afterward revealed the electric panel in this part of the building with all of the breakers turned off was a sub-panel and the 440 volt line feeding it from another panel in another part of the warehouse ran through the ceiling of the fire room. The metal conduit had melted leading to a short.

Lessons Learned: Live fire training can be done safely and successfully using NFPA 1403 as a guide, however it is still dangerous and training officers must not become complacent. Our training staff was diligent in preparations, but in retrospect, we found ourselves thinking "residential" in a commercial occupancy. All of the breakers were off and the supply to this part of the building had been turned off by city utilities personnel, but due to redundant power supplies, live wires were found in an unexpected area. After this incident, numerous fire fighters relayed similar experiences with live wires thought to be dead in commercial and residential occupancies on emergency scenes. If we had the power off in the entire occupancy, we would have averted this close call.

To prevent this from reoccurring:
- Follow NFPA 1403 to the letter. This takes tremendous planning and resources and cannot be thrown together.
- Establish a training plan with a specific goal other than "let's get it as hot as we can" and stick to it.
- Remember the differences between occupancy and building construction types that may affect fire behavior, staffing needs, hose lays, water supply, etc.

lines, tools, RIC, safety personnel, and even a walk-through of the structure are prepared ahead of time.

The emergency plan needs to include every aspect of the *planned* emergency response. It should also include plans for unexpected emergencies as well. Elements include the following:

- Rapid intervention crews should be fully staffed with experienced personnel. A federal two-in/two-out rule is standard, but outside of that, there must be enough staff on the RIC to enter, locate, rescue, and remove trapped or lost fire fighters. In permanent props, the danger of collapse or failure is almost negligible, so staffing can be less than it would be for an acquired structure. Regardless, the rapid intervention crew must be properly prepared for immediate deployment.
- The building evacuation plan must be created and known by all personnel. The building evacuation plan needs to include an evacuation signal that can be heard by all participants during interior evolutions. The signal must be tested and demonstrated. Aerosol air horns or a megaphone that feeds directly into the interior may be necessary. As part of the building evacuation plan, participants should be instructed to report to a predetermined location (rally point) for roll call or **personnel accountability report (PAR)** if evacuation of the structure is signaled. Instructors should immediately report any personnel unaccounted for to the instructor-in-charge, who will deploy the RIC. Keep in mind that recruit fire fighters may not have participated in PAR drills before.
- The possibility of losing one of the two water supplies needs to be addressed in the plan. Training evolutions need to stop and personnel need to be relocated to the exterior while the supply is reestablished or replaced.

> **Safety Tips**
>
> Every fire fighter, regardless of tenure, should be trained to constantly identify hazards and alternative escape routes during interior fire suppression operations, including training exercises. Live fire training in any structure should include directions for a secondary means of egress in case of an unexpected fire condition change. While it is not an NFPA 1403 requirement, it is a good idea for each fire fighter to identify two means of egress from each area, prior to any evolution.

> **Live Fire Tips**
>
> Lightning detectors are a useful tool to have on hand at live fire training events. They can provide advanced notice of lightning, give warning to shut down operations before lightning arrives, or alert the user to safe conditions after the threat of lightning has passed. Lightning detectors can vary greatly in price and in functionality, but a great tool to use is your head. Hearing thunder is an indication that lightning is close enough to shut down operations. Shelter should be sought immediately. Activities should not resume until the threat of lightning has completely passed.

As part of the emergency plan any time the instructor-in-charge determines that fuel, fire, or any other condition poses a potential hazard, the training evolution shall be stopped immediately. If an exercise is stopped, it should only be restarted once the hazard identified has been resolved and after the Go/No Go sequence. The **Go/No Go sequence** is a verification method by radio that ensures that all positions are ready to initiate operations. By using this defined sequence of actions, everyone is separately accountable for all teams and participants they can see, not just their own. This gives the instructor-in-charge and the safety officer more eyes around the site, checking for preparation and ensuring that each team is ready for ignition.

Nonexercise Emergencies

There may be times when other emergencies occur. Anything from on-scene problems to a neighborhood emergency can require the attention of emergency personnel. If there is ever a question of divided attention, the evolution needs to stop. Often, there are enough unengaged personnel to handle such emergencies. However, the instructor-in-charge must make the determination of whether or not to proceed. The safety officer can play a critical role in that decision-making process.

Weather

The weather can be a determining factor in the decision to cancel or continue with training. The preburn plan should include measures to shelter from the sun, for rehydration, and for cooling.

Storms and other weather systems often come with winds that can drastically change fire dynamics, heavy rain that can hamper operations, and lightning that will curtail training. Acceptable weather parameters of the plan should include wind direction and/or speed that would preclude safe operations, and "red flag" conditions that would preclude open burning, such as high winds and low humidity. The weather forecast, preferably the National Weather Service Point forecast, needs to be checked for expected untoward weather conditions that could cause problems. This should also include a final check for possible changes in weather conditions prior to ignition.

Lightning is a training ground weather hazard that exists everywhere, and can be present in rain or shine in some cases. Many training centers and agencies are now utilizing lightning detection equipment.

List of Training Evolutions

In the preburn plan, a list of the training evolutions to be conducted needs to be developed. The procedures for those evolutions should include instructors and support personnel needed, and the number of students for each. Depending on the site being used, the availability of instructors, and the number of students, this list will vary. The list of training evolutions should be developed following the development of the objectives.

Order of Operations

The **order of operations** is the steps, in sequential order, on how to conduct each live fire training evolution. The order of operations is determined after the list of the training evolutions is developed and should be noted in the preburn plan.

On the day of the training evolution, such an order includes the following:

1. Set up according to the site plan, unless conditions dictate a change. It is a good idea to have several copies of the site plan that show only the building and not the locations where the functional areas are to be placed. Revisions can then be made to the functional areas when needed. Any changes *must* be reviewed in the briefing of the live fire training instructors and participants.
2. Conduct a briefing of all instructors with clear objectives and a consensual understanding. It is important to brief the instructors first before the preburn briefing, because if there is anything overlooked or unclear, they will usually catch it before the next step, the preburn briefing.
3. A preburn briefing is required, along with a walk-through of the entire structure, prior to any ignition of the burn set. The instructor-in-charge and safety officer conduct a walk-through with all of the instructors first, and then with the students. It may be necessary to demonstrate an evolution if there is any question or clarity needed. All instructors must have a clear understanding of the learning objectives for the training evolution. Part of the preburn briefing is to point out the locations of burn sets, review the sequence of burns with the rooms numbered, review the exit markings, ventilation points, and the primary and secondary means of egress. All participants need be aware of the operations area, where all PPE is to be worn. All participants must also be aware of the fireground (warm zone) where a helmet may be the only piece required.
4. Assignments are given and personnel must report to their positions. At this time, all lines are flowed to the proper pressure. Once all lines have been flowed and the pressure is verified, the instructors and students are in place for a Go/No Go sequence. The Go/No Go sequence is the same

process that NASA follows during the launch of a spacecraft. The Go/No Go sequence is stopped if one thing is not right.

5. The Go/No Go sequence is the last step of the order of operations before ignition **Table 5-1**. The verbal confirmation by radio communication ensures that all participants are ready for action. By radio, the instructor-in-charge says, "All personnel stand by for a Go/No Go." Start with the support crew, followed by the attack crew, with pump operators checked before the entry crews. Once the pump operators confirm that they are ready, the instructor-in-charge of each crew is checked, and with each confirmation, it progresses to the next crew. The order is typically as follows:

- All positions share the responsibility and can stop the process. Any position seeing less than 100 percent preparedness needs to give a No Go report. An example of a No Go would be the safety officer observing students still getting their equipment on. Even if all of the positions had advised Go, the safety officer must stop the process.
- Prior to the safety officer giving a Go signal, he or she must inspect the structure to make sure it is clear of any occupants and that personnel are ready. This includes protective clothing and breathing apparatus inspection.
- After the safety officer advises a Go, the instructor-in-charge would declare, "We have a Go for ignition."
- Ignition is the last step in the Go/No Go sequence. As noted earlier, the utilization of specific terminology is vital. One of the most important is the terminology to identify the instructor responsible for lighting the fire and the directive to light the fire. The order for ignition should be along the lines of "Ignition, you have a go," or another very clear directive that everybody knows ahead of time. Ignition will then advise command when there is "fire in the hole," or a similar, predetermined phrase. For Class A fires in permanent live fire training structures or acquired structures, the safety officer will need to make sure the torch is returned to the staging area.
- Anytime operations are shut down, a Go/No Go sequence should take place before continuing the live fire training evolution. An example of this would be after a burn room that is being used is no longer usable, it is necessary to now move to another burn room. The fire in the room being moved away from must be fully extinguished. Remember, only one fire at a time is allowable in an acquired structure. This would also be done after an extended break, like after lunch, or after putting the fire out to check for structure integrity.

Emergency Medical Plan

NFPA 1403 requires that emergency medical services (EMS) be available on-site to handle injuries. For acquired structures, basic life support emergency medical services with transport capabilities shall be available on site. The highest level of prehospital care (i.e., paramedic) with transport capability is preferred. Response times need to be considered, especially for rural training fires. At the very least, basic life support equipment with trained personnel needs to be on-site.

The ability for a local fire department to have an advances life support transport unit on scene varies tremendously. Costs can be involved, especially when the fire department is not the local EMS provider. The local EMS provider needs to be apprised of the location and time of the evolution.

Planning needs to include identification of a landing zone with global positioning satellite (GPS) coordinates for helicopter transport (Medevac). Coordination with the provider is necessary to meet their landing zone and planning requirements, to reduce the possibility of confusion or error in the time of need. The property owner of the intended landing zone must be aware of the intent to use it. Some of the more appealing sites, such as large school fields, can create havoc with the facility and may not actually be available at the time of need.

Written reports, including patient care records, shall be filled out and submitted on all injuries and on all medical aid rendered.

Communications Plan

The communications plan will provide for coordination among the instructor-in-charge, the interior and exterior divisions, the safety officer, and external requests for assistance. Prior to beginning any evolution, all interior crews and command staff throughout the entire structure and fireground need to be equipped with working two-way radio communications. Operations need to be conducted on a dedicated radio channel that will not be used for dispatching, or for any other use, during live fire evolutions.

During a training fire that took the lives of two fire fighters, communications became a critical issue when units responding to the instructor-in-charge's request for assistance interfered with vital on-scene communications. The communications plan needs to include the ability for the instructor-in-charge to communicate with the dispatch center and incoming units on a different radio channel, and a physically different radio, than on-scene operations. In case of an emergency, the plan should include an aide for the instructor-in-charge to operate on two radio channels as needed. Radio communication capabilities need to be confirmed during the planning stage for on-scene radio-to-radio and from on-scene to dispatch.

Staffing and Organization

Instructor positions must be filled by qualified and competent people. Some states have training and even certification requirements for live fire training instructors, but in areas where such requirements are not in place, the instructors must be qualified by some process by the authority having jurisdiction (AHJ). This process needs to be compliant to NFPA 1403 and instructor standards. It is also a good idea to utilize a recognized and credentialed program, such as the International Society of Fire

Table 5-1 The Go/No Go Sequence

From the instructor-in-charge/command	Reply by team or function
Staging: Go/No Go?	Staging: Go
Rehab/Medical: Go/No Go?	Rehab/Medical: Go
Engine 1: Go/No Go?	Engine 1: Go
Engine 2: Go/No Go?	Engine 2: Go
Entry: Go/No Go?	Entry: Go
RIC: Go/No Go?	RIC: Go
Backup: Go/No Go?	Backup: Go
Attack: Go/No Go?	Attack: Go
Safety: Go/No Go?	Safety: Go

Service Instructors' (ISFSI) Live Fire Trainer program, if a state or provincial program is not in place.

The instructor-in-charge needs to assign instructors that meet the above criteria to the following roles:

- One instructor to each functional crew, which shall not exceed five students
- One instructor to each backup hose line
- Additional personnel to backup lines to provide mobility
- One additional instructor for each additional functional assignment

It is important that this selection process follows an organized structure that is known and understood by all participants. Instructors and staff need to be assigned to one of three functional groups: command staff, attack group, and support group. Instructors may swap assignments within their group at anytime, as long as the instructor-in-charge is advised of the change. The accountability system must then be updated with these changes.

Groups are to be rotated at the discretion of the instructor-in-charge. It is important to note that out of concern for divided responsibility, it is generally best for the instructor-in-charge to also be the incident commander. The command structure should follow the normal National Incident Management System (NIMS) model. Following a strict incident command structure will only help maintain a higher level of safety and accountability Figure 5-3.

The instructor-in-charge has the overall responsibility to coordinate the drill ground activities to ensure correct levels of safety, whether the exercise involves a permanent prop or an acquired structure. The instructor-in-charge must work closely with the safety officer to be sure that operations are conducted properly. By using the Go/No Go sequence, *all operational positions are empowered and given responsibility to watch for safety issues.*

Along with other considerations, personnel assignments will be based on the number of evolutions, the number of participants involved, the experience of the participants, and the desired objectives.

Safety Officer

Just as in fireground operations, the role of the safety officer is of the highest importance and the selection of the individual should be based on their abilities and knowledge of live fire training and the roles and duties of the incident safety officer. The assignment of a safety officer is required by NFPA 1403.

Unlike the incident safety officer on the fireground, the safety officer for training has the ability to see and assess the structure before the fire starts. The safety officer should be involved in the planning and preparation of the structure.

Like emergency fireground operations, the safety officer shall have the authority, regardless of rank, to intervene and control any aspect of the operations when, in his or her judgment, a potential or actual danger, accident, or unsafe condition exists.

The safety officer shall provide for the safety of all persons on the scene, including students, instructors, visitors, and spectators. It is important to note that the scope of the safety officer's duties includes ensuring the safety of nonparticipants. The responsibilities of the safety officer shall include, but shall not be limited to, the following:

1. Prevention of unsafe acts
2. Elimination of unsafe conditions

The safety officer shall not be assigned other duties that interfere with safety responsibilities. The role of safety officer requires constant vigilance and considerable mobility to be able to observe operations and to maintain constant scrutiny of structural conditions. There are responsibilities that may be deemed necessary by the safety officer that may require the use of additional safety personnel. These persons can be located strategically within the structure to react to any unplanned or threatening situation or condition, or to review other geographical or functional assignments.

The safety officer is tasked with enforcing the requirements to wear SCBA and protective clothing appropriate for

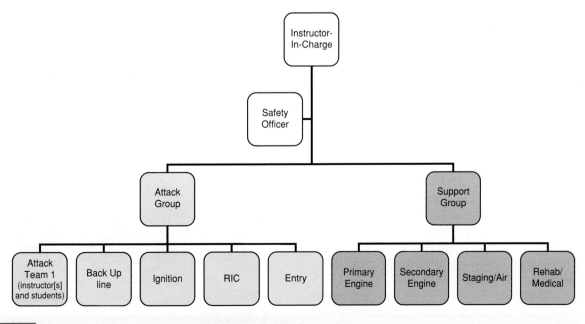

Figure 5-3 A sample incident command organization for structural live fire exercise.

Live Fire Tips

Sample incident command organization for live fire evolutions

- The **command staff** shall consist of the instructor-in-charge and safety officer.
- The **attack group** shall consist of the attack instructor, backup instructor, rapid intervention crew (RIC), and the entry officer.
- The **support group** shall consist of the primary engine, secondary engine, rehab/medical, and staging/air supply.
- The **instructor-in-charge** has overall responsibility and authority for managing and coordinating the drill, ensuring that all procedures are completed in accordance with NFPA 1403. Generally, the instructor-in-charge will also be the incident commander.
- The **safety officer** shall have the authority, regardless of rank, to intervene and control any aspect of the operations when, in his or her judgment, a potential or actual danger, accident, or unsafe condition exists.
- The **attack instructor** supervises the attack team and directs the fire attack. The attack team shall not exceed five students.
- The **backup instructor** supervises the backup team, which monitors conditions during the live burn. The backup team shall not exceed five students. As the term infers, this line is ready to be deployed, and is normally outside so it can be rapidly relocated to back up interior crews, in case of unforeseen fire spread, water supply problems to the attack lines, or other issues.
- The **rapid intervention crew** (RIC) should monitor the interior and exterior conditions and should not be made up of students, but rather instructors. This crew must be sufficiently staffed, equipped, and prepared to make entry at any point to remove a lost or entrapped fire fighter or crew. This works the same way in real emergencies, except that during training, the crew is familiar with the structure, so they can plan the rescues and be better prepared.
- The **entry officer** controls the entry point to the structure for accountability. The entry officer must maintain an entry board to hold the personnel accountability tags (PAT), PASSPORT, or other accountability system used, of students and instructors working inside to keep an accurate head count inside and out **Figure 5-4**. The entry officer also monitors exterior conditions and performs a last check on all PPE. The entry officer reports to the safety officer and is part of that division.
- The **ignition officer** ignites the fire as part of the fire control team after the decision to ignite has been made by the instructor-in-charge and safety officer. The ignition officer shall not be a student. The ignition officer ignites, maintains, and controls the fire and fuel with a hose line available after assuring the area is clear of personnel. Ignition can be accomplished using propane lighters, fusees, kitchen-type matches, and similar devices which must be removed from the area after ignition. The fire control team is utilized any time a fire is ignited and consists of a minimum of two members, one of which is the ignition officer or assigned ignition person. A member other than the ignition officer observes the ignition officer ignite and maintain the fire and watches for adverse conditions. Members must be in full PPE and SCBA when performing their functions.
- The **primary engine/pumper** establishes and maintains a continuous water supply from the primary water source. It operates to provide required water flow and discharge pressures to the attack line and all other lines as needed.
- The **secondary engine/pumper** establishes and maintains a continuous water supply from the secondary water source. It operates to provide required water flow and discharge pressures to the backup line and all other lines as needed.
- The **rehab/medical officer** assesses the established location for conditions such as positions upwind, uphill, and with easy access to EMS. The rehab officer ensures that baseline vital signs are taken on all students and instructors before the drill. The rehab officer also ensures that all students arrive at rehab from attack positions for vitals, hydration, cooling, and rest. After rehabilitation, the students are then sent to staging for reassignment. The rehab officer also manages patient care and is the contact person with local EMS.
- The **staging/air supply officer** assesses the established location for inhibiting conditions (upwind, uphill). The staging officer also maintains adequate air supply reserve, monitors all PPE, and maintains at least one crew in a state of readiness if assistance is needed in addition to the RIC and other team assignments.

the environment. No person should be exposed to smoke, toxic vapors or fumes, products of combustion, or other contaminated atmospheres. The use of SCBA must be enforced for all participants, and the instructors are often the most difficult to influence to comply. SCBA must be used any time participants, or anyone for that matter, are operating in an atmosphere that might be any of the following:

- In an atmosphere that is oxygen deficient or contaminated by products of combustion, or both
- In an atmosphere that is suspected of being oxygen deficient or contaminated by products of combustion, or both

Figure 5-4 An accountability board.

- In any atmosphere that can become oxygen deficient, contaminated, or both
- Below ground level

It is one of the major duties of the safety officer to ensure that these rules are strictly followed to avoid unnecessary injuries or fatalities.

The safety officer will often need to be assisted by several deputy safety officers to help maintain visual contact with operations and the structure that is a distance from him or her. For acquired structures, this is even more critical and is discussed further in Chapter 6. In permanent props, deputy safety officers still may be necessary for entry checks and to monitor student progress through the prop. In Class A props, the deputy safety officers may be needed to monitor the fire conditions. The safety officer has many important responsibilities during live fire training and it is smart to assign additional personnel to assist them when it is needed.

While NFPA 1403 places the overall responsibility for the rehabilitation of personnel with the instructor-in-charge, the safety officer oversees this process more directly. Appropriate health monitoring, time for rest and rehabilitation, food and fluid replenishment, and relief from climatic conditions must be provided to participants.

Visitors and spectators can be problematic, and vary from invited VIPs to media, neighbors, and even fire fighter families. Access should be clearly limited to avoid heat, smoke, trip hazards from hoselines or equipment, and from getting in the way of participants. Anyone the fire department wants there, with the permission of the instructor-in-charge, should be escorted and with proper protective clothing for the "cold" (safer) areas they may be in.

Protective Clothing and Self Contained Breathing Apparatus

The safety officer needs to ensure that all students, instructors, safety personnel, and any other participants wear their equipment appropriately. NFPA 1403 expands on this to say that all participants must wear all protective clothing and equipment "according to manufacturer's instructions." This is to be checked by an instructor or safety officer prior to entry into the structure. Therefore, safety officers must be familiar with inspection and the requirements of protective clothing, SCBA, PASS inspections, and uniform apparel. This portion of NFPA 1403 becomes very problematic and can, if followed verbatim, require the safety officer and his or her staff to read the labels on each piece of participant gear to determine if the gear is compliant to the standard. This would include turnout coat, pants, hood, boots, helmet, uniform pants and shirt, and SCBA with integrated or separate PASS. When conducting training with only your own fire department, this task should be less of a problem if your department is in compliance with NFPA 1971, *Standard on Protective Ensembles for Structural Fire Fighting and Proximity Fire Fighting,* NFPA 1981, *Standard on Open-Circuit Self-Contained Breathing Apparatus (SCBA) for Emergency Services,* NFPA 1982, *Standard on Personal Alert Safety Systems (PASS),* and finally NFPA 1975, *Standard on Station/Work Uniforms for Fire and Emergency Services.* All of the aforementioned gear should be compliant in manufacture, but their condition still needs to be checked, and the pants and coats should be of the same brand and ensemble type. With high-backed pants versus bib-cut pants and standard cut, and coats with similar differences, it is important to make sure that the pants match the coat's style and manufacture.

When multiple agencies or individuals from different agencies are allowed to participate in training, a more thorough inspection is necessary. In this case, it would be wise to have the outside agencies provide documentation to the host agency that all of the protective clothing and SCBA have been checked and are compliant. The safety officer and staff should understand that each participant is not going to bring the written manufacturer's instructions for each item. Still, due diligence is necessary to verify that the protective clothing appears to be in good condition. The safety officer should be checking the following:

- Turnout coat and pants
  - They should be of the same ensemble.
  - Outer shell: There should be no obvious signs of contamination, heat discoloration, burns, or charring, tears, holes, fraying, or weakening of material.
  - The NFPA compliance tag should be checked.
  - Ensure proper fit on participants.
  - Liner: There should be no thermal damage, delaminating of the moisture barrier, tears, heat discoloration, holes, or fraying.
  - Hardware: Ensure that snaps, zippers, Velcro, or other closures are functional.
- Hood
  - There should be no tears, holes, fraying.
  - Ensure proper fit on participants. The hood should not be stretched out.
  - The NFPA compliance tag should be checked.
- Helmet
  - Shell: There should be no obvious contamination, cracks, holes, weakened material, burns, or charring.
  - Liner: Check for thermal damage, and damage to the impact shell.
  - Check the ear flaps for functionality.
  - The hardware should be functional and properly adjusted for good fit.
  - The NFPA compliance sticker should be checked.
  - Check the strap to make sure it is in good condition and functions properly.
- Gloves
  - Outer shell: There should be no obvious contamination, burns, or charring, and no tears, holes, or fraying.
  - Liner: There should be no moisture-barrier delaminating, tears, holes, or fraying.
  - Ensure proper glove-to-coat interface.
  - The NFPA compliance sticker should be checked.
  - Ensure proper fit on participants.
- Boots
  - There should be no contamination, tears, holes, fraying, weakened material, burns, or charring.
  - The liner should show no thermal damage.
  - There should be no moisture-barrier delaminating or tears, holes, or fraying.
  - Hardware: Ensure that snaps, zippers, Velcro, or other closures are functional.
  - The NFPA compliance tag should be checked.
  - Ensure proper fit on participants.

Incident Report

Green Cove Springs, Florida - 1990

In Green Cove Springs, Florida, a live training exercise was set up near the end of a fire fighter training course. The course lead to state certification, and was taught by a state-certified fire training center operated by a local school district. Fire fighters from a municipal and county fire department joined the students for the final live exercises. A picnic for the students and their families was held before the fire and the families remained for the training fire.

The wood-framed, two-story house was almost 100 years old and had previous fire damage to it. The 24-person class with two state certified instructors and one noncertified instructor set up four 1¾" (44.5 mm) hose lines, wyed off of two 2½" (63.5 mm) hose lines. Two additional 2½" (63.5 mm) hose lines with nozzles were positioned, with the pumper supplied by a 5" (127 mm) large-diameter hose (LDH) line. Ground ladders were set on three sides as emergency egress from the second floor. The stairs leading to the second floor were 36" (914.5 mm) wide.

An initial burn was conducted in a second floor bedroom, measuring 12' × 19' (3.7 × 5.8 m) with 9' (2.7 m) ceilings. There were two doors to the room: an entry door and one that connected to an adjoining room. There were two windows, one of which looked out over the front porch roof. A 1¾" (44.5 mm) hose line was brought up and into the bedroom by the first crew. The exercises upstairs were primarily for observing fire behavior, and the actual suppression exercises were to follow afterwards in the rooms downstairs.

Approximately a half-gallon of diesel fuel was poured on a mattress that was lying on its side against a wall about 4' (1.2 m) from the entry door. It was ignited with a road flare by a noncertified instructor. The first instructor took the students into the room to observe the fire spread as he explained the fire's behavior. The students then knocked the fire down under the direction of the instructor. The instructor flipped the mattress over for reuse. The first crew and the instructor exited down the stairs.

Another certified instructor then led a second crew up the stairs. He swapped SCBAs with the other instructor that had just exited because there was a failure with his own. As they advanced the 1¾" (44.5 mm) hose line, it lost pressure because the supplying 2½" (63.5 mm) hose line burst. The 2½" (63.5 mm) hose line was quickly changed out, and the crew continued to the fire room.

Once in the fire room, the students were met with deteriorating fire conditions. Apparently, the nozzle was pulled from a student's hands. Conditions became untenable. One student crashed headfirst through the window onto the front porch roof. Another student and the instructor exited through another bedroom window down one of the ground ladders. The last student exited though the bedroom door and down the stairs. An instructor outside ordered a safety line to enter and extinguish the fire after observing the student escape though the front window. The first, noncertified, instructor now became the incident commander. He requested EMS and additional fire units.

The instructor inside apparently did not have his gloves on and received third degree burns to both of his hands, right arm, and shoulder. He lost parts of two fingers due to the burn injuries. Two students received second degree burns on their upper torsos, one with third degree burns to his back. All were flown to medical facilities by EMS helicopters.

Richard Knoff, Deputy Chief of Clay County Fire Rescue said this of the incident:

> At first we thought there might be a problem when we saw heavy black smoke coming from the bedroom windows upstairs. I called the lead instructor who had the students inside but I did not get an answer. I tried again and then we saw a student bail out headfirst through a 2nd floor window, landing on the porch roof. I called for additional fire units and EMS by radio, and instructed a crew to get the fire fighter down from the porch roof. Then, I deployed the RIC and went in with them. We got inside and everybody was already out. Looking at the room afterwards, you could not imagine how anyone was hurt. There was no heavy charring, no significant damage to their SCBA or bunker gear. Only damage to their helmet shields.

Post-Incident Analysis — Green Cove Springs, Florida

NFPA 1403 Noncompliant

Flashover and fire spread unexpected (4.3.9)

Fire "set" was close to primary entry/egress of fire room (4.4.16)

Student–instructor ratio was 8:1 (4.5.2)

Noncertified instructor (4.5.1)

No backup line (4.4.6.2)

NFPA 1403 Compliant

Ground ladders were set as secondary means of egress (4.2.12.1)

- SCBA
  - Cylinder(s): There should be no physical damage, contamination, or thermal damage. The hydrostatic test should be current.
  - Harness: There should be no physical damage, fraying of straps, contamination, or thermal damage. The hardware should be complete and functioning.
  - Regulator and hoses should be intact and functional, and there should be no physical damage affecting use or contamination.
  - Face piece should be intact, lens visibility should be good, straps and headpiece should be intact, not frayed, and should seal properly.
  - PASS device
  - Should be NFPA compliant.

Staff and Participant Rotation

Live fire training is physically demanding on the instructors as well as the students. NFPA requires that instructors rotate through assignments keeping the instructor in condition to conduct live fire training. There are a number of methods for instructor and student rotation within the structure. Due to the interior crews' close proximity to the fire, the idea is to rotate instructor and student positions to keep them from overheating. This is especially true for Class A burn buildings and acquired structures, where the ignition officer and personnel assigned to monitor and restock the fire will be exposed to greater heat over a longer period of time. Rotations need to be planned for and "choreographed" to prevent delays between entries. Any delays that occur while a fire is burning can lead to an increase in the heat of a permanent prop or can make the entire acquired structure dangerous.

Crews of recruit students should be kept as small as possible. Four students to a crew is generally the most that can be supervised at one time, but three is more desirable. NFPA 1403 states there shall be no more than five students per instructor. Many agencies use one instructor and one safety officer for each attack hose line team for recruit student training. This is done so that the instructor or safety officer is always within reach of all of the students and they are able to closely monitor their activities.

When working with nonrecruit fire fighters, the rotation is different. The instructor has no more than five students at a time and he or she can maintain integrity with that crew throughout the entire evolution. The instructor and crew will move as one team from one assignment to the next. Remember, many communities will only have two to three fire fighters on an attack line, and some may only have one fire fighter and a company officer. When planning the evolutions, it is beneficial to try and reflect the actual staffing that will be present at the burn.

Personal Protective Equipment Use

Individual fire departments should already have policies on the use of PPE and SCBA, and related safety concerns. At this time, it may be necessary to review the requirement to wear and use the equipment while within the operations area. The use of scene tape, or barrier tape, reinforces this safety practice, especially for recruit fire fighters Figure 5-5. At permanent live fire training structures, a painted line with signs can serve the same purpose.

Figure 5-5 The operations area where PPE and SCBA are required should be clearly defined.

Agency Notification Checklist

It is a good idea to keep a checklist to confirm the training with *any* agency that was previously notified of the evolutions. Notification is required prior to starting any evolution for any agency that plays a critical role in the training. These agencies may include law enforcement, emergency medical services, environmental organizations, local utilities, and any other locally required notifications. It is strongly suggested to keep contact names and numbers of all points of contact on the checklist.

Demobilization Plan

Through the use of a demobilization plan, personnel, equipment, and fire apparatus can be put back into service in an organized way. The department's policies should generally assist in the development of the plan, but in the absence of such policies, it should be determined ahead of time if apparatus and equipment can be returned at less than 100 percent. Any apparatus that may be needed to respond to an emergency needs to have the SCBA serviced, booster tank filled, and preconnected hose lines repacked and ready. This includes apparatus that may be used directly following the training evolutions and those that may need to be called for emergencies during the evolutions. Part of the demobilization plan can include bringing fresh hose to repack, personnel assigned to fill SCBA cylinders, refueling the apparatus, and so on.

Any participant who is known or suspected of sustaining an injury during the training should have the appropriate documentation completed before they depart the drill site.

Using the Preburn Plan

The preburn plan should be used prior to actual live fire training evolutions. The perfect time to review the preburn plan is during the preburn briefing session, in which all facets of each evolution are discussed and assignments for all crews are given.

Prior to the evolutions, all participants shall be required to conduct a walk-through of the acquired structure, to have a knowledge of and familiarity with the layout of the structure and to facilitate any necessary evacuation. Emergency procedures in the plan will be reviewed, as will the emergency egress routes.

During the preburn briefing, it is not necessary to disclose the actual location of simulated victims, but it is required to discuss the use of them if they will be included as part of the evolution. Participants need to be advised that there will not be any "live" victims, and any simulated victims will not be wearing protective clothing. No person(s) shall play the role of a victim. It is important to note that NFPA 1403 does not prohibit mannequins being clothed in fire fighter PPE, but out of concern for any confusion that may occur with a fire fighter down, it is highly recommended that this not be done.

Post Live Fire Training Tasks

After the fire is out, a postevolution debriefing should be done. The postevolution debriefing can review aspects of the evolutions that could have been done better, as well as those part of the training evolutions that went smoothly. Ask all participants for their input. Participants will likely be tired, but it is a good idea to get their input while the events are fresh in their minds.

Before clearing the training site, the fire department needs to be careful not to leave burning debris that could cause fire or smoke problems for neighbors. It may be necessary to leave a fire watch in place on acquired structures.

Live fire training records must also be kept in accordance with NFPA 1403. Regardless of the type of structure used in training, an accounting of the activities conducted, a list of instructors and their assignments, a list of all other participants, documentation of unusual conditions encountered, a list of injuries incurred and treatment rendered, any changes or deterioration of the structure, and documentation of the condition of the structure and surrounding areas after the training exercise, must all be kept on record. Documenting these items and saving a record of each will only help you, the instructor, in the end. If something goes wrong, even after the fact, you will have these records to go back to.

Wrap-Up

Chief Concepts

- The planning process is one of the most important parts of setting up live fire training evolutions.
- The importance of planning and preparation can not be over emphasized, to safely achieve the learning objectives and maximize the value of all the time and effort put forth to conduct and participate in the training.
- The items that should be included as part of any preburn plan include:
  - Learning objectives
  - Participants
  - Water supply needs
  - Apparatus needs
  - Building plan
  - Site plan
  - Parking and areas of operation
  - Emergency plans
  - Weather
  - List of training evolutions
  - Order of operations
  - Emergency medical plan
  - Communication plan
  - Staffing and organization
  - Safety officer
  - Staff and participant rotation
  - Personal protective equipment (PPE) use
  - Agency notification checklist
  - Demobilization plan

Hot Terms

Fire control team Comprised of a minimum of two members including the ignition officer or assigned ignition person and utilized any time a fire is ignited. Observes the ignition officer, ignites and maintains the fire, and watches for adverse conditions.

Go/No Go sequence A verbal confirmation via radio communication that each and every participant is ready for action in the live fire environment.

Order of operations The sequence of steps to conduct a procedure. In this context, the steps are in proper sequence to conduct the live fire evolution, but order of operations could also refer to any emergency scene operation. Most commonly refers to the sequence a mathematical equation is solved.

Personnel accountability report (PAR) A verification by person in charge of each crew or team that all of their assigned personnel are accounted for.

Preburn plan A briefing session conducted for all participants of live fire training in which all facets of each evolution to be conducted are discussed and assignments for all crews participating in the training sessions are given.

Live Fire Training Instructor in Action

As the instructor-in-charge of planning a live fire training evolution, you are posed with many difficult tasks that must be secured before the evolutions can take place. This process can be as exhausting as the actual operations, and will generally take a great deal of time. This is the first time you will be the instructor-in-charge of an event like this, therefore, your chief determines that he wants to work with you for the planning of the event. How do you respond when he asks you the following questions:

1. Who has the direct responsibility for the safety of the participants and instructors?
   A. Fire chief
   B. Instructor-in-charge
   C. Training instructor
   D. Safety officer

2. Identify emergencies, not fire related, that can require the training exercise to be stopped or suspended.
   A. Lightning
   B. Medical emergency of a student or instructor
   C. Off-site emergency requiring personnel or equipment at the training site
   D. All of the above

3. Which position in the organizational structure of a live fire training exercise can call a "No Go" and stop ignition?
   A. Only the safety officer has that authority.
   B. Only the safety officer and the instructor-in-charge have that ability.
   C. Only key positions actually asked in the roll call can stop the evolution.
   D. Anyone observing less than 100 percent preparedness can stop the evolution.

4. How many preburn walk-throughs are required?
   A. The standard does not specify
   B. None
   C. Two
   D. As many as are needed

5. Which of the following are *not* true by NFPA 1403 and this text about the use of "victims" during a live fire exercise?
   A. A properly equipped and trained instructor can serve as a victim.
   B. The location of the victim has to be disclosed.
   C. The use of a victim has to be disclosed.
   D. The use of a live victim has to be disclosed.

Acquired Structures

CHAPTER 6

NFPA 1403 Standard

4.2.1 All required permits to conduct live fire training evolutions shall be obtained. [p 103]

4.2.2 The permits specified in this chapter shall be provided to outside, contract, or other separate training agencies by the authority having jurisdiction (AHJ) upon the request of those agencies. [p 103]

4.7.2 The decision to ignite the training fire shall be made by the instructor-in-charge in coordination with the safety officer. [p 117]

4.7.3 The fire shall be ignited by the ignition officer. [p 116]

4.11.1 The instructor-in-charge and the safety officer shall determine the rate and duration of waterflow necessary for each individual live fire training evolution, including the water necessary for control and extinguishment of the training fire, the water supply necessary for backup line(s) to protect personnel, and any water needed to protect exposed property. [pp 104–105]

4.11.2 Each hose line and backup line(s) shall be capable of delivering a minimum of 95 gpm (360 L/min). [p 104]

4.11.3 Backup line(s) shall be provided to ensure protection for personnel on training attack lines. [pp 104–105]

4.12.6* Fuel materials shall be used only in the amounts necessary to create the desired fire size. [p 113]

4.12.7 The fuel load shall be limited to avoid conditions that could cause an uncontrolled flashover or backdraft. [p 113]

4.12.8* The instructor-in-charge and the safety officer shall assess the selected fire room environment for factors that can affect the growth, development, and spread of fire. [p 113]

4.12.9* The instructor-in-charge and the safety officer shall document fuel loading, including all of the following: [p 113]
(1) Fuel material
(2) Wall and floor coverings and ceiling materials
(3) Type of construction of the structure, including type of roof and combustible void spaces
(4) Dimensions of the room

4.12.10* The training exercise shall be stopped immediately when the instructor-in-charge or the safety officer determines through ongoing assessment that the combustible nature of the environment represents a potential hazard. [p 103]

4.12.10.1 An exercise stopped as a result of an assessed hazard according to 4.12.10 shall continue only when actions have been taken to reduce the hazard. [p 117]

4.12.11.9 All possible sources of ignition, other than those that are under the direct supervision of the ignition officer, shall be removed from the operations area.[pp 113, 117]

4.14.1 All spectators shall be restricted to an area outside the operations area perimeter established by the safety officer. [pp 114–115]

4.14.2 Control measures shall be posted to indicate the perimeter of the operations area. [pp 114–115]

4.14.3 Visitors who are allowed within the operations area perimeter shall be escorted at all times. [pp 114–115]

4.14.4 Visitors who are allowed within the operations area perimeter shall be equipped with and shall wear appropriate protective clothing. [pp 114–115]

4.14.5 Control measures shall be established to keep pedestrian traffic in the vicinity of the training site clear of the operations area of the live burn. [pp 114–115]

4.15.1 A preburn plan shall be prepared and shall be utilized during the preburn briefing sessions. [p 103]

4.15.2 Prior to conducting actual live fire training evolutions, a preburn briefing session shall be conducted by the instructor-in-charge with the safety officer for all participants. [p 116]

4.15.8 Property adjacent to the training site that could be affected by the smoke from the live fire training evolution, such as railroads, airports or heliports, and nursing homes, hospitals, or other similar facilities, shall be identified. [p 102]

4.15.9 Streets or highways in the vicinity of the training site shall be surveyed for potential effects from live fire training evolutions. [p 102]

4.15.9.1* Safeguards shall be taken to eliminate possible hazards to motorists. [p 102]

5.1.1* Any acquired structure that is considered for a structural fire training exercise shall be prepared for the live fire training evolution. [pp 106–114]

5.1.1.1 Buildings that cannot be made safe as required by this chapter shall not be utilized for interior live fire training evolutions. [p 102]

5.1.2 Adjacent buildings or property that might become involved shall be protected or removed. [p 102]

5.1.3* Preparation shall include application for and receipt of required permits and permissions. [p 103]

5.1.4* Ownership of the acquired structure shall be determined prior to its acceptance by the AHJ. [p 103]

5.1.5 Evidence of clear title shall be required for all structures acquired for live fire training evolutions. [p 103]

5.1.6* Written permission shall be secured from the owner of the structure in order for the fire department to conduct live fire training evolutions within the acquired structure. [p 103]

5.1.7 A clear description of the anticipated condition of the acquired structure at the completion of the evolution(s) and the method of returning the property to the owner shall be put in writing and shall be acknowledged by the owner of the structure. [p 103]

5.1.8* Proof of insurance cancellation or a signed statement of nonexistence of insurance shall be provided by the owner of the structure prior to acceptance for use of the acquired structure by the AHJ. [p 103]

5.1.9 The permits specified in this chapter shall be provided to outside, contract, or other separate training agencies by the AHJ upon the request of those agencies. [p 103]

5.1.10 A search of the acquired structure shall be conducted to ensure that no unauthorized persons, animals, or objects are in the acquired structure immediately prior to ignition. [p 116]

5.1.11 No person(s) shall play the role of a victim inside the acquired structure. [pp 102–103]

5.1.12 Only one fire at a time shall be permitted within an acquired structure. [p 113]

5.2.2* All hazardous storage conditions shall be removed from the structure or neutralized in such a manner as to not present a safety problem during use of the structure for live fire training evolutions. [p 109]

5.2.3 Closed containers and highly combustible materials shall be removed from the structure. [p 112]

5.2.3.1 Oil tanks and similar closed vessels that cannot be removed shall be vented to prevent an explosion or overpressure rupture. [pp 108, 112]

5.2.3.2 Any hazardous or combustible atmosphere within the tank or vessel shall be rendered inert. [p 112]

5.2.4 All hazardous structural conditions shall be removed or repaired so as to not present a safety problem during use of the structure for live fire training evolutions. [p 112]

5.2.4.1 Floor openings shall be covered to be made structurally sound. [p 111]

5.2.4.2 Missing stair treads and rails shall be repaired or replaced. [pp 108, 113]

5.2.4.3 Dangerous portions of any chimney shall be removed. [p 109]

5.2.4.4 Holes in walls and ceilings shall be patched. [p 111]

5.2.4.5* Roof ventilation openings that are normally closed but can be opened in the event of an emergency shall be permitted to be utilized. [p 109]

5.2.4.6* Low-density combustible fiberboard and other highly combustible interior finishes shall be removed. [p 111]

5.2.4.7* Extraordinary weight above the training area shall be removed. [p 112]

5.2.5* All hazardous environmental conditions shall be removed before live fire training evolutions are conducted in the structure. [p 109]

5.2.5.1 All forms of asbestos deemed hazardous shall be removed by an approved manner and documentation provided to the AHJ. [p 108]

5.2.6 Debris creating or contributing to unsafe conditions shall be removed. [p 109]

5.2.7 Any toxic weeds, insect hives, or vermin that could present a potential hazard shall be removed. [p 106]

5.2.8 Trees, brush, and surrounding vegetation that create a hazard to participants shall be removed. [p 106]

5.2.9 Combustible materials, other than those intended for the live fire training evolution, shall be removed or stored in a protected area to preclude accidental ignition. [p 109]

5.3.1 Utilities shall be disconnected. [p 108]

5.3.2 Utility services adjacent to the live burn site shall be removed or protected. [p 108]

5.4.1 Exits from the acquired structure shall be identified and evaluated prior to each training burn. [pp 106–108]

5.4.2 Participants of the live fire training shall be made aware of exits from the acquired structure prior to each training burn. [p 113]

5.4.3 Fires shall not be located in any designated exit paths. [p 117]

9.1.2* For acquired structures, records pertaining to the structure shall be completed. [p 103]

9.1.3 Upon completion of the training session, an acquired structure shall be formally turned over to the control of the property owner. [p 121]

9.1.3.1 The turnover process shall include the completion of a standard form indicating the transfer of authority for the acquired structure. [p 121]

Additional NFPA Standards

NFPA 1142 *Standard on Water Supplies for Suburban and Rural Firefighting*

Knowledge Objectives

After studying this chapter, you will be able to:
- Define an acquired structure according to NFPA 1403.
- Describe how to perform an initial evaluation on an acquired structure.
- Identify the responsibilities of the owner of the acquired structure.
- Identify the information that must be included in the preburn plan for an acquired structure.
- Describe how to ensure the water supply is adequate for training evolutions.
- Describe how to prepare an acquired structure for live fire training.
- Describe how to secure access to the site of the acquired structure.
- Identify the equipment and supplies required to prepare an acquired structure for live fire training.
- Identify the precautions to take to ensure proper entry and egress for the acquired structure.
- Identify the structural elements to evaluate when preparing an acquired structure for live fire training.
- Describe the steps of preparing the interior of an acquired structure for live fire training.
- Describe how to ensure the safety of the occupants of the homes and businesses adjacent to the acquired structure.
- List the equipment and supplies required for live fire training in an acquired structure.
- Describe how fuel for ignition is prepared safely.
- List the steps to follow prior to the ignition phase.
- Describe how to ensure safety in preparing for and during the final controlled burn.
- Describe tasks involved in the overhaul phase.
- Describe the benefits of a postevolution debriefing.

Skill Objectives

After studying this chapter, you will be able to:
- Inspect and prepare an acquired structure for live fire training.

You Are the Live Fire Training Instructor

As the training officer for your fire department, the chief has advised you of a citizen's offer to donate a structure to use for live fire training. You will be the instructor-in-charge of the live fire training evolution, and as such you will be responsible for personnel safety, protection of adjacent properties, and meeting the objectives of the training. How can you ensure the following:

1. Is the structure prepared for the evolution?
2. Are the fuel materials identified and do they have known burning characteristics?
3. What are the functional positions that need to be filled and how do you fill these with competent and knowledgeable personnel?

Introduction

NFPA 1403, *Standard on Live Fire Training Evolutions*, defines an **acquired structure** as one that is obtained by the authority having jurisdiction from a property owner for the purpose of conducting live fire training evolutions. There are a number of considerations to take into account when evaluating possible structures for live fire training exercises. This chapter will assist you, as the live fire training instructor, in readying a structure and conducting a safe and successful training within an acquired structure.

Initial Evaluation

One of the first questions asked when evaluating a structure for live fire training evolutions will be, "Why does the individual want to let the fire department burn down their building?" One reason may be that the building has structural issues. If this is the case, the building may have lost value to the owner, which also would mean that it has no value to the fire department. The condition of such structures can pose significant dangers to the participants in training. Any acquired structure that is up for live fire training consideration must be prepared for the training evolution. Remember, NFPA 1403 states, "Buildings that cannot be made safe as required by this chapter shall not be utilized for interior live fire training evolutions."

Live Fire Tips

There is an old adage in the fire service, "We Train to Fight Fires—We Don't Fight Fires to Train."

Concerns that might immediately rule out a particular structure, or require significant mitigation, should be addressed at the initial visit to the site. Many of the initial observations can be done during the first visit if you take the time to walk in and around the structure. In no particular priority, some of these concerns may be as follows:

- Proximity to other structures and exposures, such as utilities, infrastructure, sheds, trees, and heavy vegetation on and adjacent to the property that could be in danger of fire spread.
- Adjacent properties that could adversely be affected by smoke produced during evolutions such as schools, childcare facilities, hospitals or nursing homes, or possible business disruptions. The owners of these properties need to be notified of the date and time of the planned training sessions, so that they can take the appropriate precautions.
- Adjacent properties where even the presence of fire and fire apparatus could be an issue, causing concern or disruption such as schools, childcare facilities, hospitals or nursing homes, or possible business disturbances.
- Locations where local transportation could be affected, such as busy streets, highways, railroads, or airports. These interruptions could be smoke obscuring a roadway, a railroad, interfering with airport operations, or secondary issues like traffic disruptions due to passers by "checking out the scene" of smoke and fire apparatus.
- Site access restrictions, and parking for apparatus and equipment.
- Obvious structural integrity, such as a sagging roof or floors, or cracks in brick or masonry walls.
- A structure that has already had a fire. This does not automatically preclude its use, but does require determining how much damage was sustained.

- Does the structure provide for the fire department's needs? What does the fire department and its fire fighters gain from burning the structure? Will the interior configuration allow the fire department to conduct the desired evolutions in a safe manner? Are hallways wide enough for operations? Are there "mantraps" or other hazardous features or fixtures?
- Other discernable issues and concerns such as past occupancy, or contamination to the building or site, either by construction materials or by other means.

Owner Responsibilities

Based on a favorable initial evaluation, the owner needs to be advised of his or her responsibilities and of the costs involved. Careful record keeping is extremely important at this stage and fire departments are encouraged to use checklists, such as the examples included in the appendix of this text. Make sure all of those documents are easily accessible and readily available at the site when conducting the exercises. It is very common for the environmental agency or other regulatory agencies to request copies of the documents, especially the permits and environmental inspections. The authority having jurisdiction (AHJ) will need to provide any permits or permissions to outside, contract, or other training agencies upon request. However, before the fire department spends any time or money on the project, the following needs to be definitively determined:

- Evidence of ownership and clear title.
- Proof of fire insurance cancellation, or a signed statement of the nonexistence of insurance.
- Written permission from the owner detailing what the fire department can or will do with the structure.
- A document signed by both parties that details the responsibility to obtain the necessary permits and inspections, the cost of asbestos removal, and any other environmentally prohibited materials or preparation costs. It is up to the fire department to determine if the donor or the fire department will bear the costs or share them. Beyond the requirements of NFPA 1403, many environmental agencies will require inspections and the removal of certain materials in a prescribed manner.
- A document signed by both parties with a very detailed description of the anticipated condition of the structure at the completion of the evolution(s). This needs to include which party will pay for any remaining demolition, the removal of debris, and the responsibility of securing the property until the site is considered safe. The attorney for the fire department may want this document to be formal, notarized, or otherwise executed.
- Plans for lunch and appropriate cold drinks for the participants can be organized with the property owners.

Preburn Plan

The preburn plan for an acquired structure needs to include details in addition to those common to a permanent training facility. The plan needs to be developed and approved by the instructor-in-charge, and then approved by the fire department's proper chain of command. This plan must include all features of the training structure and must also include the following:

- Prior to the beginning of evolutions, all participants, including the instructors, need to participate in a thorough walk-through of the acquired structure in order to understand the expected evolutions and to have a good understanding on the primary and emergency egress routes. Emergency procedures in the plan need to be reviewed with all participants.
- Final controlled burn plan, includes final check of building for personnel, equipment, conditions, PAR, and audible "burn down" signal to all personnel
- Demobilization plan (release of personnel, returning of equipment, placing fire apparatus back into service, etc.)
- Transfer of property back to owner

Emergency Plans

A training fire is unlike an emergency response. We can plan, gather personnel, layout equipment and hose lines, and tour the building before a planned fire starts. However, when working with an acquired structure, unexpected fire spread and other problems can occur. Unlike a fire in a permanent live fire training structure, in this case, the structure itself is actually on fire. The role of the rapid intervention crew (which must now meet NFPA 1407) and the safety personnel, and purpose of the walk-through all become more critical.

Despite the weeks of planning leading up to a training exercise, fire is unpredictable and can still spread into voids, concealed spaces, and exit paths. Like a working structure fire, all personnel must be vigilant for such fire spread. When fire is discovered in any of these areas, training operations should cease, as the fire should now be treated as an uncontrolled fire. Planning must determine how unforeseeable emergencies like this will be handled. The backup hose line may take over suppression while recruits are relocated to the exterior. However, personnel need to know that operations have shifted from training to suppression. The instructor-in-charge needs to advise all personnel of the shift from training to suppression and get an acknowledgement of that shift, just like when shifting from offensive to defensive operations at a structure fire. NFPA 1403 requires that any time the instructor-in-charge determines that any condition represents a potential hazard, the training evolution shall be stopped immediately. If the fire acts in an unexpected manner for whatever reason, or problems are encountered with students, hose lines, water supply, or anything that threatens the safety of the participants, withdraw the personnel and resolve the situation. In certain situations, this will require careful analysis of the building in order to ensure it is safe to continue. Crews on the outside may need to be reassigned with new orders to extinguish the fire, until the building is assessed for safety. If the fire can be extinguished and the area made safe, sometimes ceilings can be secured and operations started again. Be very cautious with fires getting into attic and large void spaces, as this is a recurring problem in acquired structures.

If an evolution is stopped, it should only be restarted once the hazard identified has been resolved. The safety officer and any assistants need to confer with the instructor-in-charge and, if in agreement, the evolution should only continue after the "Go/No Go" sequence is initiated.

■ Water Supply

When it comes to water supply for live fire training, instructors have varying opinions on how much water should be available, what the water source(s) need to be, and how to maintain the water. On top of this, there are many times when instructors get nervous over the idea that they must do calculations to determine the amount of water needed at an evolution. The intent of NFPA 1403 is to assure that training is conducted in a safe environment. This means that if the attack line should lose its ability to fight fire due to a mechanical or other emergency, the backup line must still have pressure and enough water to put out the fire and allow the attack team to escape safely.

To this end, the instructor-in-charge is required to determine the amount of water needed for the live fire evolution as well as the amount needed for any unforeseen emergency. This requires that the instructor be able to determine the amount of water needed, identify the source of the water, and assure maintenance of the water supply for both attack and backup lines.

The instructor-in-charge shall determine the rate and duration of waterflow necessary for each individual live fire training evolution, including the water necessary for control and extinguishment of the training fire, the supply necessary for backup lines to protect personnel, and any water needed to protect exposed property. To determine the amount of water, NFPA 1403 refers the instructor to NFPA 1142, *Standard on Water Supplies for Suburban and Rural Firefighting*, which determines the minimum requirements for water supplies for structural firefighting. In areas where there is not a reliable municipal water resource, the instructor will need to be familiar with the water calculations in NFPA 1142. NFPA 1403 also requires a minimum reserve of additional water in the amount of 50 percent of the fire flow demand determined to handle exposure protection or unforeseen situations.

Determining the Required Water Supply

The **fire flow rate** is the amount of water pumped per minute (in gallons per minute or liters per minute) needed to extinguish a fire. The **minimum water supply requirement** is the total amount of water, not flow, required for a given structure based on its size, construction, and proximity to other exposures. In order to begin determining how much water will be needed, the instructor must know the occupancy hazard, type of construction, structure dimensions (length, width, and height), and any exposures that exist.

Remember that the fire flow rate and the minimum water supply needed for the evolution are two different calculations. For a nozzle that flows a minimum of 95 gallons per minute (gpm) (360 Liters per minute), which is the minimum allowed for live fire training, use the National Fire Academy's Fire Flow Rate formula, as follows:

$$\text{Fire flow rate (FFR)} = (l \times w) \div 3$$

Where: l = length of room/structure

w = width of room/structure

Per NFPA 1142, the minimum water supply (MWS) requirement formula is equal to the total volume (TV) of the structure divided by the Occupancy Hazard Classification (OHC) number multiplied by the Construction Classification (CC) number. If there is an exposure hazard (EH), the result is then multiplied by 1.5. This formula is expressed as follows:

$$\text{MWS} = (TV \div OHC) \times CC \times EH$$

The Occupancy Hazard Classification numbers and the Construction Classification numbers also come from NFPA 1142 and are as follows:

The **Occupancy Hazard Classification** numbers are a set of predetermined factors between 3 and 7 that represent the hazard levels of certain combustible materials.

Occupancy Hazard Classification:

3. **Severe hazard occupancies:** Explosives and pyrotechnics manufacturing and storage, flammable liquid spraying
4. **High hazard occupancies:** Warehouses, building materials storage, department stores, exhibition halls, auditoriums, theaters, upholstering with plastic foams
5. **Moderate:** Quantity or combustibility of contents is expected to develop moderate rates of spread and heat release
6. **Low:** Quantity or combustibility of contents is expected to develop relatively low rates of spread and heat release
7. **Light:** (Dwellings) Quantity or combustibility of contents is expected to develop relatively light rates of spread and heat release

The **Construction Classification** numbers are a set of predetermined factors between 0.5 and 1.5 that relate to the type of building construction of an acquired structure:

0.5 Type I Construction (Fire resistant)

0.75 Type II Construction not qualifying for Type I (Noncombustible)

1.0 Type III Construction: (Ordinary) exterior noncombustible or interior of wood

0.75 Type IV Construction: Heavy timber exterior and interior

1.5 Type V Construction (Wood frame)

As a rule of thumb and according to NFPA 1403, attack lines, backup lines, and rapid intervention crew hose lines should each be capable of flowing a minimum of 95 gpm (360 L/min), and two exposure lines should be capable of flowing 200 gpm (758 L/min) each. If all lines were operating at once, this would require a minimum of 700 gpm (2653 L/min). When calculating fire flow, all possible lines must be considered.

Water Supply Source

Now that the minimum water supply needs for the live fire evolution have been determined, the instructor needs to determine the source of the water. The instructor needs to consider water sources based on the location of the training. Rural water supply and urban water supply sources have different concerns, however the intent is to have reliable and valid separate water sources. When deciding the reliability of the water source, the instructor in charge needs to ask, "If my primary attack line water source is lost, or the pumper drafting should malfunction, will the backup line continue to have an adequate water source?"

If two pumpers are drafting from the same water source such as a river or pond, the instructor in charge must assure that there is enough water to supply both pumpers and both the attack and backup lines. If folding tanks are being used, two separate pumpers should be used to supply the attack and

Live Fire Tips

Example 1

A 1-story, wood-framed dwelling, measuring 20' × 20' × 10' (6.1 m × 6.1 m × 3.05 m), with a standard 4' (1.2 m) from attic floor to ridgepole and one exposure, requires what minimum water supply?

$$MWS = (TV \div OHC) \times CC \times EH$$

First calculate the total volume (TV) of the structure. This is done by multiplying length × width × height. For structures with a pitched roof, the height is equal to the wall height + half the height of the pitch. In this example, the height is 10' + ($1/2 \times 4'$) = 12'

$$TV = (20 \times 20 \times 12)$$
$$= 4800 \text{ ft}^3$$
$$OHC = 7 \text{ (light)}$$
$$CC = 1.5 \text{ (wood-framed)}$$
$$EH = 1.5$$

Therefore:
$$MWS = (4800 \div 7) \times 1.5 \times 1.5$$
$$= 685.7 \times 1.5 \times 1.5$$
$$= 1542.8 \text{ or } 1543 \text{ gallons}$$

Using these calculations, the minimum water supply needed is 1543 gallons, but in this case there is an exception. According to NFPA 1142, the minimum water supply required for any structure without exposure hazards is 2000 gallons (7570 L) and for structures with exposure hazards it is 3000 gallons (11,355 L). According to NFPA 1142, an exposure hazard is any structure within 50 ft (15.2 m) of another building and 100 ft² (9.3 m²) or larger in area. So in this case, the minimum water supply is 3000 gallons (11,355 L).

Now that the required minimum water supply has been calculated, what is the minimum fire flow rate (FFR) required for this structure?

$$FFR = (l \times w) \div 3$$
$$= (20 \times 20) \div 3$$
$$= 400 \div 3$$
$$= 133 \text{ gpm}$$

133 gallons per minute (8.4 L/sec) would be the minimum fire flow rate required for 100 percent involvement of the structure. Round down from 133 gpm to 125 gpm (7.9 L/sec) because of nozzle settings. If this example were of a two-story house, we would have to calculate an additional fire flow rate at 25 percent per floor up to 5 floors (see Example 2).

Live Fire Tips

Example 2

A 2-story, wood-frame dwelling is 50' × 24' (15.2 m × 7.3 m). Each story is 8' (2.4 m) high with a pitched roof that is 8' (2.4 m) from attic floor to ridgepole. What is the approximate minimum water supply (MWS) needed?

$$MWS = (TV \div OHC) \times CC$$

Calculate the total volume (TV) of the structure, which is equal to length × width × height. In this case, the height is equal to:

$$\text{height} = 8' + 8' + (1/2 \times 8')$$
$$= 8' + 8' + 4'$$
$$= 20'$$

$$TV = (50 \times 24 \times 20)$$
$$= 24{,}000 \text{ ft}^3$$
$$MWS = (24{,}000 \div 7) \times 1.5$$
$$= 3428.6 \times 1.5$$
$$= 5142.9 \text{ or } 5143 \text{ gallons}$$

The minimum water supply required for this structure is 5143 gallons (19,468 L).

What is the minimum fire flow rate (FFR) required for this structure?

$$FFR = (l \times w) \div 3$$
$$= (50 \times 24) \div 3$$
$$= 1200 \div 3$$
$$= 400 \text{ gpm}$$

400 gallons per minute (25.2 L/sec) would be the minimum fire flow rate required for 100 percent involvement of the structure, however since this is a 2-story house, we must add an additional 25 percent of the calculated FFR for each floor, up to 5 floors. Therefore:

$$FFR = 0.25 \times 400$$
$$= 100 \text{ gpm}$$
$$100 \text{ gpm} + 400 \text{ gpm} = 500 \text{ gpm total}$$

The total fire flow rate would be 500 gpm (31.5 L/sec) and the minimum water supply would be 5143 gallons (19,468 L).

Live Fire Tips

A technique that will assist with water supply and live fire training is to use different colored hoses to ensure that instructors and students know which lines support which function. For example, the attack line could be red, the backup line could be yellow, and the rapid intervention crew line could be green. If there is not enough colored hose to run from the engine, a different colored hose could be used at the nozzle only. There should be no confusion as to which line is being utilized for what function. The exposure lines should be clearly identified by divisions, as in exposure A/B, which would cover two sides of the structure, or exposure A and exposure C where each line is covering only one side of the structure.

---

backup lines. If possible, an added safety would be to have two dump tanks to supply the two pumpers. Regardless, the water source has to be enough, and two separate apparatus should be used to supply the lines.

If the training area is in an area with a municipal water source, the instructor needs to ask other questions such as, "If the main pipeline should rupture, are both hydrants affected?" and "Do municipal water sources have backup generators in case of loss of power?" If a municipal water supply system is used, two pumpers on two different hydrants should be used. This assures that should the attack line lose its water, the backup line should still have enough to protect the attack line and allow for escape. If the instructor-in-charge feels that there needs to be additional safety precautions, portable water tanks could be used at one of the hydrants or a pumper could be used to assure there is additional water source in an emergency.

Initial Preparation

After determining that a structure is usable, and the owner and fire department agree to terms, a more detailed inspection of the structure and property should be performed followed by preparation. The instructor-in-charge is responsible for ensuring the acquired structure is prepped to meet NFPA 1403. Although there is no rule or requirement that states this inspection be separate from the initial evaluation, it is often better to do so for organizational purposes.

Access

Access to the training site can be a problem, just as it can be in a hostile fire situation. However, unlike emergency responses, access to the training site can be planned for.

- **Access to and from the training area:** Locate a staging area for apparatus and parking for other vehicles that will not be used on the immediate scene. Remember to locate an area for in-service apparatus that will be considered available for response, so that they do not get blocked by parked vehicles, hose lines, or spectators.
- **Access to the property:** Fences or walls that limit access to the site may have to be removed. Make sure to keep areas wide open for unexpected emergencies.
- **Access around the property:** Trees, brush, and surrounding vegetation that create a hazard to participants or limit access to the scene must cut back or removed. High grass and weeds need to be cut clear in the operations area so any hidden hazards can be exposed. These include toxic weeds, insect hives, or vermin that could present a potential hazard. This removal may require professional assistance, especially with large bee colonies and similar hazards.
- **Underground dangers:** Septic tanks or other buried tanks must be identified and clearly marked and/or barricaded to ensure apparatus do not park above underground tanks. Drain fields that the owner expects to reuse need to be avoided. Both fire fighters and fire apparatus have been known to fall into septic tanks. Mark areas to be avoided **Figure 6-1**.
- **Nonparticipant access:** The press, spectators, law enforcement, and emergency medical services (EMS) will all be present on the day of training. Be prepared for this influx of spectators and plan ahead of time for parking and the area from which they will observe.

Equipment and Supplies for Preparing the Building

Depending on the structure, and according to NFPA 1403, the following items are needed to prepare the structure:

- A fully equipped engine
- Hammers and nails
- Cordless drills, battery chargers, and screws
- Cordless saws, battery chargers, and blades
- Generator(s) and extension cords
- Portable lights
- Step ladders

Figure 6-1 Septic tanks and other hazards should be marked for safety.

- Round point and flat shovels
- Brooms
- Chain saws and proper fuel
- Vent saw and proper fuel
- Weed eaters
- Bolt cutters, wire cutters, and tin snips
- Premade vent opening covers
- Utility ropes
- Barrier tape
- Water flow test kit or pilot tube and gage
- Measuring wheel with pad of paper
- Four sheets of ¾" (19 mm) CDX grade plywood (to cover openings in attic spaces)
- Six 2" × 4" × 8" (50.8 mm × 101.6 mm × 2.4 m) wooden studs (to use as braces)
- Type X Gypsum wall board

Some training centers use large bins to keep forms, markers, accountability supplies, signage, plans, reports, and other materials that are needed for each burn, in good order and stored conveniently between uses **Figure 6-2**. Use a printed inventory of what should be in each bin. The bins are good for acquired structures or permanent live fire training structures.

Entry and Egress for the Structure

All entry and egress routes must be planned for ahead of time and must be known to all participants. These routes must also be monitored during the evolutions to ensure they are clear in the event of an emergency. In addition, the primary doorways must being clear for normal access. When unexpected problems arise, fire fighters should always have a short

Figure 6-2 Large, plastic bins can be used to keep materials in order.

Figure 6-3 Hose hinges keep window coverings in place, but are easily knocked down.

travel distance to the outside. Likewise, instructors and rapid intervention crews (RIC) should be able to rapidly access the interior crews' status in a moment's notice. In selecting what rooms are to be used for the live burns, there must be at least two separate means of egress available for each room. Any room with limited access should not be used for live fire training purposes, unless a door can be cut out. Fires cannot be located in any exit paths, so that participants have a direct path to the outside.

The following are precautions to take when selecting rooms and preparing the structure for live fire training:

- Trees, brush, and surrounding vegetation that create a hazard to participants or limit entry and egress shall be removed.
- Interior and exterior doorways need to be easily accessible, with no obstructions, and no drop-offs that could cause injury. Locking mechanisms should be disabled or removed. Pneumatic or other types of door closers should be removed, as should storm doors, screen doors, and secondary security doors.
- Windows need to be easily accessible for emergency egress, and when windows are covered for the training evolutions, these coverings must be removable from inside or out by hand, without tools. Consider using hose hinges, or other methods Figure 6-3 . Any glass, hardware, horizontal or vertical cross pieces, window air conditioners and fans, or anything else that could hinder

Figure 6-4 A window can be made into a doorway by cutting and removing a section of the wall below the window.

a fire fighter from exiting should be removed. There must also be an exterior escape path that is not blocked by plants or other obstructions. If the window sill is high or is too small, cut the wall below it to make it accessible as an exit, if there are not sufficient means of egress for the evolution planned Figure 6-4 .

- Make doors where needed. Sometimes long hallways or large rooms make for long escape routes. Keep escape routes short and clear, and cut through walls if needed.
- Stairs, porches, and railings need to be in serviceable condition and not hazardous to exiting fire fighters.

Exterior Preparation

Asphalt and Asbestos

In the initial assessment, the exterior was assessed for sagging or other visual signs of potential structural damage. Now the actual exterior surface covering must be evaluated. Although there are a number of environmental concerns, the most common concerns are asphalt and asbestos. Be sure to know the applicable requirements in your jurisdiction for permitting, testing, removing, and disposing of such substances. Most areas will have requirements for asbestos testing. Many homes and buildings built before 1985 will have asbestos either in the roofing, flooring, insulation around the heater and plumbing, or possibly in exterior or interior wall surfaces. Starting in the 1920s, the National Board of Fire Underwriters recommended that homeowners use asbestos siding and roofing instead of the less fire-resistant wood materials. By 1979, the US government outlawed the use of asbestos in many building products, because of its physiological effects on humans. This ban included asbestos cement siding shingles; however, they are still found in structures today.

Siding is generally not an issue, unless it is asphalt or asbestos. Composite siding made of asphalt-impregnated fiberboard, with the surface granules similar to asphalt roof shingles, was popular on less expensive wood-frame houses built before 1950. The asphalt siding often has a pattern that makes it look like stone or brick. Also, cement asbestos board was used for lap siding and wall shingles. Asphalt shingles, paper shingles, or tarpaper roofing will most often have to be removed due to environmental requirements. It is important to note that products containing asbestos cannot be identified by sight alone, and further testing may be needed.

Houses built before 1980 should be suspected of having asbestos. Asbestos can be found in many places in a building, such as on furnaces, boilers, hot water pipes, ceiling tiles, drywall, roofing felts and shingles, and exterior siding shingles.

Asbestos siding was manufactured in a wide range of colors and patterns, but it does have some characteristics that may help identify it. Asbestos siding shingles are usually 12" × 24" (304.8 mm × 609.6 mm). They may have grooves or a wood-grain pattern pressed into the cement, or they may be smooth. Each tile usually has two or three nail holes at the bottom of each shingle. Another popular type of asbestos siding came in 27 ½" (12.7 mm) corrugated sheets of various lengths. These sheets were used in the same way that corrugated metal sheeting was used. It can be recognized by the corrugation, but these sheets were seldom used in home construction.

Utilities

Once the building construction is secure and all measures have been taken to mitigate any dangers, the utilities must be secured. This requires going beyond the household or building service disconnects and shut-offs. There have been unfortunate experiences during live fire exercises where the electrical service

Figure 6-5 Ensure that the electrical service lines are disconnected from the pole to the weather head.

lines going to the weather head were found to still be charged. Electrical service lines must be disconnected from the pole to the weather head **Figure 6-5**. Natural gas must also be shut off at the distribution line. In both cases, this disconnect must be done by the service provider, and that provider should be the point of contact if there are any questions. Liquid propane gas (LPG), fuel oil, and other such tanks must be moved away from the building. Any tank that cannot be removed must be rendered inert, most often by filling it with sand. In addition, tanks that cannot be removed shall be vented to prevent an explosion or rupture. Check with the service provider if any doubt exists.

Marking the Building

The sides of the building should be marked according to local protocol. This procedure is done to eliminate confusion and to reinforce the procedure with all of the participants. The accepted protocol is generally for the front of the building to be marked as side A, with sides B, C, and D designated in a clockwise fashion. However, there are local and regional variants, so you should check with the authority having jurisdiction **Figure 6-6**.

Ventilation

Roof ventilation is another aspect of exterior preparation that needs to be readied. This process will have to wait until the interior fire locations are finalized. Emergency ventilation must be planned for in order to limit fire spread and improve interior conditions. Neither the primary nor secondary egress points should be used for normal room ventilation. The ventilation endangering personnel operating inside and outside. Apart from the collapse concern, chimneys can allow fire spread into attics and upper floors during house fires. Although this is an infrequent problem, it should be considered. Outside chimneys should be removed before training takes place.

Additional Hazards

Any toxic weeds, insect hives, or vermin that could present a hazard to fire fighters must also be removed from the area. The last part of exterior preparation should be to remove any exposures or combustible materials outside, including storage sheds, detached garages, and materials from demolition and such. These items must either be removed or protected from unintended ignition or fire spread.

Interior Preparation

Common Hazards

Most of the preparation efforts will be spent inside, as the interior will be where the majority of operations take place. Most often, buildings donated for live fire training purposes have not been occupied for some time. Some buildings may have been used by trespassers for shelter or for illicit use. Hazards from these nontraditional uses can be present in the form of drug paraphernalia, broken glass, weapons, and even infectious clothing or bedding. Caution should be exercised during the preparation stage to protect personnel from these hazards, as well as from unseen structural dangers. It is important to use personal protective equipment, especially safety shoes, helmets, gloves, eye protection, and if using power equipment, hearing protection, etc.

A systematic manner of preparation is necessary to accomplish the desired level of security. First and foremost, determine the building's utility status. Next, before interior preparation begins, check for environmental hazards, such as insects in or around the structure that could have an effect on the building's preparation, or the evolution itself. Environmental hazards include contamination from past storage in homes and in businesses. Some such contamination includes pesticides and other chemicals, depending on the building's previous use. Illegal drug manufacturing has become an issue in many areas, and the conditions associated with these locations can dangerously contaminate a structure. Unfortunately, with abandoned structures, vagrants or drug users may have occupied them and left behind biological issues in addition to dangerous drug paraphernalia, broken glass, and trash. It may be wise to include bio-hazard protection such as latex gloves, masks, bags, and needle containers. Even storage found in normal homes can provide for dangerous contamination from household pesticides, fertilizers, paints, and other materials. It may simply be safer to remove contaminated wood shelving or flooring. Businesses can be more difficult when it comes to environmental concerns, and it is important to determine what the building was previously used for. Consider checking past inspection reports and preburn plans.

One practice currently used is to spray the inside of the building with a bleach solution using a pump sprayer, starting from the farthest point from the exit and working towards the exit. This will kill off any insects and disinfect the area prior to removing any items. The suggested concentration is two cups

Figure 6-6 The sides of the building should be physically marked according to local protocol.

Figure 6-7 Roof ventilation openings are made during preparation, not during the actual fire. Hinges, pivot, and cable allow remote release from the ground.

opening must be placed in such a way as to draw the products of combustion *from the means of egress*, not towards the means of egress. This gives the fire fighters a way out while the products of combustion are drawn away from them. When deciding where to put the ventilation openings, keep in mind fire behavior, the layout of the building, the burn room(s), hallways, and the primary and secondary means of egress. Existing roof openings that are normally closed, but could be opened in the event of an emergency, can be utilized. The ventilation opening must be cut during the preparation stage and not while any live fire evolutions are taking place **Figure 6-7**. Do not place fire fighters on the roof during a live fire training evolution. A pivot point and cable can be used to secure the covering so that it can be opened and closed from the ground as needed.

Chimneys

Chimneys need to be checked for stability to make sure they will not fail during interior operations. Once the structure is compromised, it becomes more likely that the chimney will collapse,

Near Miss REPORT

Report Number: 9-919
Date: 10/14/2009

Event Description: After completing four live burn evolutions in an acquired structure, a PPV fan was placed near the basement doors to vent the structure. All the evolutions were basement fires and were compliant with the requirements of NFPA 1403.

After ventilating for about 15 minutes, myself and a crew of two went into the structure to retrieve the metal drum and remove the 1-¾ hose that was the safety line for the ignition team. We had full structural gear without SCBA.

Once we entered the basement, we found the drum still burning and impinging onto the ceiling joists. Extinguishment was quickly made with the handline and we found no extension.

Lessons Learned: Decision-making and situational awareness are of utmost importance. The decision to remove the hose team without fully extinguishing the class A barrel could have led to an actual uncontrolled fire in the basement.

The safety officer and the instructor assigned to the basement both left without verifying extinguishment.

Lesson learned – make sure the fire is out before leaving.

---

of bleach per gallon of water (0.24 L of bleach to 3.785 L of water) in a sprayer. Be sure to use a mask when spraying the structure. Spray the structure one or two days prior to building preparation to allow for drying. Some other suggested solutions are as follows:

- 1 tablespoon of regular bleach per gallon of water, for Staph and *E.coli*.
- ¾ cup of bleach per gallon will kill Feline Parvovirus and Canine Parvovirus.
- 1 ¾ cup bleach per gallon will kill *Mycobacterium bovis* (Tuberculosis).

Furniture

Furniture needs to be removed during the interior preparation. It may be tempting to leave furniture in the structure that appears to be wood, but do so with great caution. Most furniture today is not made of solid wood, but rather pressboard with an exterior laminate to look like real wood. Only Class A materials with known burning characteristics may be used for the burn sets (materials to fuel the fire). Until recently, NFPA 1403 did not specifically prohibit the use of burning furniture, as it did the use of flammable or combustible liquids. Previous editions of NFPA 1403 required that the fuels utilized have known burning characteristics and be as controllable as possible. With furniture, it is not always feasible to determine the construction materials used. Many of the commonly used products give off considerably more heat, smoke, and toxins than would be expected.

Live Fire Tips

A bleach solution will kill the following:
- Bacteria: *Staphylococcus aureus* (Staph), *Salmonella choleraesuis*, *Pseudomonas aeruginosa*
- *Streptococcus pyogenes* (Strep), *E.coli*, *Shigella dysenteriae*
- Fungi: *Tricholphyton mentagrophytes* (causes athlete's foot), *Candida albicans* (a yeast)
- Viruses: Rhinovirus Type 37 (a type of virus that can cause colds), Influenza A, Hepatitis A virus
- Rotavirus, Respiratory Syncytial Virus (RSV), HIV-1, Herpes simplex Type 2, Rubella virus
- Adenopvirus Type 2, Cytomegalovirus

Safety Tips

During interior clean up, dust masks and eye protection are necessary. Hand protection is also strongly recommended when removing bedding, clothing, and similar items.

Safety Tips

Per NFPA 1403, ordinary combustibles such as clean wooden pallets, pine excelsior, and hay and straw not chemically treated are allowable fuels. Clean wooden pallets means that they are free from any noticeable spilled material that may have soaked into the wood such as oils, pesticides, or other material that may cause an unforeseen condition or create a hazard.

NFPA 1403 specifically prohibits the use of materials found on-site where the fire department cannot verify the environmental or health hazards associated with the materials, such as exposure to chemicals not readily apparent. Unknown chemicals pose dangers, not just in the obvious way of inhaled chemicals, but also through contact. An unknown contaminant can get on protective clothing and could be later inhaled or contacted. For this reason, unknown materials are prohibited, along with known materials such as pressure-treated wood, rubber, and plastics. Further, straw or hay that is known to be treated with pesticides or harmful chemicals is not allowed for the same concerns.

Flooring

After the furniture has been removed, the rugs, carpeting, padding, and tack strips can be removed. This will expose the structure's actual floor. All combustible material must be removed, especially linoleum, which is a solid petroleum product **Figure 6-8**. Once linoleum is heated, it can act like flammable liquid pouring on the floor. Sheet vinyl (including the backing or underlayment), vinyl tile, and vinyl adhesive may all contain asbestos and must be removed. Remember, personnel will be crawling inside of the structure, so broken glass, debris, carpet tacking strips, vinyl flooring thresholds, nails, and other sharp objects are all crawling hazards and need to be removed. Holes in the flooring can lead to unexpected fire spread. Any holes in the floors need to be covered in a manner that will not cause harm, and must be able to bear the weight of fire fighters. Openings in the floor or ceiling created because of equipment being removed or other renovations, should be evaluated for reducing structural stability.

Walls and Ceilings

Walls and ceilings need to be checked for low-density combustible fiberboard or other combustible interior finishes. Low-density combustible fiberboard has contributed to the deaths of many fire fighters, and was a major factor in fire spread in the fires at Our Lady of Angels School (Chicago, Illinois), Hartford Hospital (Hartford, Connecticut) and Opemiska Social Club (Chapais, Quebec). Be very careful with unconventional interior finishes such as burlap, carpeting, artificial turf, and other treatments as they may cause rapid fire spread and unexpected smoke production, along with greater toxicity. Ceiling fans and large light fixtures also need to be removed so that they do not fall on participants during live fire exercises. Any holes that may allow fire to travel into the concealed void spaces must be covered.

Windows and Doors

Windows, as previously mentioned, need to be available for emergency egress. Window openings can be covered to keep smoke in and control flame spread. Be careful not to seal the windows airtight, because this can inadvertently create flashover conditions. Depending on the size of the windows, a small space at the bottom of the window can be left open to allow for ventilation to reduce flashover concerns. A small opening can also be made at floor level to allow for the introduction of air. If using a chainsaw, the width of the chainsaw bar will work, and either a three- or four-sided opening should suffice.

Interior doors must also be made safe. Remove any hardware that may snag or catch on personal protective equipment (PPE) **Figure 6-9**. Either remove or secure doors that need to remain open. It is recommended to clearly mark doors

Figure 6-8 For safety, expose the structure's actual floor by removing all additional layers of flooring.

Figure 6-9 Any hardware that may snag or catch on PPE must be removed.

that are not exits, such as closets or bathrooms without usable windows, and consider covering them so that a fire fighter under adverse conditions is not confused by the door frame and door **Figure 6-10**.

Kitchens

Cabinets and kitchen appliances are considered fixed contents and their removal will vary depending on what they are made of. Some live fire training instructors may want to leave in kitchen cabinets or bathroom cabinets for the live fire evolution. Depending on whether they are solid wood or composite-board, a burn test may be needed to determine the burn characteristics. If in doubt of the type of material, it is always best to remove it.

Commercial fixtures, such as large coolers or freezers, warrant careful review. Such commercial fixtures also frequently have large voids behind or around them to allow a service person to access them for repair. Such voids need to be considered. Closed containers, including water heaters and air conditioner compressors need to either be vented or removed **Figure 6-11**. Cans of products, especially aerosol cans and smaller items, need to be removed.

Oil Tanks

Oil tanks and similar closed vessels that cannot be removed from the structure must be vented to prevent an explosion or overpressure rupture. Enforce strict safety practices when ventilating these tanks. Also, any hazardous or combustible atmosphere within the tank or vessel shall be rendered inert to prevent an unexpected explosion. All hazardous structural conditions shall be removed or repaired so as to not present a safety problem during use of the structure for live fire training evolutions.

Attics

The attic space needs to be inspected for hazardous contents. Air handlers, hot water heaters, gas heaters, storage, and other items could be a collapse hazard and some items could contribute to fireload or adverse conditions. The attic space should be accessible so that a fire could be controlled with ventilation during evolutions, if necessary. Place a piece of plywood over the opening on the ceiling and on the roof, securing them both just enough to hold them in place so that they may be removed when needed. Determination of when to open the space will depend on the layout of the building, the burn room, hallways, and the primary means of egress.

Live Fire Tips

A wood-frame house with lightweight truss construction was going to be used for a live fire exercise. When personnel checked the attic nothing was found amiss, but the safety officer noticed a wall covered in drywall that was set back about ten feet from the opening. Climbing into the attic, the safety officer found a full 200-gallon (757-liter) water tank that had been used when the house had solar panels. The tank had a weight hazard of over 1600 pounds (726 kg) and was situated directly over the hallway, which was the main egress for the burn rooms.

No matter how experienced or how many times instructors prepare a structure, when in any instructional or leadership position, do not take it for granted that everything has been done properly. Never assume!

Safety Tips

Weight above the training area can include contents on the floors above the fire, and on the roof. Anything that poses a potential hazard to participants needs to be removed.

Figure 6-10 Clearly mark doors that cannot be used as exits.

Figure 6-11 Vent or remove closed containers that could rupture when heated.

Figure 6-12 Fluorescent spray paint can be used to mark the way out.

> **Live Fire Tips**
>
> Variables leading to flashover:
> 1. The heat release characteristics of materials used as primary fuels
> 2. The preheating of combustibles
> 3. The combustibility of wall and ceiling materials
> 4. The room's size and geometry (e.g., ceiling height, openings to rooms)
> 5. The arrangement of the initial materials to be ignited, particularly the proximity to walls and ceilings
> 6. The size and location of ventilation openings, both exterior and interior

Exits

Primary exits and exit routes need to be clearly marked and evaluated before each live burn. There are various ways to show exit routes, including the use of a light rope (illuminated rope-like cord), a strobe light, or other light at the primary exit point. Bright fluorescent paint on the floor can also help indicate the way out **Figure 6-12**. Some live fire training instructors suggest marking the exit pathways at baseboard level or on the floor next to the wall area to prevent the paint from being worn off by crews crawling with hose lines. Participants in the training evolution need to be made aware of the exits and any markings used prior to each evolution.

Stairs and railings must be made safe. Any broken or missing trends need to be replaced, the stairs must be clear of trip hazards, and the weight-bearing capability must be checked. The handrails and balusters also need to be secure.

Burn Locations and Fuel Loads

Each burn room must be prepared using Class A materials only. Only one burn room can be used at a time, and only one fire set is allowed per room. Identifying the exact burn locations is critical in the preparation of the interior. The final burn locations are decided by the instructor-in-charge and safety officer.

The instructor-in-charge shall document the fuel loading, including all of the following:
- Furnishings
- Wall and floor coverings and ceiling materials
- Type of construction of the structure, including type of roof and combustible void spaces
- Dimensions of the room

Using all of the information above, the instructor-in-charge needs to carefully evaluate each fire room and burn location. The instructor-in-charge must check for factors that can affect the growth, development, and spread of fire. The instructor-in-charge must also evaluate for flashover danger and must map out the anticipated fire spread.

An excessive fuel load can help to create unusually dangerous fire behavior. When preparing the burn rooms, use only what is necessary to meet the training objectives and avoid conditions that could cause a flashover or backdraft. In several training incidents that resulted in injury or death, those running the drill did not anticipate flashover or unexpected fire spread based on the amount of fuel being used.

To avoid such tragedies, specify the fuel loading precautions during the planning stage. This should include the number of pallets to be burned, how much hay or excelsior to be initially used, and how much burn material can be added to the fire and when.

To save time, the pallets can be positioned at the burn set locations during the preparation of the interior. Hay or excelsior should be placed in the burn sets on the day of the evolution, especially if the structure is not secure. You do not want an intruder lighting the fires early. Also remember that you do not want the spare pallets and hay to become interior exposures and allow fire to spread to them. All other possible sources of ignition need to be removed from the operations area, except for those under the immediate supervision of the ignition officer.

Considerations for locating the exact burn locations include plotting the expected avenues of fire spread and the time factors for expected build-up of the fire. Carefully analyze the room's characteristics, ventilation, and openings in order to plot anticipated and "worst case" fire spread. Ventilation must be planned. Do not place the fire fighters between the fire and the ventilation opening because the fire may travel towards them. Remember that if the entry door is the only source of natural ventilation for the fire, there is a possibility that the fire will engulf the participants when it is in need of fresh air. Openings and voids within the structure can result in sudden and unexpected vertical fire spread and can trap participants by cutting off exit routes. Even worse, this fire spread can result in the unexpected weakening of the structure, which can lead to collapse. The instructor-in-charge and the safety officer need to ensure that the plans for primary and secondary exit paths do not conflict with the expected avenues of fire spread.

Burn Set

The burn set should be physically located in a corner of the room. This will vary based on where the fire stream will be directed into the room. Try not to place the burn set directly across from the door where the nozzle will be operating from, as it tends to allow personnel to operate from the hallway and not enter the room, and does not utilize the space in the room efficiently. Also, the burn set cannot be located in any designated exit path.

Lining the Room

There are several ways to extend the use of the rooms used for the live fire evolutions. One way is to "line" the room with an additional layer of dry wall for greater fire resistance. Another method is to use a "fire box" hearth. Both of these methods can also be used together.

To line the room, attach 5/8" (15.9 mm) Type X Gypsum wall board to the ceiling and walls adjacent to where the fire set will be. Type X Gypsum wall board has great fire resistance rating, and noncombustible fibers are added to the wall board to maintain greater strength and greater resistance to heat transfer during fire exposure. Put the covering on the ceiling first and then the walls, in order to provide a better seal and prevent fire spread into the attic area Figure 6-13 . Taping and "mudding" the drywall can extend its use even longer.

Hearths or fireboxes can also be used in conjunction with a "lined room," or by themselves Figure 6-14 . Hearths are metal and are used to protect the ceiling and walls from direct flame. They can be used during initial burns and then removed. There is a good chance that they may not be recovered, but when used with or without the drywall lining, they can be an effective way to provide for more evolutions in a single room.

■ Preparing the Neighborhood

It is important to be aware of the homes and businesses in the areas surrounding the training site. Properties adjacent to the burn site that could be affected by smoke or fire were identified in the initial evaluation. The owners of those identified properties need to be informed of the dates and times of the training evolutions, and any problems will need to be addressed at this time. Signs should be in place to advise people of the training session, and smoke advisory signs should also be posted. Traffic assistance from law enforcement may be necessary and should be planned for in advance.

Neighbors should be treated with courtesy and respect. Be sure to apologize for any inconveniences and try to answer all questions, to ease concern. A little planning can go far with neighborly relations. Set up a viewing area for them and let them tour the site before the fire. Explain all tasks at hand and what they should expect during the training. Make sure they know the following:

- Keep animals away.
- Close windows and shut off air conditioners.
- Bring in laundry from outside.
- Park vehicles inside the garage.

Spectators

Any live fire training evolution will inevitably draw spectators. Planning for the training must include spectator parking and the designation of spectator areas. Fire lines and warning signs need to be used to keep any pedestrians clear of the operations area, also known as the hot zone. These lines may need to be staffed for enforcement. All spectators need to be kept away from operations, where they can be hurt or get in the way. Some spectators may

Figure 6-14 Hearths or fireboxes can protect the ceiling to allow for more live fire evolutions.

Figure 6-13 Greater fire resistance can be added by lining the burn room with drywall.

Live Fire Tips

The locations of the burn sets should be all of the following:
- Determined by the instructor-in-charge with input from the safety officer and senior instructors
- With due regard to preventing flashover and fire spread
- The burn set cannot be located in any primary or secondary designated exit path.

Figure 6-15 Escort members of the media to a location outside of the operations area where they will have a view of the action and remain safe.

use cameras or video recorders, so the training session must be conducted in a professional manner. This may warrant a small reminder to the participants about proper conduct.

Media
The media will oftentimes be present and a viewing area must be established ahead of time **Figure 6-15**. Reporters should not be allowed in a structure during an actual fire. NFPA 1403 requires visitors allowed within the operations area perimeter to be escorted at all times. It also mandates that they should be equipped with and properly wear complete protective clothing.

Live Fire Training Evolution Preparation

After the initial preparation of the acquired structure, there are steps that still must be taken before conducting a live fire training evolution. The actions taken or not taken during this phase will decide the outcome of the training evolutions. Will it be a rewarding and organized training evolution or is it set up for failure from the start?

Equipment and Supplies
At a minimum, the equipment needed on the day of burn should include the following:
- Two fully equipped pumpers
- Sufficient portable radios with spare batteries and chargers for each crew, instructor position, etc. Include spares.
- Portable generator(s) with extension cords and surge suppressors (for radio chargers, cooling fans, lights, etc.)
- Accountability boards (command, entry, divisions/groups)
- Supplies to support local accountability procedures, such as personnel accountability tags, unit identification pads, Velcro, permanent markers, grease pencils, tape writer, razor knife or scissors, etc.
- Incident command vests
- Igniter (propane)
- Spray paint (a florescent color is recommended for interior painting)
- Two anemometers (wind speed meters) for command and rehab
- Thermal imaging camera(s)
- Four instant-up canopies (command, entry, rehab/medical, staging)
- Folding tables
- Folding chairs
- Six or more water coolers (command, entry, rehab/medical (3), and staging), depending on the weather. Participants should be encouraged to bring their own water as well.
- Two basic life support (BLS) kits including blood pressure cuff, stethoscope, tympanic thermometer, trauma dressings, burn sheets, and any other additional BLS supplies
- Automated external defibrillator(s)
- Documentation folder, containing:
  - Copy of NFPA 1403
  - Copy of model SOG/SOP
  - Copy of National Weather Service point forecast
- Copy of the preburn plan inclusive of the site plan, floor plan, communication plan, evacuation plan, medical plan, and demobilization plan
- Appropriate blank and executed forms
- Optional equipment:
  - Video and still cameras
  - Binoculars
  - Stopwatch
  - Scene tape to mark the operations area (the area where PPE is required), and the fire scene, which is the wide area surrounding the area of operations, in which no unauthorized personnel are allowed.
  - Traffic advisory signs, cones, and barricades; the local public works department can be a good resource.

Preparing for Ignition
The ignition officer is now a member of the fire control team, and must work in very close concert with the safety officer and instructor-in-charge to ensure that the objectives of the training fire are met. Significant time and effort have been expended thus far, so it is very important that the ignition officer follow the proper order of operations, and not deviate from the plan regarding fuel loading, burn set locations, or safety protocols.

As previously mentioned, use only Class A materials with known burn characteristics in the burn sets. Class A materials include hay, excelsior, baled cardboard, and clean wooden pallets. Any unidentified materials that could burn in unexpected ways, or could create health or environmental issues shall not be used. Pressure-treated wood, rubber, and plastic are also not allowed to be used. Remember, only one fire at a time in an acquired building.

Excelsior is a product made of wood slivers that is used in packaging, cushioning, and in stuffed animals. Excelsior will act somewhat like hay when burned, but will give off more **British thermal units (Btu)** as compared to the same amount of hay. A British thermal unit is a unit of energy, that is approximately the amount of energy needed to heat 1 pound of water, 1 degree Fahrenheit. Because excelsior has different properties than hay, you should practice first to determine the amount needed. This should always be done prior to and outside of the acquired building. Once again, it is necessary to know the burn characteristics of the materials being used.

Hay is usually the easiest to use when preparing a burn set. Clean, dry, feed-grade hay should always be used in order to limit the presence of herbicides and pesticides in the hay. A half a bale of hay is normally sufficient in most burn sets.

A maximum of five wooden pallets should be used with half a bale of hay. This should be sufficient for rooms that are approximately 10' × 10' (3 m × 3 m) or larger. Smaller rooms may only require a quarter or a third of a bale of hay with only three or four wooden pallets. One pallet should be set on the floor and two on each side leaning into each other to form a triangle, leaving an opening in the middle for the hay **Figure 6-16**. Short legs made of 2" × 4" (51 mm × 102 mm) blocks can be added to lift the bottom pallet anywhere from 6" to 12" (152 mm to 305 mm) off of the floor. This may allow for better air circulation.

The hay should not be placed into the burn sets until the day of the live fire training evolution. When putting the hay into the burn set, fluff it by pulling it apart from the bale, making sure it is not compressed. This will make for proper burning. Bunches of hay haphazardly placed in the burn set does not work, because it does not allow for proper air circulation.

Figure 6-16 One method for configuring the burn set is to lean two pallets against each other, forming a triangle, and place hay or excelsior in the middle.

Safety Tips

Only one fire at a time in acquired structures!

Operations

Setup

Setup according to the site plan unless conditions, such as changing wind direction, dictate a change. A change in wind direction or force can postpone an exercise when smoke or heat could affect exposures. Any deviation from the plan *must* be included in the briefing of the instructors and participants.

Preburn Briefing and Walk-through

A preburn briefing and a walk-through of the structure are required prior to any thought of igniting the burn set. The instructor-in-charge and safety officer conduct the walk-through with all of the instructors and students. It may be necessary to demonstrate an evolution if there is any question of how something is to be performed, or to give clarity to the instructions. Two separate walk-throughs should be conducted, first with the instructors and then with the students. All instructors must have a clear understanding of the objectives for the evolution. If a room and contents fire is planned with fire being allowed to come out of the room, not to exceed four to six feet (1.2 m to 1.8 m), then there should not be flames engulfing the entire structure or traveling down a hallway for 10 to 15 feet (3 m to 4.6 m). Part of the preburn briefing is to point out the burn sets, sequence of burns, and the burn order of the rooms with the rooms numbered. Exit markings, ventilation points, and the primary and secondary means of egress should all be addressed at this time. All participants need to know the operations area where all PPE is to be worn and the fireground (warm zone) where a minimum of a helmet only may be required.

Go/No-Go

Prior to the safety officer giving a "go" in the Go/No Go sequence, he or she must inspect the structure to make sure it is clear of any occupants, unauthorized persons, animals, or objects. The possibility of vagrants being found in buildings always exists, even after the structure is prepared for burning. Ensure that the participants, students, and instructors are ready. This includes a check of all protective clothing and breathing apparatus.

Safety Officer

After the safety officer advises a "go," the instructor-in-charge should declare, "We have a go for ignition." The ignition officer would ignite under the supervision of the safety officer with a hose line in for protection. NFPA 1403 does not allow a student to be the ignition officer. This position needs to be filled by an experienced live fire instructor or under the immediate supervision of one.

In all situations there must be one safety officer, but especially in acquired structures, that position may need "deputy" safety officers to do the following:

- Monitor the entry point, ensuring a head count on all participants is done upon entry and exit. This must include instructors, other safety personnel, ignition officer, etc.
- Monitor interior participants for safe entry and egress.

- Monitor fire conditions between entry crews (reminder: nobody operates alone inside).
- Monitor structural conditions inside and out, watching for fire spread, changes in smoke production, structural integrity, etc. Provide visual monitoring for areas outside the safety officer's field of vision.

Igniting the Fire

The decision to ignite the training fire is made by the instructor-in-charge in coordination with the safety officer, and should follow the Go/No Go sequence. The fire is then physically lit by the fire control team (the ignition officer and at least one other firefighter [not a student]) with a charged hoseline present. The fire control team and all personnel inside during the ignition process must wear full protective clothing, including SCBA. If there are any other possible sources of ignition, other than what is being watched and used by the fire control team those sources must be removed from the operations area immediately.

In actual practice, often the instructor-in-charge and the safety officer will check the room and the fuel loading and arrangement before the training starts. Remember that no one should be alone inside.

The actual ignition of the burn sets can be done in several ways. Many training agencies use propane torches, which generally result in an easy ignition **Figure 6-17**. If using propane torches, the ignition officer is responsible for removing the torch from the structure after ignition.

Road flares also work well for ignition, but caution should be taken as these flares work by burning metal powder that may drip and lead to additional ignition sites. If you use a road flare, the remaining flare should be removed from the structure and properly extinguished.

Fire Behavior Considerations

Reports by the National Institute for Occupational Safety and Health (NIOSH) and other agencies that have investigated injury and fatal training fires, indicate that not recognizing changes in fire conditions has been a common issue.

Fire behavior, and the many aspects it covers, is often overlooked. The ignition temperature of the materials, the amount of heat given off for the volume of material burning (or anticipated to burn), the arrangement of the materials, the amount of fuel compared to the size of the room, the geometry of the room, natural ventilation pathways, the amount of water needed to absorb the heat, the impact of radiant heat, the combustible gases in the smoke being produced, and the flashpoint for that smoke during the fire, are all considerations.

Issues with using furniture and other materials made of synthetics include the increased amount of heat given off, the increased amount of combustible gases with lower ignition temperatures given off, and the greater amounts of toxic chemicals released in the air.

Carbon monoxide is the most prevalent by-product of combustion, and has a relatively low ignition temperature (approximately 1148°F [620°C]), and is considered highly toxic.

Live Fire Tips

Remember:
- Ignition will only begin with the approval of the instructor-in-charge in coordination with the safety officer following the Go/No Go sequence.
- Only one fire at a time shall be permitted within an acquired structure.
- Fires shall not be located in any designated exit paths.
- Fuel materials shall be used only in the amounts necessary to create the desired fire size.
- The fuel load shall be limited to avoid conditions that could cause a flashover or backdraft.
- No unidentified materials, such as debris found in or around the structure will be used as they could burn in unanticipated ways, or create environmental or health hazards.
- No pressure-treated wood, rubber, or plastic should be burned.
- No flammable or combustible liquids should be used to assist ignition.
- Wooden pallets must be clean from any spilled material that may have soaked into the wood such as oils, pesticides, or other material that may cause an unforeseen condition or create a hazard.
- Hay needs to be dry, clean, and feed-grade to prevent using hay that has been subjected to pesticides or other unknown material.
- Nobody operates inside alone.
- Watch for debris from pallets and other materials that may hinder the access or egress of fire fighters, and be sure to remove it prior to the beginning of the next training exercises.
- Any time the instructor-in-charge determines that fuel, fire, or any other condition represents a potential hazard, the training evolution shall be stopped immediately. If an exercise is stopped, it should only be restarted once the hazard identified has been resolved and after the Go/No Go sequence.

Figure 6-17 A propane torch can be used for interior ignition.

Incident Report

Baltimore, Maryland - 2007

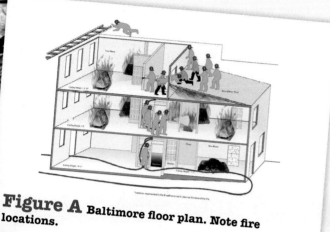

Figure A Baltimore floor plan. Note fire locations.

The Baltimore City Fire Department was conducting a live fire training exercise in a row house following the department's Fire Fighter I program. The program had not been successfully completed by all participants, and some of the recruits did not successfully complete the physical performance requirements during training. Some of the recruits had also never participated in interior live fire training evolutions.

The involved building was a three-story row house, and was one of a series of very narrow row houses, only 11'4" (3.5 m) wide. It was a 1200 ft^2 (111.5 m^2) unit, built as a single family home. It had been vacant for almost seven years, and condemned for three years. Each unit had separate exterior walls and the side walls touched on one or two sides of the adjacent unit(s). The units were trapezoidal in shape, and this was an end unit **(Figure A)**.

The live fire exercise consisted of multiple fires on all floors. The investigation indicates that approximately 11 bales of excelsior and at least 10 wooden pallets were used for the fires. Pallets were propped against the walls in several locations, with excelsior underneath. Excelsior was also placed in openings in the ceilings and in walls. In the backroom on the first floor, an automotive tire, two full-size mattresses, one twin-size mattress, one foam rubber chair, tree branches, and other debris were piled up and burned. The exact number of fires set could not be definitely determined, but there were at least nine separate fires.

It appears that the ignition of the fires was not coordinated with the preparation of students entering. Several students were not ready when the fires were ignited, and that allowed the fires to burn unimpeded while crews made final preparations to enter.

Five separate crews were to operate simultaneously as engine and truck companies. A sixth crew was designated as the rapid intervention crew (RIC), but they were not briefed and did not have a hose line.

An adjunct instructor and four students of the first crew, designated as Engine 1, entered through the front door. All of the students had PASS devices, but the adjunct instructor did not, nor did he have a portable radio. Per the direction of the instructor-in-charge, Engine 1 proceeded to the third floor and expected a second hose line to follow. They advanced an initially uncharged 1¾" (44 mm) hose line to the centrally located stairs on the first floor. There is disagreement as to when the line was actually charged.

Another crew of recruits was to advance a hose line to the second floor from the rear door. When they entered the rear of the house, they encountered the large pile of debris burning, which inhibited ingress and egress. The fire was spreading across the ceiling, and they extinguished the fire before a delayed advance to the second floor. As with the first crew, the four students had PASS devices, the adjunct instructor did not, nor did he have a portable radio.

The Engine 1 crew advanced to the second floor and encountered heavy fire conditions. Although directed to go to the third floor, the instructor determined it was necessary to knock down the fire before proceeding upstairs to the third floor. Two of the crew's members stayed on the landing between the second and third floors to pull hose, as the remainder of the crew advanced upstairs. Conditions on the third floor became too hot and the instructor did not have a radio to advise command of their situation or of interior conditions. The instructor lifted himself up and climbed out of a high window onto the back roof of the second floor. The first student was able to lift her upper body out of the window. The instructor then grabbed her SCBA straps to pull her out the rest of the way onto the second floor roof. She was then hoisted to the third floor roof by Truck 3, where she told them that her crew needed help.

The other student was now at the window as the instructor attempted to pull her through, using her SCBA harness as he had done with the other fire fighter. She was initially able to talk to the instructor to tell him that she could not help, and that she was burning up. It appears that she still had her SCBA face mask on at this time. The instructor lost his grip and she fell back into the room, landing on her feet. When he was able to get hold of her again, she was still conscious but her mask was partially dislodged or removed. Her face had visibly started to blister. The instructor did not have a radio, but yelled for help. He lost his grip on her a second time and she fell back into the room. Shortly after regaining his grip on her for the third time, she became unresponsive. At this point, the crew jumped about six feet down from the third floor roof to help pull her out, but they were unable.

With additional assistance they were eventually able to help lift her onto the roof where a student used a portable radio to advise of the fire fighter down.

Assistance was requested of on-duty battalion, engine, and truck units that had come by to watch. The engine company engaged the fire, while the truck company entered to ensure that the students were all out.

Two members of the on-duty truck company climbed the aerial ladder to the roof and assisted in placing the victim in a Stokes Basket and carried her down the ladder. An advanced life support ambulance initiated advanced life support and transported her to a shock trauma center, where her care continued until she was pronounced dead.

Incident Report

Baltimore, Maryland - 2007

Post-Incident Analysis — Baltimore, Maryland

NFPA 1403 Noncompliant

NOTE: Near total lack of compliance to NFPA 1403. 50 issues considered to be violations of NFPA 1403 by investigative team.

- Flashover and fire spread unexpected (4.3.7)
- Walk-through was not performed (4.2.25.4)
- Multiple fires on different floors (4.4.15)
- Excessive fuel loading (4.3.5)
- Fire was beyond the training and experience of the students to participate in live burn exercises in an acquired structure (4.1.1)
- No incident safety officer (4.4.1)
- Noncompliant gear on victim. (4.4.18.2)
- Instructors without PASS* (4.4.18.5)
- Instructors not equipped with portable radios (4.4.9)
- Adjunct instructors had little to no prior instructional experience (4.5.1)

NFPA 1403 Compliant

- 22 Students, 11 instructors (4.5.2)
- EMS (ALS transport) on standby on scene (4.4.11)

Other Contributing Factors

- Students not informed of emergency plans/procedures
- Rapid intervention crew staffed by students and unequipped or prepared

Flashover

An instructor must be able recognize the signs of flashover and take appropriate action. Some warning signs of flashover, like "fingers of flame" or "daggers of flame" may not be visible or noticed without vigilance. Rapid heat build-up forcing the fire fighters lower and lower, coupled with dark, black smoke that banks to the floor, is one indicator of an impending flashover. When these fire conditions exist, action must be taken immediately. This includes the following:

1. Ventilate to prevent the build-up of super-heated gases.
2. Apply short bursts of water to cool the gases at the ceiling. Be sure to not disrupt the thermal balance!
3. If conditions do not immediately improve, it is time to exit the structure.
4. When water is applied to the area without immediate results, it is time to exit the structure.

There may not be much time, so the decision must be made quickly to get the participants out of the structure. Once everyone is outside and accounted for, regroup and get the fire under control. As we know, fire is an ever-changing phenomenon as it moves through its various stages of development. Interior instructors as well as those on the outside must constantly monitor for these ever-changing conditions.

Final Controlled Burn

After the interior evolutions are complete, and prior to the aquired structure being burned completely down, the fire must be brought under total control. This means putting all of the fire out. This allows for an orderly personnel accountability report (PAR) process, to ensure everyone is accounted for. This is also the time to check the interior for tools or equipment, reposition hose lines for defensive operations, break for rehab, and start stowing equipment that will not be used during the burn down.

During this phase, there is a tendency to drop some of the precautionary procedures utilized earlier during interior suppression operations. Personnel are tired, and it may be getting late, so it must be stressed to remain vigilant.

Great caution must be exercised in the preparation for the "final burn down." Walls and ceilings can be breached to allow for fire spread and fuel loading can be heavy. Positive pressure ventilation (PPV) fans can be employed to "push" fire throughout a building by what would be improper placement and use during a hostile fire.

Staffed, charged hose lines and a rapid intervention crew need to be in place to protect personnel preparing for the final ignition. The incident management structure needs to be fully in place. The final controlled burn should only begin after a completed Go/No Go sequence is done following all of the previously discussed guidelines. Prior to ignition, a pre-established "final burn down" signal, other than the emergency evacuation signal, should be sounded. This may be accomplished by one long horn blast or by a radio transmission. Everyone must be advised of this being the final burn down.

> **Safety Tips**
>
> Any time the instructor-in-charge determines that fuel, fire, or any other condition represents a potential hazard, the training evolution shall be stopped immediately. If an evolution is stopped, it should only be restarted after the Go/No Go sequence, once the hazard identified has been resolved.

In the unusual scenario that that the building will not be burned down after training, plans for the immediate demolition of the structure should be in place. If for some reason the building will not be immediately demolished, the building will need to be checked to verify the fire is totally extinguished so there is not a rekindle. The site may also need to be secured from unwanted persons.

Overhaul

The structure needs to be reevaluated after the final burn down, to determine what measures need to take place to render the site safe. Metal roofing, chimneys, large beams, and parts of walls that remain could be dangerous to children or scavengers. Arrangements need to be made for heavy equipment to knock over remaining walls or chimneys, and to make the site safe.

It should be expected that scavengers will want to go through the debris to collect salvageable metals and materials, and the curious will want to explore the remains. Signs and barrier or scene tape can warn adults of the danger, but younger children may not understand the warning signs or the danger present. The added step of installing construction or barrier fencing will protect the property owner and the fire department until the debris is removed. Often local government will have the fencing that they use on-hand for large events and construction projects, or it can be included in the agreement with the property owner that they provide it. It is even available for rental.

Postevolution Debriefing

Once the burn down phase and overhaul are completed, conduct a postevolution debriefing. The postevolution debriefing will help those in charge learn if anything could have been done better. Encourage all participants to give their input. Even though everybody will probably be tired, it is best to take get this information while it is still fresh in their minds.

Before leaving the site, the fire department needs to be careful not to leave burning debris that could cause fire or smoke problems to neighbors. If there is any doubt, it may be necessary to leave a fire watch in place. At this point, the structure is turned back over to the property owner with a signed transfer of authority form.

Wrap-Up

Chief Concepts

- The instructor-in-charge needs to identify whether or not the proposed acquired structure will fit the department's training needs.
- A preburn plan must be utilized with all live fire training, including acquired structures.
- Developing an emergency plan will help ensure the safest training.
- The instructor-in-charge is required to determine the amount of water needed for the live fire evolution as well as the amount needed for any unforeseen emergency. NFPA 1142, *Standard on Water Supplies for Suburban and Rural Firefighting*, should be consulted.
- Once a location is finalized, check with the authority having jurisdiction for what paperwork needs to be completed.
- An initial preparation should be performed to check the property itself, surrounding properties that may be affected, access to the structure, and any other dangers that may exist.
- All entry and egress routes must be planned for and made clear to all participants.
- Preparation of the exterior should including items such as assessing for any remaining structural damage, checking for asbestos, ensuring utility shut-off, determining roof ventilation, chimney removal, toxic materials, and weed removal.
- Most of the preparation will be spent on the inside with tasks such as mitigating environmental dangers, spraying for insects and diseases, removing furniture, removing flooring, preparing the windows, eliminating snag hazards, etc.
- It is important to be aware of the homes and businesses in the areas surrounding the training site. Notify the owners of those about the dates and times of the evolutions, and address any issues or concerns at that time.
- Use only Class A materials with known burn characteristics in the burn sets for live fire training evolutions with acquired structures.
- The decision to ignite the training fire shall be made by the instructor-in-charge in coordination with the safety officer, and should follow the Go/No Go sequence. The ignition officer and the safety officer shall both be present when actual ignition occurs, and a backup line must be utilized.
- The final controlled burn, including the expected remains and the possibility of a fire watch, should be planned for ahead of time.

Hot Terms

Acquired structure A building or structure acquired by the authority having jurisdiction from a property owner for the purpose of conducting live fire training evolutions

British thermal units (Btu) A unit of measurement indicating the amount of heat required to heat one pound of water 1° Fahrenheit (F) at sea level.

Construction Classification A set of predetermined factors that are used to help determine the minimum water supply based on the type of building construction of an acquired structure.

Fire flow rate The amount of water pumped per minute (gallons per minute or liters per minute) for a fire. There are several different formulas that are commonly used to calculate this.

Minimum water supply requirement The total amount of water, not flow, required for a given structure, based on its size, construction, and proximity to other structures or properties that could be damaged (exposures).

Occupancy Hazard Classification A set of predetermined factors that are used to help determine the minimum water supply based on the hazard levels of certain combustible materials.

References

Stittleburg, P. (1998) Fire Chief, *When a live burn can come back to haunt you.*

Florida State Fire Marshal. (2002) Firefighters' Death Report-Osceola County, Bureau of Fire and Arson Investigations.

Live Fire Training Instructor in Action

You have been called to investigate an acquired structure training fire turned deadly. You have never been a part of an investigation before, but have been involved in countless successful live fire trainings. Being a third party removed from the event, your chief though it would be appropriate that you be chosen for this task. En route to the site, you begin running through questions in your mind about your last acquired structure fire:

1. Which of the following weather conditions is required to be monitored prior to and during the live fire evolution?
   A. Lightning
   B. Wind velocity
   C. Wind direction
   D. All of the above

2. Who is responsible for determining the rate and duration of water flow necessary for the evolution?
   A. Fire chief
   B. Instructor-in-charge
   C. Training instructor
   D. Safety officer

3. Which of the following is true regarding water supply for attack lines and backup lines?
   A. They can be from the same source.
   B. They must be from two separate sources.
   C. They must both be from a municipal source.
   D. None of the above.

4. Which of the following materials is acceptable to use as a fuel during the evolution?
   A. Pressure-treated wood
   B. Rubber to produce smoke
   C. Plastics
   D. Class A materials

5. To avoid uncontrolled flashover or backdraft:
   A. use small amounts of flammable liquids to create and maintain fire size.
   B. use combustible liquids in small controlled amounts.
   C. the fuel load should be limited.
   D. only hay or straw with pesticides should be used.

6. Each hose line shall be capable of delivering a minimum of how many gallons per minute?
   A. 65 gpm
   B. 75 gpm
   C. 85 gpm
   D. 95 gpm

7. The maximum number of fires permitted at one time within an acquired structure is:
   A. one.
   B. two.
   C. three.
   D. There are no limitations.

8. The participating student-to-instructor ratio shall not be greater than:
   A. 1 to 1.
   B. 4 to 1.
   C. 5 to 1.
   D. 6 to 1.

Gas-Fired and Non-Gas-Fired Structures

CHAPTER 7

NFPA 1403 Standard

4.6.12.1 The instructors and the safety officer responsible for conducting live fire training evolutions with a gas-fueled training system or with other specialty props (such as flashover simulator) shall be trained in the complete operation of the system and the props. [p 148]

4.7.1 A fire control team shall consist of a minimum of two personnel. [p 149]

4.7.1.1 One person who is not a student or safety officer shall be designated as the "ignition officer" to ignite, maintain, and control the materials being burned. [p 149]

4.7.1.1.1 The ignition officer shall be a member of the fire control team. [p 149]

4.11.5.1 The requirements of 4.11.5 do not apply to permanently sited gas-fueled training systems. [p 143]

4.11.7 There shall be room provided around all props so that there is space for all attack line(s) as well as backup line(s) to operate freely. [p 145]

4.12.1.1 Fuel-fired buildings and props are permitted to use the appropriate fuels for the design of the building or prop. [p 131]

4.12.3.1 Combustible liquid with a flash point above 100°F (38°C) shall be permitted to be used in a live fire training structure or prop that has been specifically engineered to accommodate a defined quantity of the fuel. [pp 148–149]

4.12.5 Propane lighters, butane lighters, fusees (safety flares), kitchen-type matches, and similar devices are permitted to be used to ignite training fires if the device is removed immediately after ignition of the training fire. [p 149]

4.14.1 All spectators shall be restricted to an area outside the operations area perimeter established by the safety officer. [p 143]

4.14.2 Control measures shall be posted to indicate the perimeter of the operations area. [p 143]

4.14.3 Visitors who are allowed within the operations area perimeter shall be escorted at all times. [p 143]

4.14.4 Visitors who are allowed within the operations area perimeter shall be equipped with and shall wear appropriate protective clothing. [p 143]

4.14.5 Control measures shall be established to keep pedestrian traffic in the vicinity of the training site clear of the operations area of the live burn. [p 143]

4.15.1 A preburn plan shall be prepared and shall be utilized during the preburn briefing sessions. [pp 141–142]

4.15.2 Prior to conducting actual live fire training evolutions, a preburn briefing session shall be conducted by the instructor-in-charge with the safety officer for all participants. [p 141]

6.1.2 Live fire training structures shall be left in a safe condition upon completion of live fire training evolutions. [p 149]

6.1.3 Debris hindering the access or egress of fire fighters shall be removed prior to the beginning of the training exercises. [pp 145, 149]

6.1.4 Flammable gas fires shall not be ignited manually. [p 149]

6.2.1* Live fire training structures shall be inspected visually for damage prior to live fire training evolutions. [p 149]

6.2.1.1* Damage shall be documented and the building owner or AHJ shall be notified. [p 149]

6.2.2 Where the live fire training structure damage is severe enough to affect the safety of the participants, training shall not be permitted. [p 145]

6.2.3 All doors, windows and window shutters, railings, roof scuttles and automatic ventilators, mechanical equipment, lighting, manual or automatic sprinklers, and standpipes necessary for the live fire training evolution shall be checked and operated prior to any live fire training evolution to ensure they operate correctly. [p 145]

6.2.4 All safety devices, such as thermal sensors, combustible gas monitors, evacuation alarms, and emergency shutdown switches, shall be checked prior to any live fire training evolutions to ensure they operate correctly. [p 145]

6.2.5 The instructors shall run the training system prior to exposing students to live flames in order to ensure the correct operation of devices such as the gas valves, flame safeguard units, agent sensors, combustion fans, and ventilation fans. [p 145]

6.2.6* The structural integrity of the live fire training structure shall be evaluated and documented annually by the building owner or AHJ. [p 144]

6.2.6.1 If visible structural defects are found, such as cracks, rust, spalls, or warps in structural floors, columns, beams, walls, or metal panels, the building owner shall have a follow-up evaluation conducted by a licensed professional engineer with live fire training structure experience and expertise, or by another competent professional as determined by the building owner or AHJ. [p 144]

6.2.7* The structural integrity of the live fire training structure shall be evaluated and documented by a licensed professional engineer with live fire training structure experience and expertise, or by another competent professional as determined by the AHJ, at least once every 10 years, or more frequently if determined to be required by the evaluator. [p 144]

6.2.8* All structures constructed with calcium aluminate refractory structural concrete shall be inspected by a structural engineer with expertise in live fire training structures every 3 years. [p 144]

6.2.8.1 The structural inspection shall include removal of concrete core samples from the structure to check for delaminations within the concrete. [p 144]

6.2.9* Part of the live fire training structure evaluation shall include, at least once every 10 years, the removal and reinstallation of a representative area of thermal linings (if any) to inspect the hidden conditions behind the linings. [p 144]

7.1.2 Live fire training structures shall be left in a safe condition upon completion of live fire training evolutions. [p 149]

7.1.3 Debris hindering the access or egress of fire fighters shall be removed prior to the beginning of the training exercises. [pp 145–146, 149]

7.2.1* Live fire training structures shall be inspected visually for damage prior to live fire training evolutions. [pp 145–146, 148]

7.2.1.1* Damage shall be documented, and the building owner or AHJ shall be notified. [p 149]

7.2.2 Where the live fire training structure damage is severe enough to affect the safety of the participants, training shall not be permitted. [p 144]

7.2.3 All doors, windows and window shutters, railings, roof scuttles and automatic ventilators, mechanical equipment, lighting, manual or automatic sprinklers, and standpipes necessary for the live fire training evolution shall be checked and operated prior to any live fire training evolution to ensure they operate correctly. [p 144]

7.2.4 All safety devices, such as thermal sensors, oxygen and toxic and combustible gas monitors, evacuation alarms, and emergency shutdown switches, shall be checked prior to any live fire training evolutions to ensure they operate correctly. [p 145]

7.2.5* The structural integrity of the live fire training structure shall be evaluated and documented annually by the building owner or AHJ. [p 144]

7.2.5.1 If visible structural defects are found, such as cracks, rust, spalls, or warps in structural floors, columns, beams, walls, or metal panels, the building owner shall have a follow-up evaluation conducted by a licensed professional engineer with live fire training structure experience and expertise or by another competent professional as determined by the AHJ. [p 144]

7.2.6* The structural integrity of the live fire training structure shall be evaluated and documented by a licensed professional engineer with live fire training structure experience and expertise or by another competent professional as determined by the AHJ at least once every 5 years or more frequently if determined to be required by the evaluator. [p 144]

7.2.7* All structures constructed with calcium aluminate refractory structural concrete shall be inspected by a structural engineer with expertise in live fire training structures every 3 years. [p 144]

7.2.7.1 The structural inspection shall include removal of concrete core samples from the structure to check for delaminations within the concrete. [p 144]

7.2.8* Part of the live fire training structure evaluation shall include, once every five years, the removal and reinstallation of a representative area of thermal linings (if any) to allow inspections of the conditions hidden behind the linings. [p 144]

7.3.1 The AHJ shall develop and utilize a safe live fire training action plan when multiple sequential burn evolutions are to be conducted per day in each burn room. [pp 134–135]

7.3.2 A burn sequence matrix chart shall be developed for the burn rooms in a live fire training structure. [pp 134–135]

7.3.2.1 The burn sequence matrix chart shall include the maximum fuel loading per evolution and maximum number of sequential live fire evolutions that can be conducted per day in each burn room. [pp 134–135]

7.3.3* The burn sequence for each room shall define the maximum fuel load that can be used for the first burn and each successive burn. [pp 134–135]

7.3.4* The burn sequence matrix for each room shall also specify the maximum number of evolutions that can be safely conducted during a given training period before the room is allowed to cool. [pp 134–135]

7.3.5 The fuel loads per evolution and the maximum number of sequential evolutions in each burn room shall not be exceeded under any circumstances. [pp 134–135]

Additional NFPA Standards

NFPA 1402 *Guide to Building Fire Service Training Centers*

Knowledge Objectives

After studying this chapter, you will be able to:

- Identify and/or describe compliance issues of NFPA 1403 in general for permanent, gas-fired, and non-gas-fired live fire training structures.
- Describe the operating procedures and the burn characteristics of a given of permanent, gas-fired live fire training structure.
- Identify the common system safeguards found in gas-fired live training structures, and the specific safeguards found in the gas-fired live training structure of the authority having jurisdiction (AHJ).
- Identify and describe the operation of system emergency controls found in common gas-fired live fire training structures and for the AHJ.
- Differentiate the safety concerns and characteristics of live fire training evolutions in acquired structures, gas-fired live fire training structures, and non-gas-fired live fire training structures.
- Describe the common safety hazards and steps to mitigate them during preparation and actual operations for gas-fired and non-gas-fired live fire training structures.
- Describe the dangers and adverse effects of too much of a fuel load in a non-gas-fired live fire training structure. Identify the facility's procedures for sequential live fire burn evolutions.

Chapter 7 Gas-Fired and Non-Gas-Fired Structures

- Identify the information that must be included in the preburn plan.
- Describe how to create the emergency plan and ensure it is implemented correctly.
- Describe how to ensure the safety of spectators, media, and visitors.
- Describe how to ensure the water supply is adequate for training evolutions.
- Describe how to ensure on-site facilities.
- Describe how to maintain permanent live fire training structures.
- Describe how to prepare a gas-fired live fire training structure for a safe training evolution.
- Describe how to prepare a non-gas-fired live fire training structure for a safe training evolution.
- Describe how to ensure a safe ignition and burn with a gas-fired system.
- Describe how to ensure a safe ignition with a non-gas-fired system.

Skills Objectives

After studying this chapter, you will be able to:

- Inspect and prepare a gas-fired live fire training structure.
- Inspect and prepare a non-gas-fired live fire training structure.

You Are the Live Fire Training Instructor

Following a number of recent veteran fire fighter retirements, your fire department has had a noticeable decline in fireground performance, particularly with interior suppression operations. During an officer meeting, the chief assigns you to come up with a plan that will increase performance in suppression operations. Your department owns a non-gas-fired live fire training structure, and you also have access to gas-fired live fire training structure at a nearby regional facility.

1. What types of live fire training evolutions will you utilize to improve on-scene performance?
2. Is there a difference in what you can expect for results between the two available types of live fire training structures?

Introduction

Practical experience operating in realistic fire conditions is considered an indispensable component of fire fighter recruit training. The concern for injuries and deaths of students and instructors from highly publicized incidents, and less publicized local experiences, has instigated the move to a more controllable and safer training environment, provided by permanent live fire training structures, commonly referred to as **burn buildings**. There are two basic types: gas-fired and non-gas-fired. NFPA 1403, *Standard on Live Fire Training Evolutions* covers the specific requirements for both types of structures, and those requirements will be covered in this chapter.

A 2003 US Fire Administration special report, *Trends and Hazards in Firefighter Training* states, "Strict adherence to NFPA 1403 helps prevent serious injury or death during live fire training by ensuring that adequate safety measures are taken prior to beginning the exercise." According to that same report, from 1993 to 2003 the use of acquired structures declined and the use of permanent live fire training structures had become the norm.

The use of permanent live fire training structures greatly reduces the chance of problems that could occur in acquired structures. The chances of structural collapse, fire hidden in voids, unrecognized or unanticipated fire spread, hazardous structural materials, and coverings that can produce highly toxic smoke, are all greatly reduced or eliminated.

The realism offered by gas-fired live fire training structures varies by the quality of the structure purchased, and by the variety of the situations it offers. The newer structures offer a much higher level of realism than their predecessors, with more accurate fire conditions and better smoke obscuration. Almost all of the major civilian and military fire fighter training centers now utilize gas-fired live fire training structures, either exclusively or in concert with non-gas-fired live fire training structures.

Code Requirements

Many fire departments and training academies have locked horns with local building officials over the need for emergency lighting, exit signs, and Americans with Disabilities Act requirements when building live fire training structures. Exit signs and emergency lights become useless after melting from the hot conditions present inside. Both of these features do play a good role in such structures when placed near the floor-level, and securely mounted where they will not be knocked off the wall. However, remember that these structures are generally not habitable under the building code.

NFPA 1402, *Guide to Building Fire Service Training Centers* is the guideline to use when building a new facility, but note it is a guide and not a standard. A standard is not a law unless it is adopted by the authority having jurisdiction, or a governmental authority, but it can be used as a reference during legal proceedings, such as by the Occupational Safety and Health Administration (OSHA).

Some states, such as Florida, have code requirements for live fire training structures with the jurisdiction of the state fire marshal. Liquefied petroleum gas and compressed natural gas systems are under another state agency's authority.

Making and Enforcing the Rules

Throughout the course of this text, the need to set strict policies and procedures for live fire training has been stressed. Needless injuries and even deaths have occurred from inappropriate

Figure 7-1 The policies and procedures of the live fire training structure should be clearly posted.

courage and complacency. The facility manager must ensure that there are concise procedures in place that the instructors, as well as the students, can follow. Permanent live fire training structures are inherently safer than acquired structures, but there have been injuries in gas-fired structures and even fatalities in non-gas-fired structures. Due diligence is necessary to prevent complacency. It is very important, especially in recruit training, to keep in mind that the students are watching the instructors all the time, not just when they are instructing. Complacent behavior by instructors with PPE and SCBA will be "learned," even if not intentionally taught.

The procedures need to include everything from normal to emergency activities. Expectations need to be spelled out on how to handle student performance problems, and equally clear expectations on instructor conduct and performance must be clear.

A successful tactic used by a number of training facilities is to post the rules in the training area Figure 7-1.

Types of Live Fire Training Structures

Permanent live fire training structures are designed to withstand repeated fires without incurring the damage that a normal structure would Figure 7-2. There are two main types of live fire training structures. The **gas-fired live fire training structure** is a permanent structure where the burn sets are generally fueled by liquefied petroleum gas (LPG) or propane. The **non-gas-fired live fire training structure** is a permanent structure where the fires are fueled by Class A materials such as excelsior, hay, and pallets. Table 7-1 lists the advantages and disadvantages of acquired structures and the two main types of permanent live

Figure 7-2 A modern concrete live fire training structure.

fire training structures. Walls, ceilings, and floors are all designed not to burn, limiting fire spread. Such structures can be made of steel, concrete, or masonry construction Figure 7-3. The live fire training structure can either be a freestanding structure, part of a training tower, or even a mobile structure Figure 7-4. Several specialized companies sell components or complete structures, with or without gas-fired systems.

In the past, the most common types of live fire training structures have been of concrete or masonry construction. They have been built with extremely heat-resistive material, allowing for repetitive fires and extended use. Such repetitive use is tough on a structure; the ceiling, walls, and floor surfaces heat unevenly and expand and are then subject to the rapid

Table 7-1 **Advantages and Disadvantages of Live Fire Training Structures**

	Acquired Structure	Non-Gas-Fired Live Fire Training Structures	Gas-Fired Live Fire Training Structures
Advantages	Most realistic: • Most similar to hostile fire • Room(s) itself is on fire. Fire spread more like occupied building	• Fire behavior more realistic than gas-fired • Structure itself is not on fire • Structure does not produce toxic smoke like acquired structures do • Real smoke conditions • More realistic smoke than gas-fired systems	• Safest • Nontoxic atmosphere • Ability to immediately "shut off" fire and clear atmosphere • Structure itself is not on fire • Structure does not produce toxic smoke like acquired structures do • No real smoke. Simulated smoke dissipates rapidly outside of prop. Runoff not contaminated • Faster and easier rotation of crews, no waiting for fire to build back up or restock pallets. • Minimal setup and cleanup
Disadvantages	• Most dangerous • Difficulty in acquiring • Environmental requirements • Resources in preparation, planning	• Highest temperature • Smoke less toxic than acquired, but still hot and toxic. • Environmental concerns less than acquired, but still can be problematic • Fire spread limited to fire set location • Requires very strict site management	• Fire spread limited to burn location(s) • Expensive to build, maintain, and use, but much less expensive than replacing a non-gas-fired live fire training structure

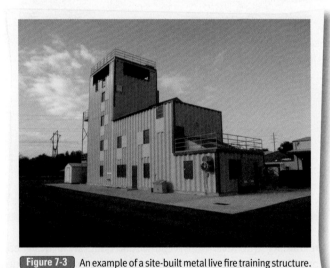

Figure 7-3 An example of a site-built metal live fire training structure.

Figure 7-4 A mobile live fire training structure.

and uneven cooling caused by hose streams Figure 7-5 . This is known as thermal shock. It is common for older concrete or masonry structures, which use Class A fuels, to be damaged from fires with too much fuel loading, or just from extensive, repeated use. Liners of insulating material became available in the 1980s to protect concrete construction, which allowed existing and new live fire training structures to extend their life spans Figure 7-6 . To extend their life span, newer facilities have begun utilizing insulation panels and replaceable, non-load-bearing walls Figure 7-7 .

Gas-Fired Live Fire Training Structures

Gas-fired live fire training structures are most often fueled with liquefied petroleum gas (LPG) or propane gas. Flammable liquids are not permitted to be burned inside. Computer-controlled systems continuously monitor the atmosphere for

Chapter 7 Gas-Fired and Non-Gas-Fired Structures

Figure 7-5 Heat damage to a permanent concrete interior ceiling. Note the exposed steel in the concrete.

Figure 7-6 Insulation panels protect the live fire training structure's integrity.

Figure 7-7 Some newer live fire training structures feature replaceable, non-load-bearing walls.

temperature, unburned gas, and oxygen levels. Should any of the monitored items approach the safety limits, the system shuts the open burning down and initiates mechanical ventilation to the area **Figure 7-8**.

With the instructor at the controls, he or she can ignite the fire, control the fire's size and intensity, add smoke, and many other options. The fires are limited to the burn scenarios inside that are made to represent different fire situations. The burn scenarios include bed fire, couch, kitchen stove or oven, flashover or rollover, attic fire, car fire (in a garage), ceiling or cockloft, electrical panel, and interior maritime and aviation fires.

In the past, gas-fired systems have been criticized for a lack of realism in their appearance and temperature. However, over the last few years considerable improvements have been made to the realistic appearance and heat production in such systems.

Current control systems utilize a control station that is removed from the fire area, which allows for activation of the systems and localized control, either by hardwire control **Figure 7-9** or wireless control **Figure 7-10**. The local controls feature a "dead man" switch that, when let go, will shut down the fire and initiate ventilation. The controls generally include changes to the fire status, smoke production, and "rollover/flashover," among others. It allows an instructor to be in the room observing operations, while at the same time having the ability to increase or decrease the severity of conditions within preset limitations. It also gives the instructor the ability to immediately shut off the fire in case of a problem. Also,

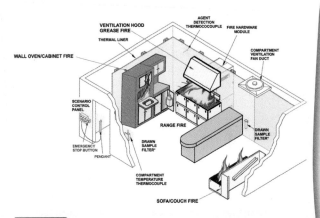

Figure 7-8 Diagram of a gas-fired, computer-controlled system.

Figure 7-9 A pendent "hardwired" control for a gas-fired prop.

Figure 7-10 A wireless control for a gas-fired prop.

gas-fired systems require less time between evolutions for set-up, since restocking and waiting for the fire to build back up are not necessary.

When working with gas-fired live fire training structures, strictly followed policies are still necessary even with the built-in safety controls. Such policies should include the following:

- **Fuel and fire locations:** Gas-fired live fire structures generally have fixed prop locations. This limits the locations of scenarios, but local policy needs to specify how the training evolutions will be conducted within the limitations and opportunities provided by the prop. Policy should state if any Class A material fires or smoke is allowed. Most manufacturers of gas-fired systems recommend against having Class A combustibles in the same area as the gas-fired props, due to possible damage to the monitoring equipment.
- **Heat:** Gas-fired systems have heat monitoring built in. A concern expressed by some instructors is the lower temperature levels that the gas-fired systems produce. Newer props appear more realistic with greater convected and radiant heat capable of damaging turnout gear, if exposed too long.
- **Simultaneous training evolutions:** Determine if simultaneous evolutions are allowed by the AHJ.
- **Staffing requirements:** Staffing needs to be clearly detailed, specific to the facility, and in accordance with the manufacturer's directions.
- **Host personnel:** It is highly recommended for training centers that allow other agencies to use their live fire training structures to require the host agency to have personnel operating the primary controls.

Non-Gas-Fired

Non-gas-fired live fire training structures offer greater safety than acquired structures, and produce higher temperatures than gas-fired live fire training structures. They tend to be less expensive to build than their gas-fired counterparts, but almost always lack the safety controls found in the gas-fired live fire training structures, such as temperature control, the ability to instantly "shut-off" the fire, and the ability to rapidly vent the area. However, some non-gas-fired live fire training structures do have automated ventilation capability, and many do have manually operated ventilation that relies on natural venting.

Many instructors prefer the non-gas-fired live fire training structures due to the more realistic fire behavior from the Class A materials. On the other hand, many administrators are very concerned about the lack of automated control that can cause injury and damage. Strictly followed policies can mitigate those concerns. Such policies should include the following:

- **Fuel**
  - Fuel materials shall be selected per NFPA 1403 and by the authority having jurisdiction (AHJ). Fuels need to be selected based on their known burning characteristics, including smoke toxicity. Pallets or other specified wood products, hay, or excelsior are allowed. Pressure-treated wood, rubber, and plastic, and straw or hay treated with pesticides or harmful chemicals shall not be used. Pallets need to be clean and not have been used to carry products that could change the burning characteristics and smoke produced.
  - Fuel load policies should specify the amount of pallets or wood products, hay, or excelsior for ignition. The majority of damage inflicted on non-gas-fired live fire training structures has been due to too much fuel loading.
- **Sequential Live Fire Burn Evolutions**
  - New to the 2012 edition of the 1403 Standard is the requirement for the AHJ to develop a burn sequence matrix chart for sequential live fire burn evolutions (see **Table 7-2** and **Table 7-3**). This recognizes that each "burn room" requires less fuel each evolution and that the room will eventually become heat saturated. Although cumbersome to determine the information for the chart, this can be a very significant contribution towards the safety of the personnel operating in the structure and for the longevity of the structure itself. Remember this is now a requirement. Specifically it includes the following:
  - The maximum fuel loading per evolution and maximum number of sequential live fire evolutions that can be conducted per day in each burn room (7.3.2.1).
    - For each room, the maximum fuel load that can be used for the first burn and each successive burn. (7.3.3)
    - Shall specify the maximum number of evolutions that can be safely conducted during a given training period before each room is allowed to cool (7.3.4).
    - The fuel loads per evolution and the maximum number of sequential evolutions in each burn room shall not be exceeded under any circumstances. (7.3.5)
    - Restocking fuel policies need to follow the facility's Burn Sequence Chart, which specifies the amount fuel that can be added (restocked) between evolutions and if any restocking is allowed during

evolutions. Any changes to the Order of Operations or the Burn Sequence Chart needs to be approved by the instructor-in-charge in consultation with the Safety Officer and Fire Control Team. The staging of additional fuel and how it will be brought into the burn areas must be specified. Do not stack pallets and/or bales in the hallway due to concern it can be knocked over and impede egress. Pallets and bales stored in an adjacent burn room that is not in use can be a problem with fire spread.

- **Heat**
    - Consider setting maximum temperature levels for the ceiling and mid-height, or other heat monitoring provisions. This needs to be done in consultation with the engineering firm that makes the non-gas-fired live fire training structure. Temperature alone is not a reliable indicator of the amount of heat present or the danger from it.
- Many non-gas-fired live fire training structures have only specific areas with thermal protection, so all ignitions must be limited to those areas.
- The local policy needs to set the **types of training evolutions** and their parameters, which are basically a script of how the evolutions are to be run.
- **Are simultaneous training evolutions allowed** by the AHJ? Concurrent training evolutions are allowed by NFPA 1403, with certain restrictions.
- **Staffing requirements** need to be clearly detailed specific to the facility, not just designated to follow NFPA 1403.
- It is recommended that training centers that allow other agencies to use their live fire training structure require the host agency to have personnel present to insure their policies are followed with regards to the above items.

Temperature Monitoring

In a non-gas-fired live fire training structure, relying too heavily on a temperature reading can be dangerous. In these settings, the value of temperature monitoring can be limited, because it is the energy of the heat that causes problems for participants, not necessarily the temperature. There are so many factors that influence the danger levels of a burn room, such as the size of the room or the number of evolutions conducted in the room. Because of weather conditions, type of structure, type of thermal lining, fuel quantity, and burn durations, temperature alone may not be sufficient. The safety officer and his or her staff, along with the ignition officer and all of the instructors, must

> **Live Fire Tip**
>
> Remember, temperature is a measure of the intensity not the amount of heat present.

Table 7-2 Class A Burn Sequence Information

Burn Sequences: The following facts will affect conditions encountered in a burn room during a live fire evolution:

(1) Larger burn rooms and rooms with higher ceilings will have more cubic feet of air than smaller burn rooms.

(2) Generally, with a given quantity of fuel, the lower the cubic footage in a room, the higher the temperatures and more rigorous the environment will be.

(3) As the number of openings in a burn room increase, the available ventilation area increases, resulting in typically lower temperatures and less severe environments.

(4) The construction of the burn room will affect how much energy the room will retain with each successive evolution. All burn rooms will retain a level of heat with each burn. The temperature and radiant heat in the burn room will increase with each additional evolution. At some point, every room will become too hot to safely conduct further training. Outside environmental conditions might also affect this.

Table 7-3 Sample Burn Sequence Chart

	Room Alpha/Bravo	Room Bravo/Charlie	Room Bravo/Charlie	Room Alpha/Delta
Burn Sequence 1	3 pallets 1 bale	3 pallets 1 bale	2 pallets 1 bale	3 pallets 1½ bales
Burn Sequence 2	2 pallets 1 bale	2 pallets 1 bale	1 pallet ½ bale	3 pallets 1 bale
Burn Sequence 3	1 pallet ½ bale	2 pallets 1 bale	End use or allow 30-minute cool down	2 pallets 1 bale
Burn Sequence 4	End use or allow 30-minute cool down	End use or allow 30-minute cool down		1 pallet ½ bale
				End use or allow 30-minute cool down

- Specify pallet and bale size and bale material (hay/straw or excelsior)
- Estimate time intervals between each room being used
- NOTE: In the above example, the rooms are not the same size, hence some will need to be cooled sooner than others

carefully monitor the interior conditions. Temperature monitoring is standard in gas-fired live fire training structures, but not common in concrete and masonry construction non-gas-fired live fire training structures. WHP, a manufacturer of metal live fire training structures, recommends the use of thermocouples placed at ceiling level and at working level to monitor temperatures in gas-fired and non-gas-fired live fire training structures.

According to High Temperature Linings, a company that designs live fire training structures and liners, a bale of hay and two to four wooden pallets will result in temperatures of about 200°F (93°C) near the floor and about 1000°F–1500°F (537°C–815°C) near the ceiling, in a standard live fire training room.

Shipping Containers

The European trend of using shipping containers for live fire training is becoming more popular in North America. Shipping containers are used with Class A fuels or are commercially equipped with gas-fired systems **Figure 7-11**. Many of these containers are even mobile. Multistory exterior props have also been assembled using shipping containers **Figure 7-12**.

Figure 7-11 Shipping containers have been successfully used for live fire training structures using Class A fuels or gas-fired systems.

Figure 7-12 Chicago's new Fire Training Facility is made of several shipping containers.

Live Fire Tips

Flashover

Flashover live fire training structures are considered separate from the other live fire training structures due to the differences in their purpose. When properly used, flashover live fire training structures provide a method to observe fire behavior in a highly controllable environment. This can make for very valuable learning that is impossible to recreate in a classroom setting. In some facilities, nozzle techniques can be taught that help reduce the chance of flashover or cool superheated gases to reduce spread and/or ignition.

There are a several different styles of structures used to demonstrate flashover. Most use Class A materials to create a "flashover" over the heads of participants, although there are gas-fired props that simulate the appearance of a flashover.

Following the 1987 deaths of two fire fighters in Sweden, attributed to a flashover, the Swedish Rescue Services Agency College in Skovde, Sweden developed a program that would train their fire fighters to recognize the indicators of impending extreme fire behavior.

The Swedes originally used shipping containers for fire training, which were built of a single level cell constructed from a 40' (12 m) container. This style evolved to include a split-level cell used for fire behavior demonstration. The Swedish split-level demonstration cell consists of two shipping containers, a smaller one raised higher than a larger one. The smaller container is used as a fire compartment and the larger container is used as an observation area occupied by the students and instructors.

The Swedish fire service training system has expanded over the years to include a single-level container used to train in nozzle technique, and a variety of other container-based structures, including the window cell for demonstration of backdraft, the large volume cell for simulation of commercial structures, and a wide range of multi-compartment cells for practicing door entry and tactical evolutions. Swedish fire service agencies use container-based structures in conjunction with live fire training structures and acquired structures in order to develop students' understanding of fire behavior and skill in nozzle technique and tactical operations.

While the use of the split level cell in the United States often involves demonstration and practice of nozzle technique, this is not common practice in Sweden. This type of cell is typically only used for observation of fire development, most commonly with the rear doors to the observation compartment open in order to keep the hot gas layer at a high level and minimize heat flux on the participants. Nozzle techniques used to cool the upper atmosphere to gain and maintain control of the fire environment are taught in a single level attack cell and multi-compartment cell.

Originally, the Swedish fire service used particle board as the primary fuel in container-based props Figure 7-13 . Today they use fiberboard, either low or high density, which is manufactured using less synthetic glue than standard particleboard. Other fire services, such as the New South Wales Fire Brigades in Australia and many fire service agencies in the United States, continue to use particle board due to its even-burning rate and level of desired smoke production.

Figure 7-13 Particle board hung by chains and along walls in the fire area.

Such systems can be built modularly, and can be initially assembled with additional modules to be added later. The standard shipping containers are steel and measure 7'8" (2.3 m) wide inside, and either 20', 40', or 45' (6.1 m, 12.2 m, or 13.7 m) long. They are either 8'6" or 9'6" (2.6 m or 2.9 m) tall.

NFPA 1403 does not specifically address the use of these containers, but such units acquired and modified either for gas-fired or non-gas-fired live fire training use need to comply with their respective NFPA 1403 chapters. There is some concern when modifying these containers, such as the effect on the structural strength if windows or doors are cut in, or if parts of the sides, roof, or bottom are removed to stack or connect shipping containers. Another concern with the protection of the shipping container's structural strength is its protection from the heat of repeated fire and thermal shock. Thermal liners may be used to protect the shipping container from thermal shock.

Figure 7-14 A simulated garage with a gas-fired car.

Features of Live Fire Training Structures

Considerable imagination and experience has brought much advancement from training towers to live fire training structures, from the four-story "cracker tin" towers with wide open floors, interior stairs, and a fire escape.

Some of the features now found in live fire training structures include the following:
- Car props Figure 7-14
- Enclosed interior stairs
- Sound systems for fireground noises, fire alarm, emergency evacuation, etc.
- More realistic layouts, not a single room. Examples include the following:
  - Hotel configurations, often with longer hallways and multiple doors Figure 7-15

Figure 7-15 A simulated apartment building or hotel.

Incident Report

Port Everglades, Florida - 2003

Figure A Maritime training prop at Port Everglades, Florida.

Note: This was the first US fatality in a permanent live fire training structure.

A live fire training exercise was being conducted in a prop constructed from shipping containers to represent a seafaring vessel **(Figure A)**. This was the first live fire experience for the students in a class leading to state certification as Fire Fighter II. The students were all new employees of the same fire department.

Three evolutions took place, with no breaks in between the evolutions to allow the all-metal structure to cool. The participants were instructed not to crawl over the metal grating and to avoid holding the handrails, since the metal structure had gotten very hot and there was a possibility of getting burned through their gloves. They were told failure to complete this evolution, or getting injured, would mean termination from the department. There was no preburn briefing or walk-through.

Some members of the third squad received burn injuries during their rotation, and an instructor left during the evolution claiming he had problems with his gloves.

The fire fighter who later died was in the fourth squad of students performing the evolution on this particular day. Five students, three instructors, and an observer with a thermal imaging camera entered the enclosed structure on the second level. As in previous evolutions, the students were sent in one-by-one with the intention of not being able to see or hear each other until arriving at the firebox. According to statements, the instructors did not monitor the students' movement or encourage their progress. Students followed the hose line through a series of three watertight hatches, and then crossed an open-grated catwalk over the gas-fueled fire in the engine room. At the end of the grated catwalk, a combustible materials fire burned in a corner. They then proceeded down a ladder and through the simulated engine room and into the "fire-box." The fire box had a raised hearth, similar to a flashover simulator, and the fire was fueled primarily by wood pallets. The students gathered in the room and took turns operating a nozzle in various patterns. They were told to avoid getting water on the fire.

The fire was knocked down, and an instructor in the burn room directed the crew to remain in the extremely hot area as the fire regained intensity. A "dead man" switch operated an open vent in the room, but it had been disabled so it remained closed.

The instructor in the fire room was quoted by several trainees as saying the environment was too hot. He instructed the group to hurry up and "get out now." One set of students turned around to exit using the same route through which they had entered the room. Now out of sequence, the safety officer led the students up the ladder and over the metal grating in a very high heat environment. Meanwhile, two instructors exited through a side door directly from the firebox to an on-grade side exit, without advising the instructor-in-charge.

There are conflicting reports as to what occurred at the exit, but it was apparent that one student was missing. It is believed that he lost track of the hose, and ended up in a chase where heated gases and smoke were venting. Handprints on the walls indicated that he was trying to find a way out. While a search for him ensued, he collapsed. After seeing the downed student's PPE, one officer entered to retrieve him, without protective clothing or SCBA, but had to abandon his attempt due to the heat. Another fire fighter wearing proper protective clothing and SCBA removed the victim to the outside. The victim's PASS device did not sound while inside; it is believed he was mobile until just before he was found.

He was unresponsive, with no pulse or respirations. After initial treatment by instructors, a medical rescue unit arrived and he was transported to a local trauma center where he was pronounced dead. He reportedly had severe burns on both hands and sloughing of the skin to both knees and hands. He was described as cyanotic from the neck up. Several members of the same crew received second-degree burns to their hands and knees while they were attempting to exit the structure. One student lost consciousness after exiting and was initially cared for by other students. He and three other students were transported to the hospital.

Two separate investigations were conducted by the state fire marshal, one for criminal violations and an administrative investigation of state training codes. No criminal charges were filed, but it went to the state attorney. The fire department was cited by the state fire marshal for numerous code violations, and the reports cited 36 specific findings including almost total failure to follow NFPA 1403.

Incident Report

Port Everglades, Florida - 2003

Post-Incident Analysis — Port Everglades, Florida

NFPA 1403 Noncompliant

- No written, pre-approved plan (5.2.12.3)(6.2.15.3)
- No previous live interior fire fighting training (5.1.1)(6.1.1)
- Safety crews did not have specific assignments (5.4.8)(6.4.8)
- No preburn walk-through for the trainees (5.2.13)(6.2.16)
- No designated safety officer (5.4.1)(6.4.1)
- No communication plan (5.4.9)(6.4.9)
- No emergency medical services on site (5.4.11)(6.4.11)
- Not all instructional personnel had specific live-fire training or instructor certification (5.5.1)(6.5.1)
- Instructors did not monitor the trainees' movement (5.5.7)(6.5.7)
- NFPA 1402 and 1403 were not complied with (both required by state code)(5.5.4)(6.5.4)
- No temperature monitoring (5.3.5)(6.3.8)
- Multiple fires inside structure burning simultaneously, including polypropylene rope and other nonorganic materials (5.3.1)(6.3.1)
- Lead instructor determined environment excessively hot, but evolution was not terminated. Also failed to identify and correct safety hazards (5.3.6)(6.3.10)

Other Contributing Factors

- No emergency plans in place
- No RIC assignment
- Command structure unknown to students and staff
- The firebox vent had been rendered unusable

NFPA 1403 Compliant

- Three to four instructors per squad of five trainees (5.5.2)(6.5.2)

- Office occupancy, either with multiple smaller rooms and doors off of a hallway, or with cubicles **Figure 7-16**
- Miami-Dade's Fire Rescue's live fire training structures include a store **Figure 7-17** complete with grocery aisles and a lunch counter with a liquid propane (LP) stovetop fire **Figure 7-18**.
- Residential configuration, often with a garage
- Noncombustible furniture appropriate to the simulated occupancy
  - Various representative types of doors and windows
  - Attic areas with low, angled ceilings and rafters **Figure 7-19**
  - Ducted smoke distribution
  - Roof vent props

Figure 7-18 A simulated lunch counter with a liquid propane stovetop.

Figure 7-16 A long hallway can simulate a hotel or office building.

Figure 7-19 A simulated attic fire.

Figure 7-17 A simulated storefront with a gas-fired kitchen.

Preburn Plan

A preburn plan needs to be developed and/or approved by the instructor-in-charge and approved by the fire department's chain of command. It should indicate all features of the training areas and structure. The general preburn plan was discussed previously in Chapter 5, Planning for Live Fire Training. The preburn plan needs to include:

- Specific objectives for the training evolution
- Who will participate
- Water supply requirements inclusive of number of training, attack, and backup lines
- Apparatus needs
- Rapid intervention crews and other emergency procedures for on-site emergencies and requests for assistance to the training site
- Incident management staffing and team organization
- Order of operations

- List of the training evolutions to be conducted and procedures for those evolutions inclusive of instructors and support personnel needed, and numbers of students
- Assignments and rotation of instructors and students
- Communications plan with emergency procedures
- A building plan with dimensions of the building, all rooms, windows, and doors with all features of the training areas inclusive of primary and emergency ingress and egress. Number each room, physically and on paper, and label which will have burn evolutions and in what sequence they will be burned. For non-gas-fired structures, fuel loading should be indicated and then confirmed and finalized on a hard copy the day of the evolutions.
- A site plan showing all aspects of the layout with measurements: the structure, command post, rehab/medical area, operations area, staging area, placement of engines, and parking area.
- An emergency medical plan.
- Areas for the staging, operating, and parking of fire apparatus involved.
- Parking for emergency vehicles not involved, so as not to interfere with fireground operations
- Parking for an ambulance or an EMS vehicle where it can quickly access participants in the evolution in case of injury, and so it can depart with regard to fireground operations, spectators, and traffic impediments
- Parking for participants and spectators
- The policy for the use of personal protective equipment (PPE) and self contained breathing apparatus (SCBA) and related safety concerns, including the requirement to wear and use within the operations area.
- Checklist to confirm with agencies already notified that the evolutions are to be conducted, including emergency medical, environmental, and any other locally required notifications. Depending on your jurisdiction, it may be necessary to notify the environmental agency, if using Class A materials.
- Checklist of requirements for other agencies, such as utilities, traffic, or EMS provider.
- Acceptable weather parameters including wind direction and/or speed that would preclude safe operations burning, or "red flag" conditions that preclude open burning. The weather forecast, preferably the National Weather Service point forecast, needs to be checked for expected untoward weather conditions that could cause problems after operations start. NFPA 1403 requires an awareness to be maintained for wind direction and velocity throughout the training.
- Demobilization plan, including the release of personnel, returning of equipment, and placing fire apparatus back into service
- Policy and plan for spectators, media, and visitors

While disclosing the location of simulated victims is not required, if the possibility of victims is part of the evolution, it will be discussed during the preburn briefing. Participants need to be advised that there will not be any "live" victims, and any simulated victims will not be wearing protective clothing similar to the participants.

Prior to the evolutions starting, all participants shall be required to conduct a walk-through of the structure in order to have a knowledge of and familiarity with the layout of structure. Emergency procedures in the plan will be reviewed as will the emergency egress routes.

Live fire training in any structure should include instruction of the participants in planning for a secondary means of egress or escape in case of an unexpected fire condition change. Prior to the evolutions starting, each fire fighter should be required to identify two means of egress or escape from each area.

Emergency Plans

Even with the high level of structural integrity offered by permanent live fire training structures, and the fire, heat, and ventilation controls offered by gas-fired live fire training structures, things can still go wrong. Permanent live fire training structures are inherently safer than acquired structures. In permanent live fire training structures, the structure itself is not on fire, and gas-fired systems offer the ability to immediately "shut off" the fire and mechanically ventilate the structure. However, there have been injuries in gas-fired live fire training structures and even fatalities in non-gas-fired live fire training structures. Due diligence is necessary to *prevent complacency*. Non-gas-fired live fire training structures tend to be hotter than acquired structures, and rehabilitation and emergency plans, such as rapid intervention crews and safety personnel, still need to be in place.

The emergency plans need to include everything that is put in place for the emergency response. Plans need to be in place for potential issues with recruits with less experience and confidence than their superiors.

Unlike an emergency response, a training fire allows for all the necessary personnel to be gathered before the fire is ignited. Hose lines, tools, rapid intervention teams, safety personnel, and even a walk-through of the structure, are all prepared prior to the evolution's beginning to ensure safety. Rapid intervention crews should be staffed with experienced personnel, depending

Safety Tips

Remember, injuries and fatalities have occurred when deviations from the plan have occurred, or when all the leadership is not aware of or not following the plan.

Live Fire Tips

It is very important, especially in recruit training, to keep in mind that the students are watching the instructors all the time, not just when they are teaching. Instructor behavior may be "learned," by students even if some behaviors are not intentionally taught. Careless behavior has the potential to be absorbed into the sponge-like minds of a new recruit student.

on local procedures (possibly more staffed than for a normal fire.) Multiple, concurrent evolutions are allowed by NFPA 1403, and RIC staffing must reflect that.

The building evacuation plan must be known by all personnel. This needs to include an evacuation signal that can be heard by all participants while inside during interior live fire training evolutions. The signal needs to be tested and demonstrated. The use of aerosol air horns or a megaphone directly into the interior may be needed. As part of the evacuation plan, participants involved in the live fire training evolutions should be instructed to report to a predetermined location for roll call if evacuation of the structure is signaled. Instructors should immediately report any personnel not accounted for to the instructor-in-charge. Remember that in many training programs the recruits will not have participated in personnel accountability report (PAR) drills. Another idea is to designate an alternate to the instructor, for each crew or team, to conduct and report a PAR, should the instructor be engaged in a rescue or other emergency operation.

Emergencies can occur either on site or off site that would redirect attention to that emergency. Everything from on-scene problems to neighborhood emergencies can occur. Any time there is question of divided attention, the training evolution needs to stop. Often there are enough unengaged personnel to handle such issues, but the instructor-in-charge must make the determination whether to proceed or not. The safety officer can play a critical role in that decision-making process. Weather has already been mentioned, but impending storms can raise winds that will change the fire dynamics, heavy rain can hamper operations, and, of course, lightning will curtail training. In the demobilization plan, a topic should be emergency reassignment, when units at the training evolution are needed for emergencies elsewhere. Part of this needs to be apparatus status, often covered in local procedures.

Spectators, Media, and Visitors

Drill grounds tend to be fenced in, and offer a higher level of security to the site than an acquired structure burn. However, the responsibility of nonparticipants still falls to the instructor-in-charge and the safety officer. Spectators need to be kept away from operations where they can not be hurt or get in the way. Specific viewing areas need to be established with fire lines to be clearly identified. This area needs to be established by the safety officer. Any visitors who are allowed within the operations area perimeter shall be escorted at all times and must wear complete protective clothing in accordance with manufacturer's instructions and in accordance with NFPA 1403. The standard does not differentiate the level of protective clothing for visitors on the exterior. If reporters are allowed into the gas-fired live fire training structures the greatest caution must be exercised in doing so. In some jurisdictions, unauthorized entry into a structure could be a violation of safety laws, as training is required before entry into such an environment. It is best to check local protocol when doing so.

As is the case with acquired structures, control measures need to be in place to mark the perimeter of the operations area. Ropes, signs, and fire line markings are all recommended by NFPA 1403.

Water Supply

The instructor-in-charge determines the rate of duration of water flow necessary for the live fire evolution, including the water necessary for control and extinguishment, the supply necessary for the attack and backup lines, and any lines needed for exposures utilizing NFPA 1142, *Standard on Water Supplies for Suburban and Rural Fire Fighting*. Each hose line shall be capable of delivering a minimum of 95 gpm (360 L/min).

Two separate water sources for the attack lines and the backup lines are necessary. A single water source is considered sufficient at training facilities where the water system has been engineered to provide adequate volume for the evolutions conducted and a backup power source and/or backup pumps are in place to ensure an uninterrupted supply in the event of a power failure or malfunction.

A minimum reserve of additional water in the amount of 50% of the fire flow demand is required for acquired structures and non-gas-fired props, to handle exposure protection or unforeseen situations. This is not required for gas-fired structures. (4.11.5.1).

On-Site Facilities

Other items are necessary for the assurance of successful operations. Restrooms (at least a Porta Potty), a hydrant, lights, security around the site, and other such features allow for better training evolutions.

An important feature seen more often now is a covered, outdoor meeting area or classroom **Figure 7-20**. The outdoor

Safety Tips

The possible loss of one of the two water supplies needs to be addressed in the emergency plan. Training evolutions need to stop and personnel relocated to the exterior while the supply is reestablished or replaced.

Figure 7-20 Outdoor classroom with lighting, ceiling fans, a large white board, and a permanent plan of the building on the right.

classroom can allow the attendees to meet and be briefed and debriefed out of the weather. A whiteboard can even be used to write and review plans and directions.

Management of Live Fire Training Structures

Instructors do not have the same responsibilities for the maintenance of a permanent live fire training structure as the management of the facility has. It is critically important that those serving as instructor-in-charge and safety officer periodically check the inspection records to make sure that they are up to date. It is important that during the inspections and operations, any defects be noted and reported.

NFPA 1403 stipulates the following frequency of the structural evaluation:

1. Once per year for live fire training structures that support more than 60 days of live fire training per year (a day of live fire training is any day during which at least one live fire training evolution has been conducted)
2. Once every two years for live fire training structures that support 31 to 60 days of live fire training per year
3. Once every three years for live fire training structures that support 30 or fewer days of live fire training per year
4. Immediately, if visible structural defects have formed, such as cracks, spalls, or warps in structural floors, columns, beams, walls, metal panels, and so on

Engineering Requirements

NFPA 1403 addresses the engineering requirements in detail. This is one of the controversial issues of the standard, due to the very small number of professional engineers (PE) working in this specialized area of work. Live fire training structures are very unique in their needs and construction, and this topic is not taught as part of a normal engineering program. NFPA 1403 requires an engineer "with live fire training structure experience and expertise," but this is not necessarily reflected in state or provincial laws that credential engineers. The intent of the standard and explanation is that the *structural* engineer has worked on at least one such project previously to gain some of the necessary experience.

When looking at projects that could at least be considered similar, such as high heat furnaces for industrial purposes, there are significant attributes that differ in how the fire service will utilize the construction technology. Examples would include large industrial furnaces that can certainly stand more heat than we want, but that are not designed for rapid cooling by hose lines (thermal shock), or the humidity caused by expanding steam that can drive it into small openings and between layers of materials or voids. Also, participants may strike the heat resistant materials, some of which become more brittle at high heat.

The engineer must have a good understanding of how the live fire training structure will be used, with repetitive heating and cooling, and how that affects the structural and nonstructural components.

Elements to Evaluate

NFPA 1403 requires that a live fire training structure's floors, walls, stairs, and other structural components be evaluated

> **Live Fire Tips**
>
> There are concerns for phenolic glues used in particle board and Orientated Strand Board (OPB) as fuels for flashover training. The concern is that phenolic glues contain isocyanates which cause irritation of skin and mucous membranes, chest tightness, and difficult breathing. Isocyanates are known to cause cancer in animals. Some effects of exposure to isocyanates are asthma and other lung problems, as well as irritation of the eyes, nose, throat, and skin. The concern primarily centers on instructors who are repeatedly exposed. The Swedes have switched to other materials such as press board (Homasote) and hardboard that are mainly held together with wax or a water-soluble paste. Also, some of the Swedish centers use fans or forced air to do an initial decontamination of the instructor before their SCBA is removed.

for safety. However, the nonstructural elements should also be evaluated at the same time. All mechanical components such as doors, windows, and vents need to be checked for operability and wear. Spray-on heat resistive linings, heat resistive panels, and such need to be inspected for damage. Following training operations, thermal linings need to be inspected for damage. If a panel is cracked, it should be removed and the area behind inspected for damage.

Heat can soak through thermal linings and reach the protected structure, without such damage being readily apparent. NFPA 1403 calls for the removal of a representative area of thermal liner to inspect the area behind it for damage every five years. Following a successful inspection and any repairs, the lining is then reinstalled. This is an expensive undertaking and NFPA 1403 does allow this to be waived if the thermal liner has *never* had a break. This means there have been no cracks, holes, breaks, or sags of insulation that would allow heat to pass though that opening to the protected structural components behind the lining, *and* that the area between the liner and structural member was monitored by thermocouples and the temperature never rose to 300°F (149°C) behind the lining.

When the above is cost prohibitive or difficult for the materials used, a concrete core can be bored from above the protected ceiling to verify the condition of the protected ceiling.

Concrete cores are required to be taken every 10 years for conventional (Portland) concrete and once every 3 years for refractory (calcium aluminate) concrete structural slabs and walls that have been exposed to temperatures in excess of 300°F (149°C) to check for hidden delamination and to test comprehensive strength.

If there is structural damage found that is severe enough to affect the safety of *any* participant, the live fire training shall not be permitted. The risk of putting participants in a live fire environment is dangerous enough without the possibility of structural problems.

Preparation

Proper preparation is essential for a safe and effective live fire training evolution.

Drillground

- Check the proximity to other structures and other exposures. Although it is doubtful that a new structure was built close to your live fire training structure, training grounds tend to collect vehicles, trailers, and other items. The area has to be ready for the live fire training.
- Identify adjacent properties that could be adversely affected by the smoke produced during the training evolutions. Even the presence of the fire and fire apparatus could be an issue, for properties such as schools, childcare facilities, hospitals or nursing homes, airports, heliports, railroads, or businesses. Likewise, verify that operations will not interfere with busy streets or any highways, railroads, or airports. Precautionary measures must be taken if interference is suspected.
- Parking for apparatus and equipment often becomes an issue, even with permanent sites. Identify and designate where nonparticipating and participating apparatus, apparatus subject to response, emergency response vehicles, and other vehicles should park. Remember, EMS must be closely located for quick access and departure.
- Parking for any press, spectators, and police should be designated.
- Fireground location for press and spectators should be designated.

Preparation and Inspection

The preparation and inspection phase of live fire training is clearly where the majority of your time will be spent. For both gas-fired and non-gas-fired live fire training structures, NFPA 1403 says, "Strict safety practices shall be applied to all structures selected for live fire training evolutions." This point cannot be stressed enough. During the planning and preparation phase, all live fire training structures must be visually inspected for damage *before* any training evolutions begin. If any damage is found, it must be documented and reported back to the owners of the structure.

The preparation and inspection for all live fire training structures includes the following:

- It is the responsibility of the instructor-in-charge to coordinate overall live fire training structure fireground activities, so as to ensure proper levels of safety. This includes the preparation stage.
- Look for signs of an obvious lack of structural integrity like a sagging roof or floors, and cracks in brick or masonry walls. Do the walls look straight or are they leaning or bulging?
- Check the doors and windows for easy operation, no obstructions, and no rubbing that could cause the door or window to stick. Hinges should be firmly attached. Not only do the primary doorways need to be clear for unimpeded travel, but emergency entry and egress must be clear and operable.
- Search the structure for unauthorized persons, animals, or objects.
- Check the stairs and railings to make sure they are stable and intact.
- The roof vent needs to be checked for operation and soundness. Due to the nature of its use and construction, it tends to wear more than other components.
- Carefully walk the floors and look for any debris that could get in the way of the training evolution, or that could potentially harm fire fighters.
- Any damage noted needs to be documented and the appropriate steps need to be taken for repair.
- If there is concern that the damage is significant enough to affect the safety of the students, training cannot be permitted.
- Devices such as automatic ventilators, mechanical equipment, lighting, manual or automatic sprinklers, and standpipes, which are necessary for the live fire training evolution, shall be checked and operated prior to any training evolution to ensure proper operation.
- All safety devices, such as thermometers, oxygen monitors, toxic and combustible gas monitors, evacuation alarms, and emergency shutdown switches need to be checked to ensure proper operation.
- Any unidentified materials, such as debris found in or around the structure, which could burn in unanticipated ways, react violently, or create environmental or health hazards must be removed from where they may unintentionally ignite or be used as fuel.
- Awareness of weather conditions, wind velocity, and wind direction is required by NFPA 1403 and is always a good idea. Ambient heat and humidity can certainly affect the participants. A final check for possible changes in weather conditions should be done immediately before actual ignition.
- Make sure all possible sources of ignition, other than those that are under the direct supervision of the person responsible for the start of the training fire, are removed from the operations area.
- Check the burn set locations or gas-fired systems to ensure there will be ample room around them for attack hose lines, as well as backup lines, to operate freely.

When training on either gas-fired or non-gas-fired live fire training structures, there are specific guidelines given by NFPA 1403 for each respective structure.

Preparation and Inspection of Gas-Fired Live Fire Training Structures

When training in gas-fired live fire training structures, NFPA 1403 provides specific guidelines for the inspection of the structure. In addition to the general guidelines that pertain to both gas-fired and non-gas-fired live fire training structures, the following must also be met to ensure the complete safety of all participants.

After the visual check of the rooms, the instructors shall run all of the gas-fired props and systems with students to ensure the correct operation of devices such as the gas valves, flame safeguard units, agent sensors, combustion fans, and ventilation fans. It is critical for all participants, instructors as well as students, to be comfortable with the controls in case of an emergency.

NFPA 1403 contains several items that the facility manager, not the instructor, has control over. The instructor, however, still needs to be aware of the following:

- The selection of liquefied petroleum gas, compressed natural gas, or butane is not the instructor's decision, as the burners for the system are set for a specific fuel types. These flammable gases shall only be permitted in structures designed for their use. The liquefied versions of these gases shall not be permitted inside the live fire training structure.
- The fuels that are utilized in live fire training evolutions shall have known burning characteristics that are as controllable as possible. While the instructor does not decide which type of fuel, it is imperative that the he or she know the burning characteristics of the fuel that will be used.
- On permanent gas-fired props, engineers have assessed the fire room environment for factors that can affect the growth, development, and spread of fire. It is important that the instructor-in-charge also be very familiar with the fire room environment for those factors. After operating at its most severe settings for several evolutions, how is the room's tenability? If a window or an exterior door is opened and wind is introduced, how will that affect interior conditions?
- Do not allow burn barrels with Class A or other materials, or other additional fire- or smoke-producing devices, to be used in a gas-fired live fire training structure without the manufacturer's and AHJ's approval. Damage to the atmospheric monitoring equipment can occur, and if located in unprotected areas, damage to the structure can occur, depending on the intensity.
- Ensure that nothing is placed on the gas-fired props, such as Class A products.
- Watch for debris from pallets and such that hinder the access or egress of fire fighters, and be sure to remove it prior to the beginning of the next training evolutions.
- Any time the instructor-in-charge determines that fuel, fire, or any other condition represents a potential hazard, the training exercise shall be stopped immediately. If an exercise is stopped, it should only be restarted once the hazard identified has been resolved. The Go/No Go sequence will be restarted only after the instructor-in-charge and safety officer determine that it is safe to do so.

Preparation and Inspection of Non-Gas-Fired Live Fire Training Structures

NFPA 1403 gives specific details about the inspection of non-gas-fired live fire training structures.

In preparation for live fire training, a careful visual inspection of the structure shall be made by the instructors and the safety officer to determine signs of damage or stress to the floors, walls, stairs, and other structural components.

NFPA 1403 requires several of the same steps for non-gas-fired live fire training structures as for acquired structures. Some of these include the following:

- Property adjacent to the training site that could be affected by the smoke from the live fire training evolution, such as railroads, airports or heliports, and nursing homes, hospitals, or other similar facilities, shall be identified.
- The persons in charge of the properties described above shall be informed of the date and time of the training evolution.
- Streets or highways in the vicinity of the training site shall be surveyed for potential effects from live fire training evolutions, and safeguards shall be taken to eliminate any possible hazard to motorists.
- The instructor-in-charge shall document fuel loading, including all of the following:
  1. Furnishings
  2. Wall and floor coverings and ceiling materials
  3. Type of construction of the structure, including type of roof and combustible void spaces
  4. Dimensions of room

As with acquired structures, the importance of identifying the exact burn locations is critical, and the final decision is made by the instructor-in-charge and safety officer, hopefully with considerable input from the participating senior instructors. With their input, the instructor-in-charge needs to carefully evaluate each fire room and burn location for factors that can affect the growth, development, and spread of fire. Fire spread beyond the room of origin is generally not an issue in non-gas-fired live fire training structures, except when fueling takes place too close by. Flashover is also generally not be an issue unless too much fuel is used. However, if materials other than pallets and hay or excelsior are allowed, the fire conditions can become more intense.

Fuel Load

An excessive fuel load can contribute to conditions that create unusually dangerous fire behavior. Only use the amount of fuel necessary to meet the learning objectives. Avoid conditions that could cause flashover or backdraft, which increases the risk to personnel to an unacceptable level and could also cause damage to the non-gas-fired live fire training structure. Local policy needs to be very specific regarding ignition and restocking of the fires. For example, a policy may state that no more than three pallets of a specific size and half a bale of excelsior should be used in a particular room **Figure 7-21**. Be on the look out for any debris that may hinder the access or egress of fire fighters. Remove any obstacles prior to the beginning of the next training evolution. Even small objects that seem harmless can get caught on turnout gear and potentially cause a major catastrophe.

Safety

Injuries and fatalities have occurred during past live fire training because of a deviation from the preburn plan, especially

---

Safety Tips

Any time the instructor-in-charge determines that fuel, fire, or any other condition represents a potential hazard, the training exercise shall be stopped immediately. The evolution should only be restarted once the hazard has been resolved. The Go/No Go sequence must be restarted as well.

Chapter 7 Gas-Fired and Non-Gas-Fired Structures

Figure 7-21 Check your local policy on fuel loads.

Figure 7-22 Vertical burn racks hold several pallets, excelsior, or hay, with a lateral flame deflector.

Figure 7-23 Steel drawers or a large pan holds pallets and hay or excelsior during fires, keeping nails, staples, and debris together.

regarding improper fuel loading. Several reminders are applicable to help ensure a high level of safety:

- The instructor-in-charge and the safety officer need to ensure the primary and secondary exit paths do not conflict with the expected avenues of fire spread and burn set locations.
- The burn set should generally be placed in a corner of the room, but permanent interior burn sites and protective lining may dictate burn set locations.
- Keep in mind where the fire stream will be directed into the room. Try not to place the burn set directly across from the door where the nozzle will be operating from.
- The burn set cannot be located in any designated exit path.
- No burn room should be used that does not have at least two separate means of egress available.
- Hearths or fireboxes can be used to protect the ceiling and walls from direct flame impingement. One type of vertical burn rack holds several pallets with excelsior and has a top on it to direct flames outward, so they do not damage the ceiling and the floor Figure 7-22 . Some burn racks can be relocated between drill sessions.
- The use of a steel "drawer" or large pan where pallets and excelsior are placed is becoming more common Figure 7-23 . These pans can help enforce restrictions on fire loading, and also help to keep staples or nails and debris contained for safety and easier removal. They can be permanently installed to slide out to the outside to be dumped. They can also be fitted with a hearth cover to protect the walls and ceiling. These are most often found in modified shipping containers.
- Each room that is to be utilized for live fire training evolutions must be prepared using Class A materials only.

Multiple Fires

Unlike acquired structures, NFPA 1403 specifically allows concurrent, multiple fires in permanent non-gas-fired live fire training structures. If there are to be multiple, simultaneous fires, the utmost caution must be exercised, and in situations with multiple teams and multiple instructors, the identity of the instructor-in-charge shall be clear to all participants. There must be one instructor-in-charge and the organization must be understood by all.

If multiple concurrent fires and evolutions are to be allowed by the AHJ, it is extremely important to include several more items during the preparation stage:

- More time and care is needed to ensure the entry and egress routes are accessible at all times. Identify possible bottlenecks and choke points where heavy foot traffic may result in backup.
- Consider the effects of having multiple simultaneous fires on the overall tenability of the structure.
- In addition to the effects of having multiple fires on the regular routes for entry and egress, consider the possible effects on the emergency evacuation routes.
- Consider the need for more personnel on the rapid intervention crews (RIC) as well as the safety positions.

Safety Tips

NFPA 1403 does allow for the use of limited quantities of a combustible liquid with a flash point above 100°F or 38°C in a non-gas-fired live fire training structure, if the structure has been specifically engineered to accommodate this fuel. While the standard does not specify quantities or purpose for using combustible liquids, they should only be used to light Class A fires. This will likely be an issue with the AHJ, and many will not allow it. If the use of combustible liquids is allowed, very careful policy needs to be in place as to how and where the material is to be stored during training, how much can be used, where it will be used, who dispenses it, how it is to be used, the use of a protection hose line, and what personnel can be in the area while it is being used.

Operations

Operations in permanent live fire training structures require a lot less personnel and are a lot less labor-intensive than operations in acquired structures. However, there are still steps that must be followed to be ready to conduct a live fire training evolution. It is during this phase where a lot can go right to make the evolutions very rewarding and organized, but not taking the necessary steps can set the operation up for failure. Many of the considerations are the same as for acquired structures. Remember, before starting any operations in a permanent live fire training structure the safety officer must walk through the entire structure and check to makes sure that all doors and windows work freely, there is no debris in the structure that could cause harm, all of the burners on gas-fired props are free of debris or unauthorized materials, look for obvious integrity concerns, and finally to make sure there are no unauthorized visitors **Figure 7-24**.

Figure 7-24 Before any use, the instructors need to walk the building to check for obvious integrity concerns, operation of doors and window, debris on the floor, and all safety features being operable.

Staffing and Organization

Instructor positions need to be filled by qualified and competent people in a strong organizational structure that is known and understood by all of the participants. All instructors delivering live fire training must be qualified by the AHJ. With gas-fired props, all of the instructors must be trained in the complete operation of the system in accordance with the manufacturer and the direction of the AHJ. This needs to include the main control station, the local controls for each prop, emergency shutdowns, and emergency procedures for the gas-fired prop. The safety officer must also be trained to operate all emergency shutdown controls and valves. The command staff may be smaller, but it follows the same organizational structure. As in acquired structures, nobody operates alone, instructor or student.

Lighting the Fire

The decision to ignite a training fire shall be made by the instructor-in-charge in coordination with the safety officer in gas-fired and non-gas-fired live fire training structures **Figure 7-25**. The Go/No Go sequence should be followed to ensure that no aspect has been overlooked. Safety is crucial at all stages, but in particular, during the ignition of the fire. If all participants are not ready for action, the fire should not be lit.

Figure 7-25 Ignition occurs only after approval of the instructor-in-charge in coordination with the safety officer with a Go/No Go sequence.

Ignition of Gas-Fired Systems

Gas-fired training systems can be equipped with ignition controls that remotely ignite fires from a control room. The fires being lit remotely and not being physically lit by an ignition officer can present a safety risk to unsuspecting personnel. The fires should not be lit without an instructor visually confirming that the burn area is clear of personnel. It is important for the instructor in the control room to keep in constant communication with the instructor present in the structure. In some large-scale structures, this communication becomes even more necessary **Figure 7-26**. Regardless of the size of the structure, communication between the instructors is critical when a fire is initiated and also

Figure 7-26 Some structural fire training props can be quite large. This four story non-gas-fueled prop in Gloucestershire, England is one of the largest in the world.

throughout the training exercise. Gas-fired systems made for fire fighter training have automatic ignition systems, and are not to be ignited manually.

Some fire departments have made their own gas-fired systems. Any man-made systems must be compliant to NFPA 1403 and other applicable codes. These systems must have monitoring equipment to detect unburned gas and oxygen levels, and must use remote ignition. They cannot require manual ignition.

The ignition officer must work in very close concert with the safety officer and instructor-in-charge to ensure the learning objectives are met. Significant time and effort have been expended so far, so it is very important that the ignition officer follows the proper order of operations, and not deviate from the plan regarding fuel loading, locations, or the set safety protocols.

The decision to ignite the training fire shall be made by the instructor-in-charge in coordination with the safety officer, and should follow the Go/No Go sequence.

Ignition of Non-Gas-Fired Systems

The set up of the burn rooms and the ignition process for non-gas-fired systems follows the same procedures as live fire training in acquired structures. In the case of non-gas-fired systems, the fire is to be physically lit by a non-student ignition officer. The ignition officer shall wear full protective clothing and be accompanied by a charged hose line while he or she is monitoring the materials for burning. NFPA 1403 states that the safety officer should also be physically present during ignition. Live fire training structures to hold more heat than acquired structures, and caution needs to be exercised.

As is the case with all live fire trainings, ignition of the fire is to only begin with the approval of the instructor-in-charge in coordination with the safety officer. Fires cannot be located in any designated exit paths, to ensure a safe exit for participants. Keep an eye out for any debris from pallets and such that may hinder the access or egress of fire fighters, and be sure to remove any items prior to the beginning of the next training evolution. Only use fuel materials in the amounts needed to create the desired fire size, and avoid conditions that could cause a flashover or backdraft. Any time the instructor-in-charge determines that fuel, fire, or any other

Live Fire Tips

Additionally, the National Institute for Occupational Safety and Health (NIOSH) report, *Fire Fighter Fatality Investigation Report F2005-31*, recommends the following for permanent live fire training structures:

- Ensure that two training officers are present with a charged hose line during the ignition or refueling of a training fire in accordance with NFPA 1403.
- Determine the minimum amount of flame, heat, and/or smoke required during live fire training evolutions to perform the training while ensuring fire fighter safety.
- Use the minimum fuel load necessary to conduct live fire training.
- Have a written respiratory protection program and ensure that self contained breathing apparatus (SCBA) face pieces are properly inspected, used, and maintained.
- Have burn rooms with at least two exits.
- Avoid having basement burn rooms.

Additionally training academies should consider the following:

- Installing instrumentation within live fire training structures to record information such as heat, the effects of suppression, and the byproducts of combustion
- Installing a ventilation system within the structure
- Having a qualified engineer evaluate fuel loads, heat retention, instrumentation, and ventilation systems of live fire training structures.

condition represents a potential hazard, the training evolution shall be stopped immediately. If an evolution is stopped, it should only be restarted once the hazard identified has been resolved and after the Go/No Go sequence. In addition to these restrictions on fuel materials, unidentified materials shall not be used because they could burn in unanticipated ways, react violently, or create environmental or health hazards. Therefore, no pressure-treated wood, rubber, or plastic shall be used either. In non-gas-fired props, a very limited amount of combustible liquids are allowed, using directives from the AHJ. Wooden pallets must be clean from any spilled material that may have soaked into the wood such as oils, pesticides, or other material that may cause an unforeseen condition or create a hazard. Hay needs to be feed grade to prevent the burning of pesticides or other unknown materials.

Nobody is to operate inside the structure alone. Also no participants should ever be playing the role of a victim. If there are going to be mannequin victims inside, they are not to wear protective clothing.

Overhaul

Once the evolutions are complete, students and instructors need to be checked for any injuries sustained, as well as any damage incurred to PPE and SCBA. If there is any damage, deficiencies, or repairs needed to the facility, they must be reported. The facility needs to be left in a safe condition. Be respectful of the next crew of fire fighters who will be using the facility. Carefully cleaning up your crew's mess will allow for an easier transition for the next crew.

Wrap-Up

Chief Concepts

- The use of permanent live fire training structures greatly reduces the chance of problems that can occur in acquired structures. The chances of structural collapse, fire hidden in voids, unrecognized or unanticipated fire spread, and hazardous structural materials are all reduced or eliminated.
- Permanent live fire training structures are designed to withstand repeated fires without incurring the damage that a normal building would.
- There are two types of live fire training structures: gas-fired live fire training structures and non-gas-fired live fire training structures.
- Gas-fired live fire training structures are most often fueled with liquefied petroleum gas or propane
- Non-gas-fired live fire training structures produce higher temperatures than gas-fired structures.
- A preburn plan for live fire training in permanent live fire training structures needs to be developed and/or approved by the instructor-in-charge and the fire department's chain of command.
- The live fire training instructor is responsible for ensuring that the permanent live fire structure is properly maintained and that the structural integrity is inspected once a year.
- All permanent live fire training structures need to be inspected for hazards and prepared before training evolutions.
- From the preburn walk-through to ignition of gas-fired props to the final overhaul, the instructor-in-charge and all instructors must ensure the safety of the fire fighters participating in the training evolutions.

Hot Terms

<u>Burn building</u> A common nontechnical term for a permanent live fire training structure.

<u>Burn sequence chart</u> The data collected and policies specified by the AHJ regarding Sequential Live Fire Burn Evolutions in a specific live fire training structure.

<u>Gas-fired live fire training structure</u> A permanent live fire training structure where the fires are fueled primarily by propane or liquefied natural gas.

<u>Non-gas-fired live fire training structure</u> A permanent live fire training structure where the fires are fueled by Class A materials such as excelsior, hay, and pallets.

<u>Sequential live fire burn evolutions</u> The amount of fuel per evolution and the number of sequential evolutions between cool-down periods, per specific room in an agency's live fire training structure on one day, as determined by the AHJ following experimentation and recording temperature and other data.

<u>Ventilation-controlled fire</u> A fire in which the heat release rate or growth is controlled by the amount of air available to the fire.

References

Colletti, D. (2007). "*Live Fire Training—Are We Making A Wrong Turn—Part 1*" Firehouse.

Klein, G. (1998). *Sources of Power: How People Make Decisions*, Massachusetts Institute of Technology.

Malo, W., Casey, D., Tomlinson, E., and Deming, R. (2004). Incident Investigation of Firefighter Death During a Training Fire—Port Everglades, Florida, BFST Safety Investigative Report 03-01.

U.S. Fire Administration/Technical Report Series (2003). *Special Report: Trends and Hazards in Firefighter Training*.

Live Fire Training Instructor in Action

You have been directed by your chief to advise a board of live fire training instructors in an upcoming burn in a permanent live fire training structure. You are unclear whether the structure is gas-fired or non-gas-fired. Before heading to the board meeting, you draw up a list of questions to refresh your knowledge on both types of live fire training structures.

1. Which of the following categories does a shipping container with a burn barrel fall into?
   A. Acquired structure
   B. Gas-fired live fire training structure
   C. Non-gas-fired live fire training structure
   D. Build and burn structure

2. According to NFPA 1403, Class A materials are typically used in:
   A. flammable liquids training.
   B. gas-fired live fire training.
   C. non-gas-fired live fire training.
   D. Class A materials are strictly prohibited by NFPA 1403.

3. Gas-fired props are most often fueled with:
   A. butane.
   B. gasoline.
   C. liquefied petroleum gas.
   D. solidified petroleum gas.

4. NFPA _____ is the guideline to use when building a new live fire training facility.
   A. 1001
   B. 1402
   C. 1403
   D. 1404

5. The instructor-in-charge shall document fuel loading, including all of the following EXCEPT:
   A. furnishings
   B. furniture
   C. wall and floor coverings and ceiling materials
   D. dimensions of room

Nonstructural Training Props

CHAPTER 8

NFPA 1403 Standard

4.2.1 All required permits to conduct live fire training evolutions shall be obtained. [p 165]

4.2.2 The permits specified in this chapter shall be provided to outside, contract, or other separate training agencies by the authority having jurisdiction (AHJ) upon the request of those agencies. [p 165]

4.2.3 The runoff from live fire shall comply with the requirements of the AHJ. [p 165]

4.4.8* Additional safety personnel, as deemed necessary by the safety officer, shall be located to react to any unsafe or threatening situation or condition. [pp 166–167]

4.5* Extreme Weather. The training session shall be curtailed, postponed, or canceled, as necessary, to reduce the risk of injury or illness caused by extreme weather conditions. [p 166]

4.6.12.1 The instructors and the safety officer responsible for conducting live fire training evolutions with a gas-fueled training system or with other specialty props (such as flashover simulator) shall be trained in the complete operation of the system and the props. [pp 163, 170–172]

4.6.12.2 The training of instructors and the safety officer shall be performed by an individual authorized by the gas fueled training system and specialty prop manufacturer or by others qualified to perform this type of training. [pp 170–172]

4.7.2 The decision to ignite the training fire shall be made by the instructor-in-charge in coordination with the safety officer. [p 171]

4.7.3 The fire shall be ignited by the ignition officer. [p 171]

4.8.7* All students, instructors, safety personnel, and other personnel participating in any evolution or operation of fire suppression during the live fire training evolution shall breathe from an SCBA air supply whenever they operate under one or more of the following conditions: [p 167]

(1) In an atmosphere that is oxygen deficient or contaminated by products of combustion, or both

(2) In an atmosphere that is suspected of being oxygen deficient or contaminated by products of combustion, or both

(3) In any atmosphere that can become oxygen deficient, contaminated, or both

(4) Below ground level

4.9.1 A method of fireground communications shall be established to enable coordination among the incident commander, the interior and exterior sectors, the safety officer, and external requests for assistance. [p 167]

4.11.1 The instructor-in-charge and the safety officer shall determine the rate and duration of waterflow necessary for each individual live fire training evolution, including the water necessary for control and extinguishment of the training fire, the water supply necessary for backup line(s) to protect personnel, and any water needed to protect exposed property. [p 167]

4.11.5 A minimum reserve of additional water in the amount of 50 percent of the fire flow demand, determined in accordance with 4.11.1, shall be available to handle exposure protection or unforeseen situations. [p 167]

4.11.6* Except under the conditions of 4.11.6.1, separate water sources shall be utilized for the supply of attack lines and backup lines in order to preclude the loss of both water supply sources at the same time. [p 167]

4.11.6.1* A single water source shall be sufficient at a training center facility where the water system has been engineered to provide adequate volume for the evolutions conducted and a backup power source or backup pumps, or both, are in place to ensure an uninterrupted supply in the event of a power failure or malfunction. [p 167]

4.12.2 Pressure-treated wood, rubber, plastic, polyurethane foam, upholstered furniture, and chemically treated or pesticide-treated straw or hay shall not be used. [p 160]

4.12.4 Unidentified materials, such as debris found in or around the structure or prop that could burn in unanticipated ways, react violently, or create environmental or health hazards, shall not be used. [pp 160–161]

4.12.11.2* All props that use pressure to move fuel to the fire shall be equipped with remote fuel shutoffs outside of the safety perimeter but within sight of the prop and the entire field of attack for the prop. [p 172]

4.12.11.3 During the entire time the prop is in use, the remote shutoff shall be continuously attended by safety personnel who are trained in its operation and who have direct communications with the safety officer and instructors. [p 172]

4.12.11.4 Liquefied petroleum gas props shall be equipped with all safety features as described in NFPA 58, *Liquefied Petroleum Gas Code*, and NFPA 59, *Utility LP-Gas Plant Code*. [pp 170–172]

4.12.11.5 Where the evolution involves the failure of a safety feature, the failed part shall be located downstream from the correctly functioning safety feature. [p 163]

4.12.11.6 Where flammable or combustible liquids are used, measures shall be taken to prevent runoff from contaminating the surrounding area. [p 165]

4.12.11.6.1 There shall be oil separators for cleaning the runoff water. [p 165]

4.12.11.7* Vehicles used as props for live fire training shall have all fluid reservoirs, tanks, shock absorbers, drive shafts, and other gas-filled closed containers removed, vented, or drained prior to any ignition. [p 160]

4.12.11.8 For flammable metal fires, there shall be a sufficient quantity of the proper extinguishing agent available so that all attack crews have the required supply as well as a 150 percent reserve for use by the backup crews. [p 161]

4.15.2 Prior to conducting actual live fire training evolutions, a preburn briefing session shall be conducted by the instructor-in-charge with the safety officer for all participants. [pp 170–172]

4.15.3 All facets of each evolution to be conducted shall be discussed. [pp 170–172]

4.15.5 The location of the manikin shall not be required to be disclosed, provided that the possibility of victims is discussed in the preburn briefing. [pp 170–172]

4.15.6 Prior to conducting any live fire training, all participants shall have a knowledge of and familiarity with the prop or props being used for the evolution. [pp 170–172]

4.15.6 Property adjacent to the training site that could be affected by the smoke from the live fire training evolution, such as railroads, airports or heliports, and nursing homes, hospitals, or other similar facilities, shall be identified. [pp 165–166]

8.1.2 Props used for outside live fire training shall be designed specifically for the evolution to be performed. [pp 160, 170–172]

8.1.3 Exterior props shall be left in a safe condition upon completion of live fire training evolutions. [p 167]

8.1.4 For outside training, care shall be taken to select areas that limit the hazards to both personal safety and the environment. [pp 160, 170–172]

8.1.5 The training site shall be without obstructions that can interfere with fire-fighting operations. [p 161]

8.1.6 Where live training fires are used outside, the ground cover shall be such that it does not contribute to the fire. [pp 170–172]

8.1.7 Debris hindering the access of fire fighters shall be removed prior to the beginning of the training exercise. [p 161]

8.2.1 Exterior props shall be inspected visually for damage prior to live fire training evolutions. [pp 170–172]

8.2.1.1 Damage to exterior props shall be documented and the owner or AHJ shall be notified. [pp 170–172]

8.2.2 All safety devices and emergency shutdown switches, plus doors, shutters, vents, and other operable devices, shall be checked prior to any live fire training evolutions to ensure they operate correctly. [p 165]

8.2.3 The structural integrity of the props shall be evaluated and documented annually. [pp 170–172]

Additional NFPA Standards

NFPA 58 *Liquefied Petroleum Gas Code*
NFPA 59 *Utility LP-Gas Plant Code*
NFPA 1971 *Standard on Protective Ensembles for Structural Fire Fighting and Proximity Fire Fighting*
NFPA 1981 *Standard on Open-Circuit Self-Contained Breathing Apparatus (SCBA) for Emergency Services*
NFPA 1982 *Standard on Personal Alert Safety Systems (PASS)*
NFPA 1975 *Standard on Station Work Uniforms for Fire and Emergency Services*

Knowledge Objectives

After studying this chapter, you will be able to:
- Describe the main types of exterior props.
- Describe the differences between an acquired prop and a manufactured prop.
- Discuss the characteristics of a Class A prop.
- Describe the types of Class A props.
- Discuss the characteristics of a Class B prop.
- Identify the information that must be included in the preburn plan for an exterior prop.

- Describe how to ensure the safety of all participants during live fire training evolutions.
- Describe how to prepare an exterior prop for safe live fire training.
- Describe how to ensure a safe ignition with a Class A prop.
- Describe how to ensure a safe ignition with a Class B prop.

Skills Objectives

After studying this chapter, you will be able to:

- Inspect and prepare an exterior prop for live fire training.

You Are the Live Fire Training Instructor

You have been approached by your fire chief regarding a need for training. There has been a recent increase in the number of fires involving tractor trailers, both mobile and stationary. Local fire departments have been unsuccessful at saving the vehicles and are not sure how to properly attack the fires. Your training site has an area designed for vehicle burning, however, the largest vehicle that can be used is a large van because of access issues. The chief wants live fire training on a tractor trailer to be held immediately and has tasked you with implementing the training.

1. What questions should you ask your chief?
2. What considerations should go into your planning?
3. Will access to the site hinder your ability to create a training evolution using tractor trailers?

Introduction

There are a variety of nonstructural exterior props used to train fire fighters nationwide. These exterior props range from homemade, inexpensive props to donated vehicles. It is important to understand the differences between the types of exterior props, their unique characteristics, and their functions. Regardless of the type of exterior prop used, safety in its operation and proper familiarization are key factors. **Exterior props** are used to simulate an actual fire emergency and can either be acquired or manufactured **Figure 8-1**. An **acquired prop** is a piece of equipment, such as an automobile, that was not initially designed for burning but is used for live fire training evolutions **Figure 8-2**. A **manufactured prop** is an exterior prop that is built to resemble an emergency scenario, such as a simulated sedan, tank truck, rail car, aircraft, ship, pipeline, fuel spill, cylinders, tanks, industrial units, etc. Live fire training with acquired and manufactured props can be conducted using Class A materials, or can be designed to burn using liquefied

Figure 8-1 Exterior props can range from small to large simulators, either commercially engineered and manufactured or locally made.

Figure 8-2 An acquired prop. The gas lines are underground and the car is stabilized for safety.

petroleum gas (LPG), compressed natural gas (CNG), butane, or diesel fuel.

Choosing the right type of prop designed specifically for the evolution to be performed is essential in assuring that the learning objectives and goals are met. If you are going to use the exterior prop over and over, the authority having jurisdiction (AHJ) may want to invest resources in an exterior prop that can be used repeatedly.

Types of Props

There are typically two main types of exterior props used in live fire training. Class A props use ordinary combustibles such as wood products, cardboard, hay, straw, or excelsior as fuel. Class B props are either gas fueled, liquid fueled, or a combination of the two.

Class A Props

Exterior Class A props are not as common as exterior Class B props and are most often intended to represent a specific item on fire, such as a dumpster or other container, or a vehicle such as an automobile. Class A props will be discussed in terms of the fuels that can be used, the container or prop that holds the fuels, safety considerations of the Class A props, and visual inspections.

Class A props use ordinary combustibles for their fuel source. These ordinary combustibles can be untreated pallets, hay, straw, as well as commercially purchased excelsior. Paper and cardboard have also been used as fuel; however, these fuels can be easily scattered by wind or by the use of portable fire extinguishers.

Determining which fuel will be used in the Class A prop is one of the first steps to address. NFPA 1403 states that the fuel being used must have known burning characteristics that can be controlled and the live fire training instructor must determine the amount that will be required to meet the objectives of the training and create the desired fire size. As previously discussed, pressure-treated wood, rubber, plastic, and treated hay or straw may not be used. Material in the area around the Class A prop that could burn in unanticipated ways or create a hazard shall not be used and should be removed from the area. Once the amount of fuel has been determined, the instructor-in-charge is responsible for documenting the fuel load. The fuel load should be noted on the preburn plan and monitored throughout the evolution. The fuels can be placed directly on the ground for students to extinguish or they can be placed inside of the Class A prop (the container) of various shapes and sizes.

The Class A prop is a valuable tool in training fire fighters. The appeal of Class A props is both their use of ordinary combustibles and the lower cost compared with flammable liquids. A Class A prop can be an effective training aid, if it is used to simulate a true emergency, as it is intended.

Types of Class A Props

The Class A prop can be as simple as a household product, such as a garbage can, a manufactured prop, or can be an acquired prop, such as a vehicle. Some of the more common Class A props that are being used are metal drums, dumpsters, vehicles, and kitchen props. The purpose of the Class A prop is to contain the fuel from being scattered.

Metal drums can be cut in half lengthwise to provide two low-profile burn pans. These can be moved easily from one area to another. A grate can be placed over the top of the barrel to contain the fuels. Cleanup is fairly simple because ash and water can be dumped into a designated area. Because ordinary combustibles are used, disposal is much more simple than when fuels are used. The instructor may decide to keep the barrel intact and simply remove the lid to maintain the height. However, thought needs to be given to how the ash is to be removed and the need for drainage holes to allow water to escape. Very important safety issues with using barrels are considering what was previously stored in the barrel and what its condition is.

A dumpster may be acquired through local resources. As dumpsters get older and become damaged, agencies may simply give them to you as an acquired prop. As with barrels, the instructor or authority having jurisdiction has the responsibility to ensure that the acquired prop does not contain any hazardous materials or would be a hazard to a student training on the acquired prop. A dumpster can also be easily manufactured using sheet metal and metal braces. Depending on the type of metal, the acquired prop could be reinforced to ensure longer life.

Regardless of whether the Class A prop is acquired or manufactured, if you are burning Class A materials, such as wood and pallets, you will want to be able to occasionally clean it out. Acquired dumpsters can be retrofitted to assist with the removal of debris. This can be accomplished by cutting out a rectangular section of one of the sides towards the bottom and creating a pan that will slide in and out. After burning, the pan can be pulled out to remove any remaining debris. This will also assist in controlling loose nails on the fireground.

Vehicles can be acquired or manufactured so that Class A materials can be used as fuel. NFPA 1403 specifies the preparation of acquired vehicles to assure safety for all participants. Fluid reservoirs, tanks, shocks absorbers, drive shafts, and other gas-filled closed containers are to be removed, vented, or drained prior to any ignition. Other items to consider are bumper compression cylinders, shock absorbers, fuel tanks, drive shafts, batteries, and brake shoes (containing asbestos). The oil pan, transmission, and differential drain plugs should also be removed, and the fluids drained and disposed of properly. The preparation of the vehicle is important to ensure safety, the first priority of any live fire training evolution **Figure 8-3**. There have been documented cases of compression cylinders that have not been removed and subjected to heat which have released and penetrated turnout pants. Fluids and containers left in the vehicle can be considered additional fuels that you may not know the burning characteristics of.

It is also very important to check the trunk and interior and remove any debris from the acquired vehicle. Remember that people carry anything and everything in their vehicles. A variety of items have been found including gas containers, paint

Chapter 8 Nonstructural Training Props

Figure 8-3 A manufactured Class B automobile prop with flexible piping to allows movement in and out of a simulated garage.

Safety Tips

An additional safety concern when discussing exterior props, specifically vehicle fires, is the use of flammable metals. If the exterior prop is to simulate fires involving flammable metals, the instructor-in-charge must ensure a sufficient quantity of extinguishing agent is available so attack crews have the required supply plus 150 percent reserve for backup crews. It is very important for instructors to understand the fire behavior of the fuels and props that are being used. Vehicles may contain combustible metals; however, instructors may chose to demonstrate fire behavior of flammable metals simply by placing the product on the ground or pad. It cannot be stressed enough that the learning objectives must be clear on all evolutions and the exterior prop must simulate a fire that the participant will encounter.

Figure 8-4 These vehicle tires are made from metal.

and paint thinner, clothes, and other items which add to the fuel load and are prohibited during live fire training. There may be slight confusion where NFPA 1403 states that rubber cannot be used as a fuel. However, tires are not required to be removed. As an instructor, you must consider if there is a purpose to leaving tires on a vehicle Figure 8-4 .

There have been acquired props, such as kitchen stoves, used in training as well. The oven can be filled with hay, straw, or excelsior to simulate an oven fire. As with all training, these props need to simulate an actual incident. Stoves are not found in the middle of an open area. However, if this prop is placed under a roof with three concrete walls surrounding it, then the evolution is more realistic and can be a valuable learning tool.

Safety Considerations for Class A Props

There are safety considerations that apply to all Class A props. Safety in terms of placement of the prop, general usage guidelines, and visual inspections are important factors when using Class A props. The Class A prop has to simulate an actual emergency and safety is the first priority. If the decision is to build a Class A prop, remember that it must meet code and standard requirements and needs to be strong enough to maintain its integrity under repeated burns. It also needs to be thoroughly tested before being used in training. The testing of manufactured props must be performed in a controlled environment with safety measures in place because there are dangers of malfunction, explosion, or unexpected fire behavior that can result in injury or death. All Class A props should be tested in an environment where any unforeseen events can be controlled. Barriers between the Class A prop and any personnel need to be present for protection. The Class A prop should only be used for live fire evolutions when the agency is absolutely sure that it is stable and operating as expected.

The placement of the Class A prop is also a safety concern. Not only must there be sufficient space for both attack and backup lines to advance and maneuver effectively and safely, but topography must also be considered. The area should be flat and clear of debris. While the fuel should be contained within the Class A prop itself, flaming debris could blow away from the prop and result in wildland fire. Dry chemical extinguishers will induct air and the inert propellant will blow away paper and smaller items. This is an important item to consider when deciding on the prop's location.

Class B Props

It would be nice if every fire department close to an airport could have a Boeing 737 to conduct training on. It would also be great if an LPG truck was readily available to all for practice on advancing hose lines and extinguishing the fire. The best training would be on the actual plane or vehicle; however, this is not cost effective or feasible in most cases. Enter the Class B props. Class B props can be designed to be clean-burning and can be used repeatedly. Class B props can be reinforced to extend the life of the prop. Like live fire training structures, many of the Class B gas-fueled fires actually simulate what would be a Class A fire or a Class B liquid fuel fire. Vehicles can easily be adapted into Class B props to simulate crash rescue training.

Near Miss REPORT

Report Number: 10-723
Date: 04/21/2010

Event Description: Note: Brackets denote reviewer de-identification.

I was the instructor for a live fire exercise involving multiple departments in a regional multi-company operation. I had previously attended a "Train the Trainer" for the gas fired facility that is privately owned and operated for maritime firefighters. The facility and scenarios were intended to simulate class "A" fire behavior in the first and third component of the scenario, with the second component simulating a small fuel fed fire involving machinery. The conditions I experienced in my instructor training were significantly different from those experienced with the crews on my first day as an instructor on [date omitted].

The goal of the training was to stress the importance of ventilation when attacking a fire below grade, hydraulic ventilation, nozzle skills to control overhead byproducts of combustion when working towards the seat of a fire, class "B" firefighting techniques, and the importance and role of a back-up team.

The prop is a simulated ship made from metal shipping boxes welded together. Each fire is lit independently by the same individual so there is little chance to have more than one prop burning at a time. However, there are no fail-safe mechanisms at any of the props. If an instructor forgets to shut off a valve and moves to the next prop or turns the valve the wrong direction, there could be an unwanted buildup of gases that result in an unintended explosive ignition. Additionally, the closing of a valve does not result in an immediate shutoff of the gas. The gas supply line at any given prop will bleed the gas from the point of shut off to the prop. In one prop in particular, this distance is over 30 feet. The operating pressure of the gas is variable, but generally, it is in excess of 200 psi. This pressure, and the distance to the prop, means a significant amount of gas is released into the prop after shut off. Additionally, it is the responsibility of the instructor at the valve to shut the valve off to prevent a buildup of gases in the space. The limited visibility and the resulting steam, training smoke and cooling of the fire makes it difficult to determine if the fire has been extinguished. If the fire is extinguished, we are simply placing unburned propane into this confined space. We were told to give the prop three seconds after the last visible fire before shutting off the valve, as we may not be able to see the fire due to the firefighting activities. This practice is far from automatic, consistent, or in my opinion safe. The following sequence would have to take place:

1. Recognition of no fire
2. Reaction by counting off 3 seconds
3. Shutting off the valve
4. Excess gas in the line bled off into the space.

With the average human reaction time being nearly .5 sec, the variable in the counting of 3 seconds, and the time to bleed excess gas in the line, it is not unrealistic to conclude that unburned propane could enter the space at pressures in excess of 200psi for over five seconds in the best of circumstances. This resulted in some unexpected and unwanted fire behavior during my first day as an instructor at the training.

The potential for explosion at this prop is real, and was seen on a small scale. The first attempt of the day on [date omitted] resulted in a rapid ignition of gases that blew doors open and material flying from parts of the prop. No personnel were inside the prop at the time; however, a piece of debris struck a captain on the side of the face while he was getting crews ready to enter. No injury was reported. However, later that day, a ball of fire was pushed through the opening created by the earlier failure and resulted in a minor heat injury to the eye of the same captain.

Later that day another narrow escape occurred when firefighters began descending the stairs as the prop was being "lit off". The exterior instructor is relying on the eyes and notification by the interior instructor(s) to shut off the gas in an episode of this nature. Again, due to the reaction time and "bleed" time a fire could easily extend to and up the stairwell resulting in injury although no injury resulted.

My final experience of the day was a smaller version of what occurred after I left the training for the day. A limited buildup of gas occurred after the line was shut down and ignited throughout the space to within 2' off the floor. The ignition was only 1–2 seconds and no damage or injury resulted.

> The final incident occurred in a scenario after I had left the facility. I will discuss how I believe the incident at the end of the day occurred. I believe there are several factors that contributed to the incident to include engineering controls, or lack thereof, at the facility, the nature of the fuel being used, and the tactics and actions implemented at this training.
>
> I have discussed some of the engineering controls of the site. In short, there are no automatic shutoffs or fail safe measures built into the system. A failure by an operator of the valve to close the valve in a timely manner introduces flammable gas into a confined space.
>
> The properties of the gas affected the situation significantly. The gas is liquefied propane. The bottles are heated using warm water to increase the pressure within the bottles. The operating pressure is generally in excess of 200psi. The properties of LPG include it being a flammable gas that is heavier than air. The process requires it to travel from a heated tank through piping to the prop. As it flows from the tank, it cools and enters a space that has been heated by the ongoing fire event. This process increases the affinity of the gas to sink in the space when it is not ignited rather than stratify at a higher level as class "A" fire gases would.
>
> The tactics being used are basic class "A" fire tactics, meaning cooling the overhead gases as the team advances towards the seat of the fire. A narrow fog pattern is used in a forward progression for this action. The venturi effect of the nozzle as the crew moves forward had the unintended result of drawing gases back toward the nozzle and around the stream (away from the ignition source). In the event on [date omitted], the crew moved forward after "extinguishing the fire gases" just as they would in a class "A" fire event, cooling the overhead and working towards the seat of the fire. However, in this case they only accomplished cooling the gases below their ignition temperature, while drawing the gases away from the ignition source as gas continued to flow into the space. As the crew began hydraulically ventilating the space, (near the ignition source) the gas was brought into its explosive limits and it ignited at the pilot and flashed throughout the space. The crew was between the fire (the entire space) and the ventilation point (a place we just learned in fire dynamics we do not want to be) and attempted to react by backing up when they saw fire behind them. The ignition of the space occurred so rapidly that the crew was unable to react before they were enveloped in ignited propane gas.
>
> The only reason these individuals were not severely injured was due to proper usage of PPE and the limited gas in the structure (as it had been turned off by the instructor). The gas burned off rapidly but not before heat damage to PPE had occurred. I believe we were lucky to have a single minor injury on [date omitted] after several narrow escapes. I also believe my chief has made the correct decision in immediately ending all future training at this facility following these events, as the safety concerns cannot be effectively mitigated.
>
> **Lessons Learned:** 1. Proper use of PPE. 2. Identifying adequate fail-safe measures for gas fired props. 3. Identifying the effects of actions on the fuel type being used. 4. Identifying the characteristics of LPG and how they will be used in a facility. 5. Limited instructor cadre for high risk training events. 6. Willingness to identify poor or unsafe actions or conditions that could contribute to injury and being willing to end these actions or mitigate the conditions before continuing.

There are a variety of exterior props designed to simulate flammable liquid fires **Figure 8-5**. Creating a Class B prop for a training evolution on extinguishing a flammable liquid with fire extinguishers can be accomplished by constructing a burn pan that will hold water and then placing a layer of flammable liquid on top.

There are also manufactured props that are similar to chutes in construction, in order to simulate a flowing burning liquid. Others are designed to simulate a storage tank rupture with fire. Instructors who will be operating manufactured props should receive training on the prop by the manufacturer of the prop or their designated personnel.

It is very important that all instructors and safety officers are familiar with all components and operations of Class B props. While all excellent tools, there are additional safety concerns that need to be addressed when using Class B props.

The instructor-in-charge needs to identify the amount of flammable liquid that will be used and that will create the desired effect to practice the use of fire extinguishers.

Safety Tips

Should the instructor-in-charge decide to run a training evolution involving the simulated failure of a safety feature (i.e., a pressure relief valve that "fails" or a hand-operated valve that does not operate), additional safety steps must be followed. For example, if the evolution is designed to fight a fire and the safety shutoff valve is set up not to function, the malfunctioning safety feature must be downstream from the functioning safety feature. If the fuel flows downstream, and the students realize the malfunctioning safety feature does not work and panic, the fuel will need to be stopped as fast as possible. A correctly performing safety valve between the fuel source and "malfunctioning" valve would stop the flow and prevent any injuries to students.

The instructor-in-charge is responsible also for knowing the burning characteristics of the flammable liquid as well as documenting the fuel load that will be used in the preburn plan.

Figure 8-5 **A.** Flammable liquid "Christmas tree." **B.** A Class B prop for training evolutions extinguishing flammable liquids.

Live Fire Tips

Foam

It would not be appropriate to discuss flammable liquids training without the discussion of foam. Environmental protection is highly valued during actual emergencies and the idea should be replicated in the training environment as well. The instructor controls the training environment and therefore also controls environmental impact. One of the main concerns with using foam is that it can contain perflurochemicals (PFCs). There is concern that the PFCs in Class B foam can travel from the training area to a municipal or private well. Once they enter the environment, PFCs enter the soil and water and remain there for a long time. Several manufacturers have minimized the use of PFCs in foam, however in many cases it is not known how much PFC is in the foam being used. In turn, Class A foam has become more popular. Class A foam typically does not contain PFCs and makes the water more able to penetrate combustible material. In an emergency, the use of foam is invaluable; however, in training, the risks are a concern.

Fire fighters need to be trained on when to use foam, how to use foam equipment, and how foam aids in fire extinguishment. Some training centers have used detergent to simulate foam, which is helpful in showing how a foam blanket is generated, but it does not show how the foam actually works. It is very important to explain to students when to use Class A foam and when to use Class B foam. It is highly suggested that if training is to be done with Class B foam, that surface waters or storm sewer drains are avoided, which would allow the foam to enter and spread throughout the system. Once in the system, it does not break down and remains in the system for a long time. If possible, foams developed specifically for training should be used. These training foams are believed to have no or less PFCs than normal Class B foam. Some manufacturers make training foam that can be proportioned at 3 percent or 6 percent, is biodegradable, and does not contain any fluorochemicals or polymers that are used in firefighting foam agents.

There may be situations where using training foam is not an option. In high-risk fires, such as airport firefighting, fire fighters need to know exactly how the foam is going to react and work with the involved aircraft. In these cases, Class B foams may need to be used, and the runoff will need to be contained and disposed of properly.

Another concern with Class B foams and aqueous film forming foam (AFFF) and runoff is the contamination of oil/water separators. Foams work by forming an impenetrable layer. Oil/water separators are designed to function so that the water at the bottom of the pit can be allowed to drain. Many of these separators have chambers with drains at various heights so that the lowest water is the cleanest. This draining system may not function properly because of the nature of the foam. Also, if a layer of foam manages to get through the system, the chemicals will enter the ground-water system and can contaminate drinking water.

Safety Considerations for Class B Props

Flammable liquids by nature can be splashed outside of the vessels that are containing them. There is the possibility for the liquid to be splashed onto fire fighters' turnout gear, contaminating it, and leaving it open to ignition.

Not only will the instructor-in-charge and safety officer need to be aware of contaminated bunker gear, but they will also need to remember that the hose and other appliances may also be contaminated if they are dragged through any liquid that has splashed outside the container. It will be necessary to assure that the hose is clean before being placed back into service for any use. Gear will need to be washed thoroughly and inspected for degradation.

Live Fire Tips

Runoff needs to be contained when working with Class B props and all props should be left in a safe condition at the end of training.

Another concern is to assure that surrounding exposures are protected. Crews need to be prepared for exposure protection, should wind direction or conditions change. The parking of apparatus is also an area for concern because the amount of heat generated can damage an apparatus if it is parked too close to a Class B prop.

The instructor-in-charge and safety officer will need to be aware of state permits or regulations that need to be followed. For example, Kentucky allows for a mixture of 50 percent diesel fuel and 50 percent gasoline by volume, may be burned in a properly designed and constructed pit for purposes of fire training involving flammable liquids. In addition, instructors need to be familiar with the props they are using as well as the guidelines of the state or authority having jurisdiction (AHJ).

Preburn Plan

Student Prerequisites

Students must possess prerequisite knowledge and skills, as dictated by NFPA 1403, *Standard on Live Fire Training Evolutions*, in order to participate in a live fire training evolution involving an exterior prop. These prerequisites must be documented and verified by the instructor. If the instructor did not personally train the students, he or she must obtain written documentation that the student has met the prerequisites prior to participation. Several AHJs have established a procedure where the students must provide the written documentation as well as demonstrate their skill levels with self-contained breathing apparatus (SCBA) and personal protective equipment (PPE). Using this procedure, the instructors see their students' familiarity with their gear and PPE first hand, which are key concerns during live fire training.

Environmental Concerns

A significant concern with exterior props and any drill ground using liquid fuels is the possibility of ground contamination or runoff contaminated with fuel, foam, etc. reaching waterways or environmentally sensitive areas Figure 8-6 . NFPA 1403 now specifies that environmental concerns must be addressed and meet the requirements of the environmental authorities having jurisdiction. Environmental protection agencies may require permits and prior notification when conducting live fire exercises.

It is very important to know what is required at your facility and you should expect there will be multiple agencies (local, state/provincial, federal) with jurisdiction at your site. Training sites using liquid fuels will generally be required to have collection systems and oil/water separators Figure 8-7 . The run-off is collected and then treated to remove foam and to separate fuel from the water. The products collected from the water are transferred to go to a holding tank to be pumped off.

Depending on local laws and the immediate environmental concerns, training centers that use only gas fueled and class A fueled props may still be required to have run off protection.

Neighbors

Property owners adjacent to the area that may be affected by the smoke need to be informed of the training date and time. Some examples of these properties would be railroads, airports or heliports, nursing homes, and hospitals. Streets and highways that may be affected also need to be identified and safeguards should be put in place to eliminate potential hazards to motorists. Pedestrians around the training site should be kept clear from the operations area through the use of fire lines.

Figure 8-6 A liquid fuel burn area protects the water supply from ground contamination when using foam.

Figure 8-7 A fuel/oil separator.

Safety Tips

Ensure that all safety devices, such as thermometers, oxygen, toxic and combustible gas monitors, evacuation alarms, emergency shutdown switches, doors, shutters, vents, and other operable devices, are checked prior to any live fire training evolutions to ensure proper operation. If the safety devices are not in operating order, there is a safety issue, and the exterior prop should not be used. This also reinforces the need for instructors and the safety officer to be fully knowledgeable of the exterior prop and its operating components.

Safety Tips

When deciding to use exterior props, ensure that any exposures, utility services, or property that might become ignited and involved in fire are protected or removed. If there are combustible materials in the area that are not part of the fuel load, they should be removed from the training area.

Weather

Weather is a very important consideration when using exterior props. High winds will affect the path of flames and smoke and could put protected exposures at potential risk. Winds can also circle the flames back and surround attack team personnel. In colder regions, ice and snow are safety concerns that need to be addressed when using exterior props. The AHJ should have procedures in place to address outdoor training in extreme heat or cold situations. Ice-covered grounds increase the chances of injuries caused by slip-and-falls. If extreme weather is present, or is forecasted, it may be necessary to postpone or cancel the training in order to reduce the risk of injury or illness.

Water Supply

Instructors need to be familiar with NFPA 1403 requirements on water supply because water supply issues must also be determined for exterior props. The instructor-in-charge is responsible for determining the rate and duration of water needed to control and extinguish the training fire, as well as making sure there is enough supply for backup lines and for protection of any exposed property. Water must also be available to protect exposures and handle unforeseen situations. The supply of the attack lines and backup lines must be from separate sources. The only way it is permissible for them to be from the same source is when the training evolution is being held at a training center where the system has been engineered to provide an adequate volume, and a backup power source or backup pumps, or both, are in place to ensure an uninterrupted supply in the event of a power failure or malfunction.

Part of the water supply component of the preburn plan needs to identify the number of attack lines and backup lines needed. Also identify what type of water sources will be used, if operating from an engine to what pressure should the lines be charged, and what type of nozzles should be used. The preburn plan can act as a checklist helping to ensure that all equipment is available when it is time to conduct the training. At a minimum, each line must be capable of providing 95 gpm (360 L/min).

Parking Areas

Areas for apparatus need to be identified and must include areas for staging, operations, and parking. The parking area should be properly located so that fire apparatus and vehicles that are not a part of the training evolution do not interfere with fireground operations. If an apparatus is in service with the possibility of being dispatched, it should be parked in an area where it can facilitate a prompt response. Additional parking areas may be required if police assistance or the media are attending the training. A parking area for an ambulance or an emergency medical services (EMS) vehicle should be designated and located where it will facilitate a prompt response in the event of an injury. The areas for apparatus and other vehicles need to have both ingress and egress routes designated in case they are needed for an emergency.

Operations Area

The preburn plan should also identify the operations area. This area should be visibly marked so that spectators and anyone not participating in the training evolution remain outside of the operations area. This perimeter shall be established by the safety officer, and must be marked using ropes, signs, or fire line markings, for example. If a visitor is to be allowed inside the operations area, he or she must be escorted at all times and must wear complete protective clothing according to manufacturer's instructions and in accordance with the requirements listed in NFPA 1403. The PPE requirements for participants and supervised visitors are the same, and should follow the standard.

> **Live Fire Tips**
>
> Permits may be needed based on local requirements, and required permits may be requested and should be provided to outside, contract, or other separate training agencies.

Emergency Medical Services

NFPA 1403 states that emergency medical services shall be available onsite to handle injuries. Any injury or medical aid rendered requires a written report be filed. It is crucial to document these instances in case an investigation is launched on the department.

Safety

The use of a safety officer is required for all live fire training evolutions. Not only is appointing a safety officer a requirement, but it also gives the instructor a second pair of eyes. Nobody can be in more than one place at a time, so the safety officer acts as a mobile partner, working in conjunction with the instructor-in-charge. Should the safety officer observe anything that requires an immediate stop to the evolution, he or she has the authority to do so.

The safety officer is not only responsible for the safety of the participants, but also for that of the visitors and spectators. For this reason, he or she needs to assure that the operations area is clearly marked as well as the areas for visitors, spectators, and the media.

The safety officer should not be assigned to other duties that could interfere with safety responsibilities, because of the importance of their role. A safety officer needs to be able to walk around the operations area to ensure that the evolution is running smoothly and that no condition exists that could pose a potential risk. The safety officer also has some additional responsibilities in regards to exterior prop training, such as knowing how the exterior prop operates, its safety features, and any emergency stops, buttons, or other shutoff mechanisms.

In order to provide maximum safety for all participants, backup lines need to be in place, charged, and staffed. These lines need to have assigned personnel, and an instructor should be assigned to work with the team if they are newer recruits. Simply laying a line and charging it is not ensuring protection for your personnel. The line needs to have assigned personnel. This is also the perfect opportunity for instructors to teach and reinforce topics such as fire behavior, hazardous environment conditions to be aware of, and how to create a barrier between the source of the fire and fire fighters.

The instructor-in-charge also needs to assign an instructor to each functional crew, maintaining the minimum instructor-to-student ratio of one to five. The instructor-in-charge also needs to assign an instructor for each backup line. Additional personnel may be needed on the backup lines to assure that they can be maneuvered into place in an efficient and rapid response.

If there are more than one manually activated safety stations, an additional safety officer needs to be placed at each station. Additional safety officers may also need to be assigned strategically to ensure that the entire operations area can be observed and participants can be stopped if there are apparent hazards. Depending on the terrain, additional safety officers may be necessary simply to observe trip-and-fall hazards. During inclement weather, additional safety officers can observe changes in wind direction and the possibility of flames coming back towards the participants. If the weather is going to be an issue, and could possibly cause illness or injury to anyone, the training should be stopped or postponed until it subsides. Also, nighttime evolutions can add additional risks if the scene is not properly illuminated, so additional safety officers may be needed in these cases.

Communications

Communication is very important to safety. Efficient, functioning communications are a must to assure smooth operation during live fire training. At a *minimum*, communications must have the capability of allowing the instructor-in-charge, interior and exterior sectors, safety officer, and ignition officer to coordinate activities. Additionally, communications must be available to request external assistance, if needed. Some departments have radios that are programmed simply for training operations, which can work for the evolution. However if an emergency occurs, these radios may not be able to call for immediate assistance. It is very important that requests for help can be made immediately. No matter how much planning is done, there is always the possibility of the fire growing out of control, the fire spreading to an exposure, or a fire fighter getting injured. All of these would require additional resources that could only be reached through established communications. Additionally, the goal is to have students train as if it were a true emergency. Instructors should consider having students use communications to give size-up information, request resources, and coordinate the fire attack. This decision will be based on the knowledge and experience level of the students and what the learning objectives of the training evolution are. However, students should have communications capabilities. Different channels should be used for instructional personnel and students.

Participant Safety

Participants are not to be used as victims inside an exterior prop, as is the case for all live fire training evolutions. Most training evolutions in exterior props will not consider the use of victims.

Live Fire Tips

Most training evolutions in exterior props will not consider the use of victims. Placement and protection of the dummy will be very important.

Although this is highly practiced in live fire training structures and acquired structures, most instructors do not practice this with exterior props. As an instructor, you may want to consider having a victim next to a burning LPG tank, or having a victim located inside a vehicle that is on fire. Protecting the dummy will be very important, but the instructor will need to consider where to place the dummy for safety reasons. Fires cannot be located in any exit path.

Participants must be in full PPE and SCBA and must be inspected by the safety officer prior to entering the live fire training evolution. Again, the safety officer needs to be knowledgeable of NFPA 1971, *Standard on Protective Ensembles for Structural Fire Fighting and Proximity Fire Fighting*, as well as knowing the manufacturers' instructions on wearing the gear. All gear needs to be in serviceable condition to reduce the risk of steam burns, or worse. The safety officer also needs to be familiar with NFPA 1981, *Standard on Open-Circuit Self-Contained Breathing Apparatus (SCBA) for Emergency Services* to assure that the SCBA is in operable condition, including the PASS devices, as covered in NFPA 1982, *Standard on Personal Alert Safety Systems (PASS)*. NFPA 1975, *Standard on Station/Work Uniforms for Fire and Emergency Services* also addresses station and work uniforms, and the safety officer needs to assure that this standard is also followed.

Personal Protective Equipment

As required by the standard, all protective coats, trousers, hoods, footwear, helmets, and gloves need to meet the requirements outlined in NFPA 1971. NFPA 1403 also allows persons *not* involved in firefighting, or exposed to hazards of structural firefighting, to use helmets, meeting federal OSHA requirements. This includes personnel such as pump operators, extra support instructors not assigned to a function, and other key personnel such as an instructor in the rehab area or in the staging area.

The safety officer must assure that students, instructors, safety personnel, and other personnel involved in the evolution or fire suppression use the correct level of PPE and that it is in the condition required by the manufacturer. Furthermore, it is the safety officer's responsibility to assure that SCBA is being used in any atmosphere that is, or is suspected to be, oxygen deficient or contaminated by products of combustion. SCBA use is also required any time training is below ground level, or in any environment that can become oxygen deficient or contaminated. New fire fighters are continuously told to keep their SCBA on until the air monitoring shows it is safe to breathe. This is a good time to reinforce the practice of using air monitoring equipment. Again, if trained correctly, the fire fighters will follow these guidelines when an actual emergency occurs.

Live Fire Tips

Ensure that the exterior prop is left in a safe condition once all training evolutions are complete. This will save the next group of fire fighters time. As you would want the courtesy of coming into a clean, safe training environment, your fellow fire instructors would too.

Incident Report

Parsippany, New Jersey - 1992

At the end of a Fire Fighter I program, search and rescue drills were being conducted in a converted school bus by members of the Greystone Park Fire Department for members of several area departments.

The windows of the converted school bus were covered with welded steel plates, and the interior seats were removed. The fire department that owned the bus used it for hot smoke training, normally using kerosene-soaked wood chips in drums. It should be noted that this was noncompliant with the legal requirements of the county it was located in.

Shredded paper was added to a couch in the bus and ignited with a road flare. The fire was allowed to burn for about 10 minutes in order to produce heat and smoke in the bus. The students were told the purpose of the drill, and were instructed to enter the bus through the rear emergency exit carrying a forcible entry tool, pass the burning couch, conduct a "primary search," and then exit through the front bi-fold bus doors.

The instructor followed three students into the bus, and two more students followed behind them, leaving the back door open. After roughly two minutes, a flashover occurred in the bus. The instructor and the last two students escaped through the back door without injury. Students trying to exit through the front doors became jammed, as students on the outside pried the doors open with forcible entry tools. Two students escaped, but interior conditions were so bad that their protective clothing was smoking. At this point, the sound of the third student's low air alarm echoed through the bus. Students pulled a preconnected 1½" (38.1 mm) hose off the engine, but the hose lines did not have nozzles. Another student used the booster line and extinguished the fire from the back door. Without wearing his protective gloves or SCBA, the ignition officer entered through the front, found the trapped student against the front partition to the stairs, burned his hands when trying to remove the student, and had to exit. The lead instructor and a student then entered though the front door and tried to drag the student out through the front door, but were unable to, due to his size. Another student, in full PPE, entered though the rear door and was able to drag him out that way.

The injured student went into respiratory arrest and EMS was summoned. He was revived and transported, and received second and third degree burns to roughly 30 percent of his body, as well as respiratory burns. The other two fire fighters who were initially trapped both received second and third degree burns, one to 15 percent of his body and the other to 20 percent of his body.

The fire was investigated by the county prosecutor's office, and eventually taken to a grand jury, although a criminal indictment was not issued.

The lead investigator for the prosecutor's office said, "This training facility violated all applicable standards, such as those governing emergency ventilation, emergency lighting, fuel sources, communications, and emergency evacuation."

Post-Incident Analysis: Parsippany, New Jersey

NFPA 1403 Noncompliant

- Student-to-instructor ratio was 17:1 (7.5.2)
- No written emergency plan (7.2.24.3)
- Students not informed of emergency plans/procedures (7.2.24)
- Interior personnel without hose line (7.4.8)
- No established water supply (7.2.22)
- No incident safety officer (7.4.1)
- Flashover and fire spread unexpected (7.3.5)
- No incident commander or instructor-in-charge (7.5.4)
- Absence of ventilation to prevent flashover (7.3.6)
- Ingress and egress routes not monitored (7.2.23.5)
- No walk-through by students for familiarization, escape routes, etc. (7.2.25)
- No EMS standby (7.4.11)
- No inspection of props to ensure safety devices were operable (front vehicle passenger door) (7.2.8)

Maintenance

The structural integrity of the exterior props must be evaluated and documented annually. If the prop was purchased, the AHJ should follow the manufacturers' procedures for evaluation, which may require inspection by the manufacturer. The manufacturer also can provide guidelines on how to conduct an inspection prior to use. Usually this will include looking at the structural components to assure there are no deformities or bending of metal.

Preparation

The instructor-in-charge must inspect nonstructural, exterior props prior to utilizing them for live fire training, as is done with live fire training structures and acquired structures. If the instructor-in-charge finds concerns during the inspection, the concerns need to be documented. If the concerns will cause a potentially unsafe training condition, then the exterior prop should not be used until it is fixed.

When conducting the inspection, the instructor-in-charge must confirm that the exterior prop is actually designed to meet the learning objectives of the training evolutions. Safety is the key concern with all training and strict safety practices must be in place for all exterior props. When live fire training fires are to be held outside, the area and ground-cover need to be clear of debris, trees, brush, and any surrounding vegetation, because they can become a safety hazard for participants. This will also assist in preventing a training evolution from turning into a wildland emergency or structure fire.

The instructor-in-charge must be familiar with and understand all functions and operations of the exterior prop in order to conduct an effective inspection. This is a key point in ensuring that the exterior prop is operating normally. The training must be done by the specialty prop manufacturer or other authorized personnel. Without the proper training and knowledge of the system mechanics, a system fault may not be uncovered during the inspection. Again, any damage or issues must be documented, and training may have to be postponed until the damage can be repaired.

The area around the exterior prop should also be inspected visually. All possible sources of ignition other than those being used in the burn must be removed. Exterior props can be situated on concrete pads, on an area filled with rock, or on the ground. Exterior props placed directly on the ground can lead to fires starting in surrounding grass. If an exterior prop is located on concrete or on a rock, the flaking of concrete is expected, but large holes can become a hazard. Pieces of rock will be scattered when hit with fire streams, and can result in a safety hazard. If you are using a pan to hold a flammable liquid, the integrity of the walls of the exterior prop need to be visually inspected. When exposed to heat after numerous evolutions, the metal may deform and cause the containment of the flammable liquid to be compromised.

If LPG props are being used, the instructor-in-charge must ensure that the safety features in NFPA 58, *Liquefied Petroleum Gas Code*, and NFPA 59, *Utility LP-Gas Plant Code*, are all present. The instructor-in-charge should verify that the LPG tank has been inspected. Piping for the LPG prop, including the operation of valves, needs to be visually inspected and evaluated. Valves should operate freely and piping should be in good condition. Piping for the LPG prop can be permanent or can use flexible lines to attach to the liquefied petroleum gas source. Regardless of which type is used, the piping should have no leaks. Rust and dents to piping should be carefully examined to assure that the use of the LPG prop is not jeopardized. If damages are found, specific documentation is required. An LPG prop needs to be inspected by the manufacturer. If the AHJ has built the prop, then it must be evaluated and documented annually by a qualified person.

Acquired props are visually inspected prior to any burn. It may be necessary to replace acquired vehicles within a couple of months. The replacement will be based on several factors, such as the number of burns and the structural integrity of the vehicle. If the learning objective is to teach students how to extinguish an engine compartment fire, but the vehicle is burnt to the point that the hood can no longer be opened, then the acquired prop is not simulating a realistic condition and should be replaced. Similarly, if the doors can no longer be opened, then it is difficult to teach students how to extinguish a compartment fire. Acquired vehicles normally have to be replaced frequently, so an annual inspection may never occur. Vehicles used as acquired props may have components that could pressurize and either release flammable liquids (e.g., fuel tank) or explosively decompress with the danger of causing injury (e.g., fluid reservoirs, tanks, shock absorbers, drive shafts, other gas-filled closed containers). These components need to be removed, vented, or drained.

Prior to utilizing a Class A prop, a visual inspection must be done. While conducting visual inspections, look for signs of container failure, reassure that all combustibles are removed from the area unless they are being used for the burn, debris is removed, and there is adequate maneuvering space for all lines and personnel. Also, any containers in the fire areas must not be sealed to prevent sudden decompression. Even though Class A materials are being used, any runoff should be contained.

Prior to any burn, the Class B prop should be ignited and the instructor-in-charge and safety officer should be in agreement that it is in proper working order. If the Class B prop is not properly functioning, the training should not be conducted until it is fixed by a qualified individual.

Operations

Preburn Briefing Session

The instructor-in-charge must hold a preburn briefing session for all participants involved in the training. During this briefing, all facets of each evolution shall be discussed.

This includes the locations of the fires, the identification and location of instructors, the training evolution's learning objectives, the possibility of simulated victims, evacuation procedures,

> **Safety Tips**
>
> Acquired props need to be visually inspected prior to any training evolution. Based on the structural integrity of the vehicle, or prop, it may be necessary to replace it after a couple of months.

and participant assignments. Prior to any live fire training, all participants are required to be familiar with the exterior props being used. The preburn briefing will help facilitate this.

Lighting the Fire

If the ignition of the exterior prop is to be done by an ignition officer, and not by mechanical means, the ignition officer must not be a student. The ignition officer must wear full PPE and SCBA. This person is also designated to control the burning material, which may be as simple as keeping hay from being blown by the wind or as complex as assuring that the proper amount of Class B material is available in the Class B prop.

The exterior prop is only ignited in the presence of, and under the direct supervision of the safety officer. Prior to ignition, communications should be established to assure that all instructors and assigned functions are prepared and ready to proceed with the training, using the Go/No Go sequence. Once all participants are ready to go, the instructor-in-charge and the safety officer will make the decision to ignite the training fire.

Ignition of Class A Props

Before the burn and during the training evolution, the instructor-in-charge must continuously watch the conditions of the fire and environment to determine if the fire growth is at the desired level. If the fire changes and presents a hazard, or is not behaving as expected, he or she should stop the evolution. Once an evolution has been stopped due to a hazard, it can only be resumed after the instructor-in-charge has determined that the hazard has been reduced or eliminated.

Ignition of Class B Props

Many Class B props at training centers are actually ignited and controlled through handheld pendants. This allows the instructors and students to be at a safe distance during ignition. A safer method to ignite a Class B prop is to use a propane tank and a six- to eight-foot wand **Figure 8-8**. The wand has a handle on the end closest the propane tank that controls the amount of fuel and allows ignition to take place from a safe distance. After the Class B prop is lit, the wand can be laid down outside of the operations area, ready to ignite the next fire.

Figure 8-8 A wand and propane tank can ignite a Class B fire.

Safety Tips

It is very important to assure that a safety officer is stationed at all emergency remote shutoff valves and mechanisms **Figure 8-9**. These stations should be located outside of the operations area so that the entire area can be monitored for any issues that require immediate shutdown of the Class B prop.

Figure 8-9 A safety officer is stationed at the remote emergency shut off during the live fire training.

Figure 8-10 A control station for a Class B prop.

When considering the use of Class B props, it is important that the instructor know how to control the prop. There are numerous types of controls that exist. Before discussing these, the placement of the control area needs to be addressed. The instructor controlling the Class B prop needs to be far enough away from the evolution to assure that he or she does not only concentrate on the size of the fire, but must remain close enough to assure that he or she can see any problems that require immediate shutdown of the Class B prop.

In some locations, one central panel can control several Class B props **Figure 8-10**. Many Class B props will color code the controls so that the ignition officer knows exactly which buttons control which prop. This is an excellent tool to have, should an emergency occur. Other Class B props may simply have turn-valves, and the instructor acting as the ignition officer must know which valve operates which prop. If a Class B prop uses both liquid and vapor LPG, there may be three controls at the panel: one bringing fuel to the control panel, one for vapor, and the third for liquid Figure 8-11 .

Since Class B props vary on their controls and piping, so it is important that the ignition officer be familiar with the entire system and its operation. If an emergency occurs, the ignition officer should be able to shut down the prop without having to guess which valve to turn Figure 8-12 .

The instructor-in-charge must monitor the fire growth throughout the training evolution. Unexpected fire growth or any unusual circumstances should result in the immediate stop of the evolution. Only after the issue has been addressed and measures have been taken to reduce or eliminate the hazard, should the evolution continue.

Because flammable liquids can move and splash, flammable liquid props that use pressure to move fuel need to have remote fuel shutoffs. These valves must be continuously monitored during evolutions and should be located so that they are outside of the safety perimeter, which allows the instructor to see the entire field and training evolutions.

Safety During Evolutions

To help maintain safety during training evolutions, first do a walk-through of the attack with no fire present and no water

Safety Tips

The safety officer and all instructors need to be vigilant at keeping the primary concern on safety by ensuring that students are working as a team. Students should not be pushing or pulling each other, which could result in a fall or serious injury. Also, it cannot be stressed enough that all students and instructors need to be constantly aware of their surroundings and fire behavior and be ready to react appropriately, should an emergency occur.

flowing, which is sometimes called a **dry run** (or dry attack). The dry run is to assure that crews work together, move the hose line in a smooth fashion, and attack the exterior prop using the desired method. A second run should then be conducted with charged lines and no fire. Students will now advance the lines working as a team, and can then concentrate on the correct nozzle pattern to attack the fire. Finally, after the students have shown they have the hose line movement and nozzle patterns down, the fire can be added to the equation. Using this method will help give students confidence when it finally comes to attacking the fire.

As always, students must be monitored and supervised during the training evolution and the instructor-in-charge is responsible for all activities during the training evolution.

Instructional Technique

One final note involves instructional technique. It is important to teach students both offensive and defensive techniques when working on exterior props. Not all fires are easily extinguished, and not every fire is a win–win situation. Instructors need to establish learning objectives and training evolutions where the fire is not extinguished and students have to identify when it is time to pull out for their crew's safety, as well as their own.

Figure 8-11 A Class B prop can simulate a cylinder leak and fire.

Figure 8-12 Insure all controls are clearly and conspicuously labeled.

Wrap-Up

Chief Concepts

- Exterior props can be either acquired or manufactured.
- Class A props use ordinary combustibles such as wood products, cardboard, hay, straw, or excelsior as fuel.
- Class B props are either gas fueled, liquid fueled, or a combination of the two.
- There are a variety of Class B props designed to simulate flammable liquid fires. These props can range from a simple burn pan to a professionally manufactured chute.
- With exterior props, runoff containing hazardous materials is a concern. Training sites must take measures to ensure that liquid fuels do not contaminate the ground or ground water.
- The structural integrity of the exterior prop must be inspected once a year.
- As with live fire training structures and acquired structures, the instructor-in-charge must ensure that all exterior props are prepped to meet NFPA 1403.
- It is very important to assure that a safety officer is stationed at any and all remote emergency shutoff valves and mechanisms to monitor the training evolution for any issues that require the immediate shutdown of the Class B prop.

Hot Terms

Acquired prop A piece of equipment that was not designed for burning but is used for live fire training.

Dry run Used during initial training, an attack on a prop with no fire and no water flowing to give students confidence in hose handling techniques. Followed by second evolution flowing water, and the third evolution introducing fire.

Exterior prop A nonstructural, outdoor live fire training prop.

Manufactured prop A type of exterior prop that is built to resemble an actual emergency for the purposes of live fire training.

Wrap-Up

Reference

Wendt, G. (1994) Fire Engineering, *New Jersey Training Burn: Real-Life Lessons.* March 1994, p. 42–48.

Live Fire Training Instructor in Action

You are a candidate in a live fire training instructor class with twenty other candidates. Your agency has access to all types of exterior props, including a car prop and a van prop. These exterior props are unfamiliar to most other instructors in the classroom, as most of them only have access to non-gas-fired permanent live fire training structures. Some of the instructors are curious about the use of exterior props and begin asking your questions about their use.

1. What is the purpose of exterior props?
   A. To take the place of structural firefighting when an acquired structure is not available.
   B. To give the local fire department the opportunity to create their own props.
   C. To simulate an actual fire emergency.
   D. To acquire cars.

2. The class instructor builds a stovetop/oven made out of metal to resemble an actual kitchen appliance for training fire fighters how to attack grease fires in kitchens. According to NFPA 1403, this prop is a(n):
   A. acquired prop.
   B. homemade prop.
   C. prop that should not to be used in training.
   D. manufactured prop.

3. Which of the following do you have to know when using nonstructural props?
   A. When the prop was built or manufactured
   B. When the prop was last inspected
   C. How to operate all safety mechanisms of the prop
   D. An instructor does not need to know specific information about the prop.

4. How do you determine where to place the controls for a Class B prop?
   A. The controls need to be directly outside the hot zone.
   B. The controls should not be in visual sight of the evolution since it should be operated via communications.
   C. The controls should be inside the hot zone so the safety officer can see the evolution.
   D. The controls should be far enough away to oversee the evolution but close enough to assure immediate shutdown if a problem occurs.

5. What do you do with the runoff from a Class B prop?
   A. Runoff needs to be contained to ensure that the ground and surrounding water sources are not contaminated.
   B. Runoff is not a concern as fire training with nonstructural props is exempt from rules and regulations.
   C. The runoff does not matter, as long as there are no water sources close by.
   D. The runoff will be collected by the ground and is not a main concern.

6. How does weather affect the use of flammable liquids?
   A. As long as it is not raining, the weather will not affect training.
   B. Weather is not a factor in determining whether an evolution should be postponed.
   C. The weather should only be a concern if lightning is present.
   D. Wind conditions and severe weather may cause postponement of training.

Live Fire Training Evolutions

CHAPTER 9

NFPA 1403 Standard

9.1.4 A post-training critique session, complete with documentation, shall be conducted to evaluate student performance and to reinforce the training that was covered. [p 183]

Additional NFPA Standards

NFPA 1001 *Standard for Fire Fighter Professional Qualifications*

Knowledge Objectives

After studying this chapter, you will be able to:
- Define a live fire training evolution.
- Describe the tasks a learning objective performs within a live fire training evolution.
- List the two types of student who participate in live fire training evolutions.
- Identify the challenges and concerns in designing a live fire training evolution for a recruit fire fighter.
- Identify the challenges and concerns in designing a live fire training evolution for an experienced fire fighter.
- Identify the items discussed during the preburn briefing.
- Describe the role of the live fire training instructor in ensuring participant safety during a live fire training evolution.
- Describe the goals of the postevolution debriefing.

Skills Objectives

After studying this chapter, you will be able to:
- Develop clear learning objectives for live fire training evolutions.

You Are the Live Fire Training Instructor

You, as the live fire training instructor, have all of the necessary information and have done all of the required research needed to conduct live fire training in your area. You are anxious to get your department together and have a safe and meaningful training evolution using the acquired structure you have prepared. During a discussion with the safety officer about what goals you want to accomplish, you come up with the following questions for your team:

1. What types of training evolutions will you conduct?
2. How will you address the training challenges of the new recruit and the experienced veteran?
3. Once this training evolution is complete, what are your responsibilities?

Introduction

As an instructor of live fire training, you will have the responsibility of creating realistic and challenging evolutions for live fire training. According to NFPA 1403, *Standard on Live Fire Training Evolutions*, live fire training **evolutions** are a set of prescribed actions that result in an effective fireground activity. There are evolutions that may be appropriate for a crew of experienced veteran fire fighters, but do not match the level of training and experience of a new recruit. The planning and execution of live fire training require that the learning objectives be clearly written and communicated. Once the evolutions are completed, it is important to tie together the learning objectives with the experience and obstacles encountered in a postburn critique. The postburn critique is as important to the learning experience as any other component. This chapter will help guide you through this often difficult task of putting all of the pieces together.

Learning Objectives

Every live fire training evolution must have a clearly defined purpose. When conducting live fire training, clearly defined learning objectives for every training evolution must be established and communicated to all participants prior to beginning the drill.

Clear and measurable learning objectives help keep everyone on track and assure that the evolution has a learning purpose. Learning objectives should be written for each specific task that is expected to be performed during the evolution. "Getting experience fighting interior fires" is not a specific learning objective. "Ladder crew will formulate and implement a plan for horizontal ventilation" is a specific learning objective.

When writing learning objectives, keep a picture in mind of what the tasks should look like when they are being performed correctly. For example, if a task were to include advancing a hose line and extinguishing a fire, a sample learning objective may be as follows:

The engine crew of three will advance a 1¾" (44.5 mm) hose line to the interior of the building, and while maintaining thermal balance, effectively extinguish a room and contents of fire.

This learning objective can be broken down into more detailed tasks, including the following:

- Checking the hose stream before entering the structure
- Checking personal protective equipment (PPE) and self-contained breathing apparatus (SCBA) before entry
- "Chocking" doors as entry is made
- Ensuring proper positioning of personnel on the hose line
- Utilizing a defined search pattern
- Adapting correct movement for conditions present
- Positioning attack for best shielding and protection of personnel
- Controlling proper stream application

Each task expected to be completed at the live fire training should be written out in detail and included in the preburn plan. Task completion is measurable, making it possible to identify when the learning objective is met.

Live Fire Tips

When learning objectives are clearly written and specific, it assists the live fire training instructor in setting up the training evolution.

Live Fire Tips

The importance of adherence to local Standard Operating Procedures (SOP) cannot be overemphasized. The most effective live fire training evolutions replicate the SOPs that are used during actual fireground operations.

Live Fire Tips

For many live fire training evolutions, crew assignments of three are ideal. The three fire fighter crew should be made up of two fire fighters and one live fire training. The instructor may communicate the crew's progress and the completion of assigned learning objectives during the live fire training evolution.

Be careful not to over-assign fire fighters and create crews that are unrealistic to standard firefighting operations. Many complaints during the postevolution debriefing occur when crews are overstaffed or too many crews are assigned to work in small areas.

Types of Students in the Live Fire Training Environment

Before conducting any type of class, it is important to know who your audience is going to be. This is especially true when conducting the high-risk training evolutions associated with live fire training. Of course, we would never place a student in a live fire environment who is not adequately prepared to perform basic firefighting skills. However, should a new recruit be faced with the same level of difficulty and skill requirements as those of an experienced veteran? Would a group of experienced fire fighters in a simple search and rescue evolution with a small fire in a steel barrel question the relevance and quality of the evolution. How can an instructor ensure that all training needs are met, without putting anyone in danger or wasting anyone's time?

■ Recruit Students

In live fire training, we are posed with a challenge that does not exist in the other types of trainings. Fire service recruits are the newest and least experienced members of our organization and introducing them into the most hazardous types of training requires extreme vigilance and careful preparations by

---

Live Fire Training Sample Evolutions

Evolution: Obvious Rescue with Room and Contents Fire

OBJECTIVE #1	Initial on-scene crew will experience a degree of urgency and difficulty in performing the rescue of a victim presenting from a window. The crew will perform the rescue.
OBJECTIVE #2	Attack crews will encounter fires in two separate locations. Crews will extinguishment fires and begin hydraulic ventilation.
OBJECTIVE #3	Search crews will conduct secondary search.
OBJECTIVE #4	Ladder crew will perform vertical ventilation from roof using roof ladder and axes.
OBJECTIVE #5	Rapid intervention crew will observe exterior conditions and estimate the fire location and intensity through smoke color, density, and force.
OBJECTIVE #6	Command will practice using the Incident Command System to direct crews on interior firefighting and will maintain personnel accountability.

A higher level of fire fighter experience may be necessary to complete the simultaneous learning objectives presented in this evolution. The initial on-scene crew will find a simulated victim at a window in distress, requiring their immediate attention. Later arriving crews will be assigned to suppress the interior fire, perform hydraulic and roof ventilation, and to the rapid intervention crew. Since both fire suppression and rescue operations need to be conducted at the same time, an experienced Incident Commander will need to coordinate these operations.

A minimum of five crews made up of three fire fighters each and one Incident Commander will be required. The equipment required includes standard PPE, hose line, nozzle, search and rescue tools, ventilation tools, and one rescue dummy.

Evolution: Attached Garage Fire Extension into House

OBJECTIVE #1	The initial attack crew will find heavy fire in an attached garage extending into an occupied house. The initial attack crew will advance to confine and extinguish the fire.
OBJECTIVE #2	Ladder crew will ventilate garage windows and perform vertical ventilation over the garage.
OBJECTIVE #3	Secondary attack crew will stretch a backup line to the interior and check for fire extension.
OBJECTIVE #4	Rapid intervention crew will observe exterior conditions to determine the fire's starting point, the current location, and its future path. This size-up will assist in determining the potential for fire fighters becoming trapped by spreading fire.
OBJECTIVE #5	Command will practice using the Incident Command System to direct crews on interior firefighting and will maintain personnel accountability.

An attached garage fire with extension into the occupied structure will challenge all levels of fire fighters. Coordinated fire attack, ventilation, and a rapid intervention crew on stand-by will simulate a realistic fireground operation. The initial attack crew will attack the fire while the ladder crews are assigned ventilation. The secondary attack crew will check for fire extension.

An interior safety crew will be needed to monitor for deteriorating conditions inside the live fire training structure and they must be protected by a charged hose line and a thermal imaging camera. A suggested minimum of 13 fire fighters should be assigned to this evolution with four crews of three fire fighters and one Incident Commander. The equipment required includes standard PPE, hose lines, nozzles, and ventilation tools.

(Continues)

(Continued)

> **Evolution: Lost, Trapped Fire Fighter with Fire Conditions**
>
> OBJECTIVE #1 Rapid intervention crew will be presented with a simulated distress call from an engine crew. Rapid intervention crew will need to determine number of lost fire fighters, location, and implement a rescue plan.
>
> OBJECTIVE #2 Ladder crew will provide ventilation and coordinate with rescue operations.
>
> OBJECTIVE #3 Initial attack crew will confine and extinguish the fire while rescue operations are taking place.
>
> OBJECTIVE #4 Command will gain experience in directing crew performing fire fighter rescues. Command will order a roll call to determine the number of fire fighters missing. Command will follow SOP for requesting additional resources, etc.
>
> OBJECTIVE #5 Search crew will assume back up rescue responsibilities. Search crew will observe exterior conditions and formulate a plan to back up the rapid intervention crew.
>
> This is a high-level training evolution. A simulated rapid intervention crew deployment scenario will acquaint the participants in the high stress conditions that surround these events. Searching for lost or trapped fire fighters in a training evolution can create high levels of stress and negative reactions to that stress. If recruit fire fighters are assigned to this evolution, experienced fire fighters and live fire training instructors must be assigned with them to maintain safety and to manage any negative reactions to stress.
>
> A suggested minimum of 13 fire fighters should be assigned to this evolution with four crews of three fire fighters and one Incident Commander. The equipment required includes standard PPE, hose lines, nozzles, rapid intervention crew search and rescue tools, and ventilation tools. No live victims are to be used at anytime during this live fire training evolution.
>
> **Evolution: Fire Behavior Drill**
>
> OBJECTIVE #1 Operating from a safe distance within the structure, fire fighters will identify fire growth and spread characteristics while observing a room and contents fire.
>
> OBJECTIVE #2 Operating from a safe distance within the structure, interior crews will identify smoke travel and thermal layering using thermal imaging cameras.
>
> OBJECTIVE #3 Operating from a safe distance within this structure, fire fighters will identify the effects of fire stream application to a room and contents fire.
>
> OBJECTIVE #4 Exterior crews will observe smoke conditions from outside the structure while the fire behavior drill is conducted.
>
> This training evolution is a basic fire behavior observation drill that will allow participants to monitor fire behavior patterns and to observe the effects of fire streams on a growing fire. Exterior crews observing from remote or outside positions will benefit from watching smoke conditions during the evolution. Great care must be taken to not create hazardous interior conditions by introducing excess amounts of steam and limiting fire fighter visibility beyond safe limits. Control of exiting participants and the maintenance of a low profile position during the evolution should be considered.
>
> The required levels of PPE as stated by the AHJ must be adhered to at all times. The number of crews assigned to this evolution will be dictated by the interior conditions and size of the controlled fire. At least one crew of three fire fighters should be positioned with a charged hose line during the evolution.
>
> **Evolution: Thermal Imaging Camera Drill**
>
> OBJECTIVE #1 Search crews will identify heat patterns and imaging through the use of a thermal imaging camera while conducting a search of an acquired structure with fires burning in contained steel drums.
>
> OBJECTIVE #2 Search will demonstrate the use of a thermal imaging camera to identify simulated victims within a structure with limited visibility. Fire within the acquired will be contained to steel barrels.
>
> OBJECTIVE #3 After the failure of a thermal imaging camera which was being used to conduct a search pattern, search crews will demonstrate their ability to exit a structure with limited visibility.
>
> OBJECTIVE #4 Initial attack crews will demonstrate their ability to direct a fire stream using a thermal imaging camera in a limited visibility environment.
>
> This training evolution is designed to train fire fighters on the use of a vital piece of life safety equipment, the thermal imaging camera. Using steel drums or barrels to produce smoke and minor amounts of heat will change the environment enough to require the use of the thermal imaging camera to complete the learning objectives. Search crews will be tasked with performing a simulated search and identifying heat patterns with the thermal imaging camera and exiting the acquired structure. A malfunction simulation should also be performed requiring the search crew to shut down the thermal imaging camera and re-orientate to their location and condition without the use of the thermal imaging camera. The initial attack crews will use the thermal imaging camera to direct a fire stream to a desired location. The equipment required includes standard PPE, hose line, nozzle, search and rescue tools, and a thermal imaging camera.

all instructors involved in the live fire training. Recruit fire fighters lack the experience of working with a crew at structural fires. Remember that the natural instincts that many fire fighters have acquired over time as veteran fire fighters are not developed in the new recruit.

As recruits, these students have not mastered all of the skills necessary to be proficient fire fighters. They are still developing and comprehending their fireground skills. When introducing these newest of fire fighters to live fire training evolutions as instructors, be prepared for their reactions to be unconventional compared to experienced fire fighters. A smoke condition that would cause a veteran fire fighter to step back, slow down, and assess the situation may not get a second glance from a new fire fighter because they do not fully understand smoke and fire behavior yet. On the contrary, a situation where a window breaks suddenly for ventilation, which may be a sign of relief to experienced fire fighters, may create a state of panic in a recruit. Instructors must constantly keep their focus on the behavior of recruits during evolutions and evaluate their reactions and performance Figure 9-1 .

New recruits are still learning how to follow orders and maintain awareness of their surroundings. They may not recognize a dangerous situation or condition. For this reason, an instructor must be close at hand to keep the recruit safe, and to instruct them on what to be looking for when fighting a fire.

Live fire training instructors should rehearse potential scenarios that may occur when introducing new recruits to live fire training. Some common untoward reactions seen in new recruits during live fire training include the following:

- Becoming lost or disoriented
- Not opening the hose line when instructed to do so
- Standing up in high heat and low visibility conditions
- Self-contained breathing apparatus (SCBA) malfunctions during live fire conditions
- Apprehension, hyperventilation, and panic
- Moving in too quickly, getting too close to the fire
- Not handling tools properly

Each of these reactions poses a different challenge for the instructor. The safety of the participants is dependent on how quickly the situation is identified and rectified. Instructors should practice how they will correct the above problems prior to conducting live fire training, especially with new recruits.

Figure 9-1 An exterior prop with clear views into the interior will allow instructors to closely monitor the actions of recruits during training evolutions.

Safety Tips

When operating in low visibility, the instructor must have an awareness of where each participant is, if they are moving with confidence or unsure of themselves, and must listen to their breathing rates and, of course, what they say. Keeping discussion going and asking questions can help judge the student's status and keep them better focused.

Being able to rapidly access a student and prevent the student from causing injury takes practice.

Live Fire Training Evolutions

Live fire training is the pinnacle of a recruit's training experience in an academy. NFPA 1403 requires that prior to exposing a recruit to live fire training, they must meet the job performance requirements in NFPA 1001, *Standard for Fire Fighter Professional Qualifications*. Live fire training *should not* be the testing ground for these skills. Instead, it *should* bring together all of the skills they have successfully demonstrated up to this point in their training.

When conducting live fire training with new recruits, the difficulty level of the planned evolutions should match the students' abilities. Some common training evolutions conducted with new recruits include the following:

- Fire behavior and fire growth patterns
- Simple search patterns
- Hose line advancement and stream application
- Thermal imaging camera training
- Room and contents fire on ground-level floors
- Cold smoke drills

Another example of an excellent training evolution for recruit students is the "One Room Fire with Trapped Victims" evolution. The learning objectives for this evolution are:

OBJECTIVE #1   Initial attack crews will locate and extinguish a fire in dense smoke conditions.

OBJECTIVE #2   Search crews will execute a search and rescue of victims under dense smoke and high heat conditions. Search crews should break windows as the search progresses to understand the value of ventilation.

OBJECTIVE #3   Ladder crew will formulate and implement a plan for horizontal ventilation.

OBJECTIVE #4   Rapid intervention crew will observe exterior heat and smoke conditions and note changes in color and density as the fire progresses and is eventually extinguished.

OBJECTIVE #5   Command will practice using the Incident Command System to direct crews on interior firefighting and will maintain personnel accountability.

This training evolution is considered to be a bread and butter evolution. In this evolution, the recruit students will be assigned to either the initial attack crew or search crew. The initial attack crew will encounter a one room fire in either a live fire training structure or an acquired structure. The search crew will perform ventilation tasks and search and rescue victims in the same structure. A controlled fire will produce sufficient heat and limit visibility to challenge the recruit fire fighters. A theatrical smoke machine may assist in limiting visibility to the desired level while maintaining safety. At least one rescue dummy should placed to simulate a victim.

Using these learning objectives, a minimum of four crews of three fire fighters (two fire fighters and one live fire training instructor), and one Incident Commander are required. The equipment required includes standard personal protective equipment (PPE), hose line, nozzle, search and rescue tools, and ventilation tools. In addition, consider the use of interior safety officers protected by a charged hose line and equipped with a

> **Live Fire Tips**
>
> When working with fire fighters from another fire department, many live fire instructors require written verification that the fire fighters' prerequisite knowledge and skill levels match or exceed the objective content for the live fire training evolutions. In addition, the instructor-in-charge may conduct a quick series of skill level assessments by observing the fire fighters don PPE, and perform emergency evacuation procedures, search and rescue procedures, and ventilation procedures.

> **Safety Tips**
>
> Having a thermal imaging camera available to instructors may be a necessary safety precaution.

> **Safety Tips**
>
> Instructors must constantly keep their focus on the behavior of recruits during evolutions and evaluate their reactions and performance.

thermal imaging camera to observe the actions of the crews and to monitor conditions.

Experienced Students

As much as a recruit can be a challenge for instructors during live fire training, the veteran or experienced fire fighter poses a different set of concerns. The veteran fire fighter comes to training with significant practice and experience in the trade, which can be a double-edged sword for instructors. "Breaking" a veteran student of bad habits can be just as difficult as teaching new recruits, if not more so. Stay focused and follow every part of NFPA 1403, even though students may have performed live fire training many times throughout their careers.

Some particular concerns that experienced student fire fighters pose include a possible reluctance to follow orders, complacency in wearing PPE, and a belief that live fire trainings are controlled and therefore pose no danger. This is a recipe for a tragedy on the training ground.

Another challenge is this: twenty experienced fire fighters may have received their initial training from twenty different sources. In addition, some veteran fire fighters may have been taught practices and protocols that differ from the Standard Operating Procedures (SOPs) of the instructor's fire department.

Unauthorized "shortcuts" may be used on the fireground by some experienced fire fighters, which could pose a problem for the instructor trying to teach recruit fire fighters how to perform tasks according to local SOPs. In many cases, these shortcuts were developed to aid in the completion of tasks on the fireground, and were modified or in conflict with teachings from the academy. These practices may violate safety protocols and should not be allowed when conducting live fire training.

Experienced fire fighter students will want to be challenged during their training. This desire to "show their stuff" may lead to variances from the standard. As an instructor, do not be pressured to modify the training evolution in order to provide a more "exciting" evolution for the experienced personnel. When the instructor-in-charge has set the evolution and determined the scope of the training, it shall not be altered.

Live Fire Training Evolutions

One advantage of having a class of experienced fire fighters is being able to include training evolutions which require more skill and work to complete. Evolutions which require multiple companies to work simultaneously are as realistic as it gets when conducting live fire training. Some examples of these evolutions include a room fire with simulated victims, content fires which require vertical ventilation, attic or cockloft fires, fires in an attached garage requiring an attack from the interior of the house, or fires that require forcible entry to be performed prior to entering the building.

An example of an excellent training evolution for experienced students is the "Attic Fire" evolution. The learning objectives for this training evolution are:

OBJECTIVE #1 Initial attack crew will demonstrate ability to gain control of a fast moving attic fire.

OBJECTIVE #2 Ladder crew will experience the urgency of vertical ventilation to assist in interior extinguishment. Ladder crew will conduct vertical ventilation using axes.

OBJECTIVE #3 Search crews will accomplish a rescue while attack crew is attacking the fire. Coordination of crews working in the same room will need to be obtained.

OBJECTIVE #4 Rapid intervention crew will observe rapid fire build up in a confined area. Rapid intervention crew will also observe smoke characteristics in a fire involving structural members.

OBJECTIVE #5 Command will practice using the Incident Command System to direct crews on interior firefighting and will maintain personnel accountability.

An attic fire evolution presents numerous challenges for experienced fire fighters and requires a carefully developed plan to safely control this evolution. The initial attack crew will be assigned to attack the attic fire to gain control and stop the progress of the fire spread while the ladder crew will be assigned vertical ventilation with hand tools. The search crew will perform a primary search and a rapid intervention crew serves as support functions to the initial attack crews. A minimum of 13 fire fighters should be assigned to this evolution—four crews of three fire fighters and one Incident Commander. The equipment required includes standard PPE, hose line, nozzle, search and rescue tools, and ventilation tools.

Operations

Prior to conducting the live fire evolution, all participants and instructors must be briefed and inspected by the instructor-in-charge. A simple rule to remember is that there can be no surprises when conducting actual training. This requires that all participants walk through the building and are made aware of all facets of the structure. During this walk-through paths of egress, and windows, doors, and stairways that can be used to evacuate or retreat if necessary should be pointed out. The preburn briefing and inspection is our last opportunity to ensure the safety of our students and the success of the drill.

Before beginning each training evolution, all participants shall be briefed on their roles during the evolution during the preburn briefing. NFPA 1403 does not require the disclosure of such or where the fire will be located or where to find victims during the drill, provided that the possibility of victims is known. When the evolution begins, there should not be any confusion about what each fire fighter's task will be and what the learning objectives of the training evolution are.

The safety rules pertinent to the training shall also be reviewed prior to starting. These may include trip hazards, projections from the wall such as standpipe connections, emergency lights, or other drillground or prop hazards. This is also a good time to explain the building evacuation procedure and demonstrate the evacuation signal that will be used. Emphasize that the evacuation signal will only be used in real emergencies and shall not be used as a prop for any training evolution.

All hose lines used for the evolution shall be changed and checked prior to beginning. Assure that you have adequate flow and water supply. This exercise of testing all hose lines is also a good way to confirm that the pump operator is competent, and there are no mechanical difficulties. It is recommended that all of the hose lines to be used are flowed simultaneously. This includes hose lines to be used in the evolution, the ignition hose line, and any exposure or safety hose lines. This will ensure that sufficient flow capability is available. At this time, assure that you have adequate personnel to safely handle all of the hose lines which will be used.

Determine if the tools necessary to complete the training evolution are in place and ready, prior to starting the scenario. This includes ensuring that portable radios are working and on the proper channel. Check components such as fuel levels in saws and other tools. Also, ensure that the rapid intervention crew has all of the tools necessary to complete a rescue, if needed.

Each participant, including instructors, shall be checked before entering a live fire evolution, to assure that PPE is being worn correctly and that there are no defective pieces of equipment. There should not be any exposed skin prior to personnel entering into the immediately dangerous to life and health (IDLH) environment. Also, prior to beginning the evolution, the safety officer should inspect everyone's SCBA to confirm that they are beginning the training evolution with an adequate air supply. Inspection of PPE and SCBA must be done prior to every training evolution. It is not sufficient, or acceptable, to check all of these items at the beginning of the training day and then not evaluate them anymore.

Postevolution Debriefing

It is highly recommended that instructors get in the habit of collecting information from the participants immediately following each evolution **Figure 9-2**. This is the postevolution debriefing. The training is still fresh in participants' minds and instructors can get extremely helpful feedback for their next session of live fire training evolutions. In some cases, the students may gain confidence from the postevolution debriefing and it should reenforce what they have learned and clarify proper procedures. It may also forewarn instructors of impending problems with participants being fearful, overly aggressive, or cocky.

Use a common format to conduct the postevolution debriefing so that participants are accustomed to the dialogue. Some components of the postevolution debriefing can include a review of the learning objectives, a description of events by fire officers, and an assessment of any obstacles that may have been encountered. The postevolution debriefing should be under shelter whenever possible and out of the way of foot traffic and distractions.

During this time, focus on keeping the criticism constructive and remember that this is a learning environment. Reinforce things that went well and keep the discussion focused on the learning objectives of the training evolution. Limit the discussion when necessary to keep the training on track and within allotted time frames. It is often tempting to debate tactics and policy following an evolution. If these subjects do not add to the constructive criticism of the evolution, limit the discussion.

Finally, before ending the postevolution debriefing, confirm that there were no injuries sustained while conducting the training evolution.

Figure 9-2 The postevolution debriefing should be conducted after removing PPE and getting fluids, but as soon as possible after the training evolution.

Incident Report

Greenwood, Delaware - 2000

A volunteer fire department was conducting a live fire training exercise in a 100+-year-old house **(Figure A)**. Upon completing interior operations, personnel were preparing for the "final burn down" of the 2½ story house.

Three officers were in the attic. One officer was in full protective clothing and SCBA. He was using a sprayer can of diesel fuel to spray fuel. Two other officers, without SCBA, lit multiple fires. The fire spread rapidly and as conditions worsened, the two ignition officers exited to the top of the staircase and told the remaining officer to follow. He agreed to follow, but heat and fire conditions in the attic became untenable very rapidly. The others escaped down the stairs, but fire blocked their colleague's escape.

Multiple rescue attempts were unsuccessful, and the roof quickly collapsed into the attic.

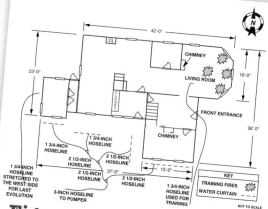

Figure A Floor plan of Greenwood, Delaware.

Post-Incident Analysis	Greenwood, Delaware

NFPA 1403 Noncompliance

Interior personnel without hose line (4.4.6)

Flashover and fire spread unexpected (4.3.9)

Flammable/combustible liquids used in large quantities (4.3.6)

Multiple fire sets (4.4.15)

Interior stairs only exit other than windows without ladders in place (4.2.12.1)

Participants not wearing PPE or SCBA (4.4.18)

Wrap-Up

Chief Concepts

- Every live fire training evolution must have a clearly defined purpose.
- Learning objectives clearly state the tasks that can be measured during the live fire training evolution.
- Instructors should rehearse potential scenarios that may occur when introducing recruit fire fighters to live fire training.
- Some common untoward reactions seen in recruit fire fighters during live fire training include:
  - Becoming lost or disoriented
  - Not opening the hose line when instructed to do so
  - Standing up in high heat and low visibility conditions
  - SCBA malfunctions during live fire conditions
  - Apprehension, hyperventilating, and panic
  - Moving in too quickly, getting too close to the fire
  - Not handling tools properly and endangering others
- Some common training evolutions conducted with recruit fire fighters include the following:
  - Fire behavior and fire growth patterns
  - Simple search patterns
  - Hose line advancement and stream application
  - Thermal imaging camera training
  - Room and contents fire on ground-level floors
  - Cold smoke drills
- Experienced fire fighters will want to be challenged during live fire training. However, do not be pressured into modifying the training evolution in order to provide more "excitement."
- Every participant in live fire training must attend the preburn briefing to review the learning objectives of the training evolution.
- The postevolution debriefing provides vital feedback on how to improve live fire training evolutions.

Hot Terms

<u>Evolution</u> According to NFPA 1403, a set of prescribed actions that result in an effective fireground activity.

Wrap-Up

Reference

Wahenitz, F & Romano, N (2001) Fire Fighter Fatality Investigation Report F2000-27, National Institute for Occupational Safety and Health.

Live Fire Training Instructor in Action

During interior operations at a live fire training evolution, a recruit fire fighter is almost seriously injured when he panics and freezes. As a veteran liver fire training instructor, you do everything possible to calm the fire fighter down and talk him out of harm's way. Thankfully, nobody is hurt in the incident and the training evolution is successful.

After the training evolution, you and a few fellow instructors sit down to review the evolution. As you begin reviewing the evolution, the image of the frozen fire fighter keeps replaying in your mind. What if something terrible had happened to him? How can you prevent this from happening again?

1. Why is it important for an instructor to know the differences between recruit and veteran students when conducting live fire training?
   A. There are higher expectations for veteran service personnel
   B. The level of difficulty will be different.
   C. Level of compliance to NFPA 1403 is different.
   D. Both A and B are correct

2. Under what conditions can veteran fire fighters be more problematic that recruit fire fighters during live fire training?
   A. Veterans tend to know "tricks of the trade" that may not be officially condoned.
   B. Veterans may be complacent in wearing PPE.
   C. Veterans may be more apt to challenge an exercise they feel is not sufficiently challenging.
   D. All of the above

3. Which of the options below is *not* an appropriate method of determining student confidence during an interior evolution?
   A. Peppering them with questions that have nothing to do with the topic to distract them
   B. Listening to their breathing rates
   C. Asking questions about what they see, hear, and feel
   D. Observing their movements and degree of confidence shown

4. Which of the following are TRUE about the preburn briefing requirements of NFPA 1403?
   A. The instructor must disclose the locations of the fire sets.
   B. The instructor does not have to disclose the possibility of victims.
   C. All participants have to do a walk-through of the building.
   D. A preburn briefing plan is only used with acquired structures.

5. Which of the below is true with regards to recruit fire fighter live fire training?
   A. It should be considered a test for recruit skills.
   B. It should be a testing tool to evaluate if a recruit "has what it takes."
   C. Fire operations must be severe enough to determine if the recruit will know what to do at a "real" fire.
   D. None of the above are correct.

An Extract From: NFPA® 1403, *Standard on Live Fire Training Evolutions*, 2012 Edition

Chapter 4: General

4.1 Application. All live fire training evolutions shall comply with this chapter and the appropriate chapter for the type of training being performed.

4.1.1 Strict safety practices shall be applied to all structures selected for live fire training evolutions.

4.2 Permits.

4.2.1 All required permits to conduct live fire training evolutions shall be obtained.

4.2.2 The permits specified in this chapter shall be provided to outside, contract, or other separate training agencies by the authority having jurisdiction (AHJ) upon the request of those agencies.

4.2.3 The runoff from live fire shall comply with the requirements of the AHJ.

4.3 Student Prerequisites.

4.3.1* Prior to being permitted to participate in live fire training evolutions, the student shall have received training to meet the minimum job performance requirements for Fire Fighter I in NFPA 1001, *Standard for Fire Fighter Professional Qualifications*, related to the following subjects:
 (1) Safety
 (2) Fire behavior
 (3) Portable extinguishers
 (4) Personal protective equipment
 (5) Ladders
 (6) Fire hose, appliances, and streams
 (7) Overhaul
 (8) Water supply
 (9) Ventilation
 (10) Forcible entry
 (11) Building construction

4.3.2* Students participating in a live fire training evolution who have received the required minimum training from other than the AHJ shall not be permitted to participate in any live fire training evolution without first presenting prior written evidence of having successfully completed the prescribed minimum training to the levels specified in 4.3.1.

4.4 Safety Officer.

4.4.1 A safety officer shall be appointed for all live fire training evolutions.

4.4.2 All live fire training instructors and safety officers shall be trained on the application of the requirements contained in this standard.

4.4.3 The safety officer shall have the authority, regardless of rank, to intervene and control any aspect of the operations when, in his or her judgment, a potential or actual danger, potential for accident, or unsafe condition exists.

4.4.4 The responsibilities of the safety officer shall include, but not be limited to, the following:
 (1) Prevention of unsafe acts
 (2) Elimination of unsafe conditions

4.4.5 The safety officer shall provide for the safety of all persons on the scene, including students, instructors, visitors, and spectators.

4.4.6 The safety officer shall not be assigned other duties that interfere with safety responsibilities.

4.4.7 The safety officer shall be knowledgeable in the operation and location of safety features available for the live fire training structure or prop, such as emergency shutoff switches, gas shutoff valves, and evacuation alarms.

4.4.8* Additional safety personnel, as deemed necessary by the safety officer, shall be located to react to any unsafe or threatening situation or condition.

4.5* Extreme Weather. The training session shall be curtailed, postponed, or canceled, as necessary, to reduce the risk of injury or illness caused by extreme weather conditions.

4.6 Instructor in Charge and Instructors.

4.6.1 The instructor-in-charge shall have received training to meet the minimum job performance requirements for Fire Instructor I in NFPA 1041, *Standard for Fire Service Instructor Professional Qualifications*.

4.6.2 The instructor-in-charge shall be responsible for full compliance with this standard.

4.6.3 It shall be the responsibility of the instructor-in-charge to coordinate overall fireground activities to ensure correct levels of safety.

4.6.4 The instructor-in-charge shall assign the following personnel:
 (1) One instructor to each functional crew, each of which shall not exceed five students
 (2) One instructor to each backup line
 (3) One additional instructor for each additional functional assignment

4.6.5 The instructor-in-charge shall provide for rest and rehabilitation of participants operating at the scene, including any necessary medical evaluation and treatment, food and fluid replenishment, and relief from climatic conditions. (See Annex D.)

4.6.5.1* Instructors shall be rotated through duty assignments.

4.6.6 All instructors shall be qualified by the AHJ to deliver live fire training.

4.6.7 Additional instructors shall be designated when factors such as extreme temperatures or large groups are present, and classes of long duration are planned.

4.6.8 Prior to the ignition of any fire, instructors shall ensure that all protective clothing and equipment specified in this chapter are being worn according to manufacturer's instructions.

4.6.9 Instructors shall take a personal accountability report (PAR) when entering and exiting the structure or prop during an actual attack evolution conducted in accordance with this standard.

4.6.10 Instructors shall monitor and supervise all assigned students during the live fire training evolution.

4.6.11 Awareness of weather conditions, wind velocity, and wind direction shall be maintained, including a final check for possible changes in weather conditions immediately before actual ignition.

4.6.12 **Training Instructors on How to Use Specialty Props**.

4.6.12.1 The instructors and the safety officer responsible for conducting live fire training evolutions with a gas-fueled training system or with other specialty props (such as flashover simulator) shall be trained in the complete operation of the system and the props.

4.6.12.2 The training of instructors and the safety officer shall be performed by an individual authorized by the gasfueled training system and specialty prop manufacturer or by others qualified to perform this type of training.

4.7 **Fire Control Team**

4.7.1 A fire control team shall consist of a minimum of two personnel.

4.7.1.1 One person who is not a student or safety officer shall be designated as the "ignition officer" to ignite, maintain, and control the materials being burned.

4.7.1.1 The ignition officer shall be a member of the fire control team.

4.7.1.2* One member of the fire control team shall be in the area to observe the ignition officer ignite and maintain the fire, and to recognize, report, and respond to any adverse conditions.

4.7.2 The decision to ignite the training fire shall be made by the instructor-in-charge in coordination with the safety officer.

4.7.3 The fire shall be ignited by the ignition officer.

4.7.4 The fire control team shall wear full personal protective clothing, including SCBA, when performing this control function.

4.7.5 A charged hose line shall be available when the fire control team is igniting or tending to any fire.

4.7.6 Fires shall not be ignited without an instructor visually confirming that the flame area is clear of personnel being trained.

4.8 **Personal Protective Clothing**.

4.8.1 All students, instructors, safety personnel, and other personnel shall wear all protective clothing and equipment specified in this chapter according to manufacturer's instructions whenever they are involved in any evolution or fire suppression operation during the live fire training evolution.

4.8.2* All participants shall be inspected by the safety officer prior to entry into a live fire training evolution to ensure that the protective clothing and SCBA are being worn correctly and are in serviceable condition.

4.8.3 Protective coats, trousers, hoods, footwear, helmets, and gloves shall have been manufactured to meet the requirements of NFPA 1971, *Standard on Protective Ensembles for Structural Fire Fighting and Proximity Fire Fighting*.

4.8.4 SCBA shall have been manufactured to meet the requirements of NFPA 1981, *Standard on Open-Circuit Self-Contained Breathing Apparatus (SCBA) for Emergency Services*.

4.8.5* Where station or work uniforms are worn by any participant, the station or work uniform shall have been manufactured to meet the requirements of NFPA 1975, *Standard on Station/Work Uniforms for Emergency Services*.

4.8.6 Personal alarm devices shall have been manufactured to meet the requirements of NFPA 1982, *Standard on Personal Alert Safety Systems (PASS)*.

4.8.7* All students, instructors, safety personnel, and other personnel participating in any evolution or operation of fire suppression during the live fire training evolution shall breathe from an SCBA air supply whenever they operate under one or more of the following conditions:

(1) In an atmosphere that is oxygen deficient or contaminated by products of combustion, or both
(2) In an atmosphere that is suspected of being oxygen deficient or contaminated by products of combustion, or both
(3) In any atmosphere that can become oxygen deficient, contaminated, or both
(4) Below ground level

4.9 Communication.

4.9.1 A method of fireground communications shall be established to enable coordination among the incident commander, the interior and exterior sectors, the safety officer, and external requests for assistance.

4.9.2* A building evacuation plan shall be established, including an evacuation signal to be demonstrated to all participants in an interior live fire training evolution.

4.10 Emergency Medical Services (EMS).

4.10.1 Basic life support (BLS) emergency medical services shall be available on site to handle injuries.

4.10.1.1 For acquired structures, BLS emergency medical services with transport capabilities shall be available on site to handle injuries.

4.10.2 A parking area for an ambulance or an emergency medical services vehicle shall be designated and located where it will facilitate a prompt response in the event of personal injury to participants in the evolution.

4.10.3 Written reports shall be completed and submitted on all injuries and on all medical aid rendered.

4.11* Water Supply.

4.11.1 The instructor-in-charge and the safety officer shall determine the rate and duration of waterflow necessary for each individual live fire training evolution, including the water necessary for control and extinguishment of the training fire, the water supply necessary for backup line(s) to protect personnel, and any water needed to protect exposed property.

4.11.2 Each hose line and backup line(s) shall be capable of delivering a minimum of 95 gpm (360 L/min).

4.11.3 Backup line(s) shall be provided to ensure protection for personnel on training attack lines.

4.11.4 The minimum water supply and delivery for the live fire training evolutions shall meet the criteria identified in NFPA 1142, *Standard on Water Supplies for Suburban and Rural Fire Fighting*.

4.11.5 A minimum reserve of additional water in the amount of 50 percent of the fire flow demand, determined in accordance with 4.11.1, shall be available to handle exposure protection or unforeseen situations.

4.11.5.1 The requirements of 4.11.5 do not apply to permanently sited gas-fueled training systems.

4.11.6* Except under the conditions of 4.11.6.1, separate water sources shall be utilized for the supply of attack lines and backup lines in order to preclude the loss of both water supply sources at the same time.

4.11.6.1* A single water source shall be sufficient at a training center facility where the water system has been engineered to provide adequate volume for the evolutions conducted and a backup power source or backup pumps, or both, are in place to ensure an uninterrupted supply in the event of a power failure or malfunction.

4.11.7 There shall be room provided around all props so that there is space for all attack line(s) as well as backup line(s) to operate freely.

4.12 Fuel Materials.

4.12.1* The fuels that are utilized in live fire training evolutions shall only be wood products.

4.12.1.1 Fuel-fired buildings and props are permitted to use the appropriate fuels for the design of the building or prop.

4.12.2 Pressure-treated wood, rubber, plastic, polyurethane foam, upholstered furniture, and chemically treated or pesticide-treated straw or hay shall not be used.

4.12.3 Flammable or combustible liquids, as defined in NFPA 30, *Flammable and Combustible Liquids Code*, shall not be used in live fire training evolutions.

4.12.3.1 Combustible liquid with a flash point above 100°F (38°C) shall be permitted to be used in a live fire training structure or prop that has been specifically engineered to accommodate a defined quantity of the fuel.

4.12.4 Unidentified materials, such as debris found in or around the structure or prop that could burn in unanticipated ways, react violently, or create environmental or health hazards, shall not be used.

4.12.5 Propane lighters, butane lighters, fusees (safety flares), kitchen-type matches, and similar devices are permitted to be used to ignite training fires if the device is removed immediately after ignition of the training fire.

4.12.6* Fuel materials shall be used only in the amounts necessary to create the desired fire size.

4.12.7 The fuel load shall be limited to avoid conditions that could cause an uncontrolled flashover or backdraft.

4.12.8* The instructor-in-charge and the safety officer shall assess the selected fire room environment for factors that can affect the growth, development, and spread of fire.

4.12.9* The instructor-in-charge and the safety officer shall document fuel loading, including all of the following:
(1) Fuel material
(2) Wall and floor coverings and ceiling materials
(3) Type of construction of the structure, including type of roof and combustible void spaces
(4) Dimensions of the room

4.12.10* The training exercise shall be stopped immediately when the instructor-in-charge or the safety officer determines through ongoing assessment that the combustible nature of the environment represents a potential hazard.

4.12.10.1 An exercise stopped as a result of an assessed hazard according to 4.12.10 shall continue only when actions have been taken to reduce the hazard.

4.12.11* The use of flammable gas, such as propane and natural gas, shall be permitted only in live fire training structures specifically designed for their use.

4.12.11.1 Liquefied versions of the gases specified in 4.12.11 shall not be permitted inside the live fire training structure.

4.12.11.2* All props that use pressure to move fuel to the fire shall be equipped with remote fuel shutoffs outside of the safety perimeter but within sight of the prop and the entire field of attack for the prop.

4.12.11.3 During the entire time the prop is in use, the remote shutoff shall be continuously attended by safety personnel who are trained in its operation and who have direct communications with the safety officer and instructors.

4.12.11.4 Liquefied petroleum gas props shall be equipped with all safety features as described in NFPA 58, *Liquefied Petroleum Gas Code*, and NFPA 59, *Utility LP-Gas Plant Code*.

4.12.11.5 Where the evolution involves the failure of a safety feature, the failed part shall be located downstream from the correctly functioning safety feature.

4.12.11.6 Where flammable or combustible liquids are used, measures shall be taken to prevent runoff from contaminating the surrounding area.

4.12.11.6.1 There shall be oil separators for cleaning the runoff water.

4.12.11.7* Vehicles used as props for live fire training shall have all fluid reservoirs, tanks, shock absorbers, drive shafts, and other gas-filled closed containers removed, vented, or drained prior to any ignition.

4.12.11.8 For flammable metal fires, there shall be a sufficient quantity of the proper extinguishing agent available so that all attack crews have the required supply as well as a 150 percent reserve for use by the backup crews.

4.12.11.9 All possible sources of ignition, other than those that are under the direct supervision of the ignition officer, shall be removed from the operations area.

4.13 Parking/Staging.

4.13.1 Areas for the staging, operating, and parking of fire apparatus that are used in the live fire training evolution shall be designated.

4.13.2 An area for parking fire apparatus and vehicles that are not a part of the evolution shall be designated so as not to interfere with fireground operations.

4.13.3 If any of the apparatus described in 4.13.2 is in service to respond to an emergency, it shall be located in an area that will facilitate a prompt response.

4.13.4 Where required or necessary, parking areas for police vehicles or for the press shall be designated.

4.13.5 Ingress and egress routes shall be designated, identified, and monitored during the training evolutions to ensure their availability in the event of an emergency.

4.14 Visitors and Spectators.

4.14.1 All spectators shall be restricted to an area outside the operations area perimeter established by the safety officer.

4.14.2 Control measures shall be posted to indicate the perimeter of the operations area.

4.14.3 Visitors who are allowed within the operations area perimeter shall be escorted at all times.

4.14.4 Visitors who are allowed within the operations area perimeter shall be equipped with and shall wear appropriate protective clothing.

4.14.5 Control measures shall be established to keep pedestrian traffic in the vicinity of the training site clear of the operations area of the live burn.

4.15 Preburn Plan/Briefing.

4.15.1 A preburn plan shall be prepared and shall be utilized during the preburn briefing sessions.

4.15.1.1 All features of the training areas shall be indicated on the preburn plan.

4.15.2 Prior to conducting actual live fire training evolutions, a preburn briefing session shall be conducted by the instructor-in-charge with the safety officer for all participants.

4.15.3 All facets of each evolution to be conducted shall be discussed.

4.15.4 Assignments shall be made for all crews participating in the training session.

4.15.5 The location of the manikin shall not be required to be disclosed, provided that the possibility of victims is discussed in the preburn briefing.

4.15.6 Prior to conducting any live fire training, all participants shall have a knowledge of and familiarity with the prop or props being used for the evolution.

4.15.7 Prior to conducting any live fire training, all participants shall be required to conduct a walk-through of the acquired structure, burn building, or prop in order to have a knowledge of and familiarity with the layout of the acquired structure, building, or prop and to facilitate any necessary evacuation.

4.15.8 Property adjacent to the training site that could be affected by the smoke from the live fire training evolution, such as railroads, airports or heliports, and nursing homes, hospitals, or other similar facilities, shall be identified.

4.15.8.1 The persons in charge of the properties described in 4.15.8 shall be informed of the date and time of the evolution.

4.15.9 Streets or highways in the vicinity of the training site shall be surveyed for potential effects from live fire training evolutions.

4.15.9.1* Safeguards shall be taken to eliminate possible hazards to motorists.

Chapter 5: Acquired Structures

5.1 Structures and Facilities.

5.1.1* Any acquired structure that is considered for a structural fire training exercise shall be prepared for the live fire training evolution.

5.1.1.1 Buildings that cannot be made safe as required by this chapter shall not be utilized for interior live fire training evolutions.

5.1.2 Adjacent buildings or property that might become involved shall be protected or removed.

5.1.3* Preparation shall include application for and receipt of required permits and permissions.

5.1.4* Ownership of the acquired structure shall be determined prior to its acceptance by the AHJ.

5.1.5 Evidence of clear title shall be required for all structures acquired for live fire training evolutions.

5.1.6* Written permission shall be secured from the owner of the structure in order for the fire department to conduct live fire training evolutions within the acquired structure.

5.1.7 A clear description of the anticipated condition of the acquired structure at the completion of the evolution(s) and the method of returning the property to the owner shall be put in writing and shall be acknowledged by the owner of the structure.

5.1.8* Proof of insurance cancellation or a signed statement of nonexistence of insurance shall be provided by the owner of the structure prior to acceptance for use of the acquired structure by the AHJ.

5.1.9 The permits specified in this chapter shall be provided to outside, contract, or other separate training agencies by the AHJ upon the request of those agencies.

5.1.10 A search of the acquired structure shall be conducted to ensure that no unauthorized persons, animals, or objects are in the acquired structure immediately prior to ignition.

5.1.11 No person(s) shall play the role of a victim inside the acquired structure.

5.1.12 Only one fire at a time shall be permitted within an acquired structure.

5.2 Hazards.

5.2.1 In preparation for live fire training, an inspection of the structure shall be made to determine that the floors, walls, stairs, and other structural components are capable of withstanding the weight of contents, participants, and accumulated water.

5.2.2* All hazardous storage conditions shall be removed from the structure or neutralized in such a manner as to not present a safety problem during use of the structure for live fire training evolutions.

5.2.3 Closed containers and highly combustible materials shall be removed from the structure.

5.2.3.1 Oil tanks and similar closed vessels that cannot be removed shall be vented to prevent an explosion or overpressure rupture.

5.2.3.2 Any hazardous or combustible atmosphere within the tank or vessel shall be rendered inert.

5.2.4 All hazardous structural conditions shall be removed or repaired so as to not present a safety problem during use of the structure for live fire training evolutions.

5.2.4.1 Floor openings shall be covered to be made structurally sound.

5.2.4.2 Missing stair treads and rails shall be repaired or replaced.

5.2.4.3 Dangerous portions of any chimney shall be removed.

5.2.4.4 Holes in walls and ceilings shall be patched.

5.2.4.5* Roof ventilation openings that are normally closed but can be opened in the event of an emergency shall be permitted to be utilized.

5.2.4.6* Low-density combustible fiberboard and other highly combustible interior finishes shall be removed.

5.2.4.7* Extraordinary weight above the training area shall be removed.

5.2.5* All hazardous environmental conditions shall be removed before live fire training evolutions are conducted in the structure.

5.2.5.1 All forms of asbestos deemed hazardous shall be removed by an approved manner and documentation provided to the AHJ.

5.2.6 Debris creating or contributing to unsafe conditions shall be removed.

5.2.7 Any toxic weeds, insect hives, or vermin that could present a potential hazard shall be removed.

5.2.8 Trees, brush, and surrounding vegetation that create a hazard to participants shall be removed.

5.2.9 Combustible materials, other than those intended for the live fire training evolution, shall be removed or stored in a protected area to preclude accidental ignition.

5.3 Utilities.

5.3.1 Utilities shall be disconnected.

5.3.2 Utility services adjacent to the live burn site shall be removed or protected.

5.4 Exits.

5.4.1 Exits from the acquired structure shall be identified and evaluated prior to each training burn.

5.4.2 Participants of the live fire training shall be made aware of exits from the acquired structure prior to each training burn.

5.4.3 Fires shall not be located in any designated exit paths.

5.5 Rapid Intervention Crew (RIC). A RIC trained in accordance with NFPA 1407, *Standard for Training Fire Service Rapid Intervention Crews*, shall be provided during a live fire training evolution.

Chapter 6: Gas-Fired Live Fire Training Structures

6.1 Structures and Facilities.

6.1.1 This section pertains to all interior spaces where gas-fired live fire training exercises occur.

6.1.2 Live fire training structures shall be left in a safe condition upon completion of live fire training evolutions.

6.1.3 Debris hindering the access or egress of fire fighters shall be removed prior to the beginning of the training exercises.

6.1.4 Flammable gas fires shall not be ignited manually.

6.2 Inspection and Testing.

6.2.1* Live fire training structures shall be inspected visually for damage prior to live fire training evolutions.

6.2.1.1* Damage shall be documented and the building owner or AHJ shall be notified.

6.2.2 Where the live fire training structure damage is severe enough to affect the safety of the participants, training shall not be permitted.

6.2.3 All doors, windows and window shutters, railings, roof scuttles and automatic ventilators, mechanical equipment, lighting, manual or automatic sprinklers, and standpipes necessary for the live fire training evolution shall be checked and operated prior to any live fire training evolution to ensure they operate correctly.

6.2.4 All safety devices, such as thermal sensors, combustible gas monitors, evacuation alarms, and emergency shutdown switches, shall be checked prior to any live fire training evolutions to ensure they operate correctly.

6.2.5 The instructors shall run the training system prior to exposing students to live flames in order to ensure the correct operation of devices such as the gas valves, flame safeguard units, agent sensors, combustion fans, and ventilation fans.

6.2.6* The structural integrity of the live fire training structure shall be evaluated and documented annually by the building owner or AHJ.

6.2.6.1 If visible structural defects are found, such as cracks, rust, spalls, or warps in structural floors, columns, beams, walls, or metal panels, the building owner shall have a follow-up evaluation conducted by a licensed professional engineer with live fire training structure experience and expertise, or by another competent professional as determined by the building owner or AHJ.

6.2.7* The structural integrity of the live fire training structure shall be evaluated and documented by a licensed professional engineer with live fire training structure experience and expertise, or by another competent professional as determined by the AHJ, at least once every 10 years, or more frequently if determined to be required by the evaluator.

6.2.8* All structures constructed with calcium aluminate refractory structural concrete shall be inspected by a structural engineer with expertise in live fire training structures every 3 years.

6.2.8.1 The structural inspection shall include removal of concrete core samples from the structure to check for delaminations within the concrete.

6.2.9* Part of the live fire training structure evaluation shall include, at least once every 10 years, the removal and reinstallation of a representative area of thermal linings (if any) to inspect the hidden conditions behind the linings.

Chapter 7: Non-Gas-Fired Live Fire Training Structures

7.1 Structures and Facilities.

7.1.1 This section pertains to all interior spaces where non-gas-fired live fire training exercises occur.

7.1.2 Live fire training structures shall be left in a safe condition upon completion of live fire training evolutions.

7.1.3 Debris hindering the access or egress of fire fighters shall be removed prior to the beginning of the training exercises.

7.2 Inspection and Testing.

7.2.1* Live fire training structures shall be inspected visually for damage prior to live fire training evolutions.

7.2.1.1* Damage shall be documented, and the building owner or AHJ shall be notified.

7.2.2 Where the live fire training structure damage is severe enough to affect the safety of the participants, training shall not be permitted.

7.2.3 All doors, windows and window shutters, railings, roof scuttles and automatic ventilators, mechanical equipment, lighting, manual or automatic sprinklers, and standpipes necessary for the live fire training evolution shall be checked and operated prior to any live fire training evolution to ensure they operate correctly.

7.2.4 All safety devices, such as thermal sensors, oxygen and toxic and combustible gas monitors, evacuation alarms, and emergency shutdown switches, shall be checked prior to any live fire training evolutions to ensure they operate correctly.

7.2.5* The structural integrity of the live fire training structure shall be evaluated and documented annually by the building owner or AHJ.

7.2.5.1 If visible structural defects are found, such as cracks, rust, spalls, or warps in structural floors, columns, beams, walls, or metal panels, the building owner shall have a follow-up evaluation conducted by a licensed professional engineer with live fire training structure experience and expertise or by another competent professional as determined by the AHJ.

7.2.6* The structural integrity of the live fire training structure shall be evaluated and documented by a licensed professional engineer with live fire training structure experience and expertise or by another competent professional as determined by the AHJ at least once every 5 years or more frequently if determined to be required by the evaluator.

7.2.7* All structures constructed with calcium aluminate refractory structural concrete shall be inspected by a structural engineer with expertise in live fire training structures every 3 years.

7.2.7.1 The structural inspection shall include removal of concrete core samples from the structure to check for delaminations within the concrete.

7.2.8* Part of the live fire training structure evaluation shall include, once every five years, the removal and reinstallation of a representative area of thermal linings (if any) to allow inspections of the conditions hidden behind the linings.

7.3 Sequential Live Fire Burn Evolutions.

7.3.1 The AHJ shall develop and utilize a safe live fire training action plan when multiple sequential burn evolutions are to be conducted per day in each burn room.

7.3.2 A burn sequence matrix chart shall be developed for the burn rooms in a live fire training structure.

7.3.2.1 The burn sequence matrix chart shall include the maximum fuel loading per evolution and maximum number of sequential live fire evolutions that can be conducted per day in each burn room.

7.3.3* The burn sequence for each room shall define the maximum fuel load that can be used for the first burn and each successive burn.

7.3.4* The burn sequence matrix for each room shall also specify the maximum number of evolutions that can be safely conducted during a given training period before the room is allowed to cool.

7.3.5 The fuel loads per evolution and the maximum number of sequential evolutions in each burn room shall not be exceeded under any circumstances.

Chapter 8: Exterior Live Fire Training Props

8.1 Props, Structures, and Facilities.

8.1.1 This section pertains to all exterior props where live fire training exercises occur.

8.1.2 Props used for outside live fire training shall be designed specifically for the evolution to be performed.

8.1.3 Exterior props shall be left in a safe condition upon completion of live fire training evolutions.

8.1.4 For outside training, care shall be taken to select areas that limit the hazards to both personal safety and the environment.

8.1.5 The training site shall be without obstructions that can interfere with fire-fighting operations.

8.1.6 Where live training fires are used outside, the ground cover shall be such that it does not contribute to the fire.

8.1.7 Debris hindering the access of fire fighters shall be removed prior to the beginning of the training exercise.

8.2 Inspection and Maintenance.

8.2.1 Exterior props shall be inspected visually for damage prior to live fire training evolutions.

8.2.1.1 Damage to exterior props shall be documented and the owner or AHJ shall be notified.

8.2.2 All safety devices and emergency shutdown switches, plus doors, shutters, vents, and other operable devices, shall be checked prior to any live fire training evolutions to ensure they operate correctly.

8.2.3 The structural integrity of the props shall be evaluated and documented annually.

Chapter 9: Reports and Records

9.1 General.

9.1.1* The following records and reports shall be maintained on all live fire training evolutions in accordance with the requirements of this standard:
  (1) An accounting of the activities conducted
  (2) A listing of instructors present and their assignments
  (3) A listing of all other participants
  (4) Documentation of unusual conditions encountered
  (5) Any injuries incurred and treatment rendered
  (6) Any changes or deterioration of the structure
  (7) Documentation of the condition of the premises and adjacent area at the conclusion of the training exercise

9.1.2* For acquired structures, records pertaining to the structure shall be completed.

9.1.3 Upon completion of the training session, an acquired structure shall be formally turned over to the control of the property owner.

9.1.3.1 The turnover process shall include the completion of a standard form indicating the transfer of authority for the acquired structure.

9.1.4 A post-training critique session, complete with documentation, shall be conducted to evaluate student performance and to reinforce the training that was covered.

APPENDIX B

NFPA® 1403 Correlation Guide

Chapter 4: General

Objectives	Corresponding Chapter(s)	Corresponding Page(s)
4.1.1	1	8-9
4.2.1	6, 8	103, 165
4.2.2	6, 8	103, 165
4.2.3	8	165
4.3.1	1	17
4.3.2	1	16-19
4.4.1	5	89-91
4.4.2	1	7
4.4.3	5	89-91
4.4.4	5	89-91
4.4.5	5	89-91
4.4.6	5	89-91
4.4.7	5	89-91
4.4.8	5, 8	91, 166-167
4.5	8	166
4.6.1	1	7
4.6.2	1	6-8
4.6.3	5	90
4.6.4	5	88, 89
4.6.5	4	64-65, 70-71, Appendix D
4.6.5.1	5	94
4.6.6	1	7
4.6.7	5	84
4.6.8	5	91, 94
4.6.9	3	48-49
4.6.10	5	84
4.6.11	5	87
4.6.12.1	7, 8	148, 163, 170-172
4.6.12.2	8	170-172
4.7.1	5, 7	90, 94, 149
4.7.1.1	5, 7	90, 94, 149
4.7.1.1.1	5, 7	90, 94, 149

Objectives	Corresponding Chapter(s)	Corresponding Page(s)
4.7.1.2	5	90, 94
4.7.2	5, 6, 8	90, 94, 117, 171
4.7.3	5, 6, 8	90, 94, 116, 171
4.7.4	5	90, 91
4.7.5	5	90, 94
4.7.6	5	90, 94
4.8.1	5	90-91, 94
4.8.2	5	90-91, 94
4.8.3	5	90-91, 94
4.8.4	5	90-91, 94
4.8.5	1	8
4.8.6	1	8
4.8.7	5, 8	90-91, 167
4.9.1	5, 8	88, 167
4.9.2	5	86-87
4.10.1	5	88
4.10.1.1	5	85
4.10.2	5	85
4.10.3	5	95
4.11.1	6, 8	104-105, 167
4.11.2	6	104
4.11.3	6	104-105
4.11.4	1	8
4.11.5	8	167
4.11.5.1	7	143
4.11.6	8	167
4.11.6.1	8	167
4.11.7	7	145
4.12.1	1	7
4.12.1.1	7	131
4.12.2	8	160
4.12.3	1	7
4.12.3.1	7	148-149
4.12.4	8	160-161

Objectives	Corresponding Chapter(s)	Corresponding Page(s)
4.12.5	7	149
4.12.6	6	113
4.12.7	6	113
4.12.8	6	113
4.12.9	6	113
4.12.10	6	103
4.12.10.1	5, 6	87, 117
4.12.11	1	7
4.12.11.1	1	7
4.12.11.2	8	172
4.12.11.3	8	172
4.12.11.4	8	170-172
4.12.11.5	8	163
4.12.11.6	8	165
4.12.11.6.1	8	165
4.12.11.7	8	160
4.12.11.8	8	161
4.12.11.9	6	113, 117
4.13.1	5	85
4.13.2	5	85
4.13.3	5	85
4.13.4	5	85
4.13.5	5	85
4.14.1	6, 7	114-115, 143
4.14.2	6, 7	114-115 143
4.14.3	6, 7	114-115 143
4.14.4	6, 7	114-115 143
4.14.5	6, 7	114-115 143
4.15.1	5, 6, 7	83-88, 103, 141-142
4.15.1.1	5	85
4.15.2	5, 6, 7, 8	87, 116, 141, 170-172
4.15.3	5, 8	87, 170-172
4.15.4	5, C, D, E	87, 222, 225, 230, 242
4.15.5	5, 8	95, 170-172
4.15.6	5, 8	87 and 170-172
4.15.7	5	87
4.15.8	6, 8	102, 165-166
4.15.8.1	5	85
4.15.9	6	102
4.15.9.1	6	102

Chapter 5: Acquired Structures

Objectives	Corresponding Chapter(s)	Corresponding Page(s)
5.1.1	6	106-114
5.1.1.1	6	102
5.1.2	6	102
5.1.3	6	103
5.1.4	6	103
5.1.5	6	103
5.1.6	6	103
5.1.7	6	103
5.1.8	6	103
5.1.9	6	103
5.1.10	6	116
5.1.11	6	102-103
5.1.12	6	113
5.2.1	5	83
5.2.2	6	109
5.2.3	6	112
5.2.3.1	6	108, 112
5.2.3.2	6	112
5.2.4	6	112
5.2.4.1	6	111
5.2.4.2	6	108, 113
5.2.4.3	6	109
5.2.4.4	6	111
5.2.4.5	6	109
5.2.4.6	6	111
5.2.4.7	6	112
5.2.5*	6	109
5.2.5.1*	6	108
5.2.6*	6	109
5.2.7	6	106
5.2.8	6	106
5.2.9	6	109
5.3.1	6	108
5.3.2	6	108
5.4.1	6	106-108
5.4.2	6	113

Objectives	Corresponding Chapter(s)	Corresponding Page(s)
5.4.2	6	117
5.5	5	86

Chapter 6: Gas-Fired Live Fire Training Structures

Objectives	Corresponding Chapter(s)	Corresponding Page(s)
6.1.1		
6.1.2	7	149
6.1.3	7	145, 149
6.1.4	7	149
6.2.1	7	149
6.2.1.1	7	149
6.2.2	7	145
6.2.3	7	145
6.2.4	7	145
6.2.5	7	145
6.2.6	7	144
6.2.6.1	7	144
6.2.7	7	144
6.2.8	7	144
6.2.8.1	7	144
6.2.9	7	144

Chapter 7: Non-Gas-Fired Live Fire Training Structures

Objectives	Corresponding Chapter(s)	Corresponding Page(s)
7.1.1		
7.1.2	7	149
7.1.3	7	145-146, 149
7.2.1	7	145-146, 148
7.2.1.1	7	149
7.2.2	7	144
7.2.3	7	144
7.2.4	7	145

Objectives	Corresponding Chapter(s)	Corresponding Page(s)
7.2.5	7	144
7.2.5.1	7	144
7.2.6	7	144
7.2.7	7	144
7.2.7.1	7	144
7.2.8	7	144
7.3.1	7	134-135
7.3.2	7	134-135
7.3.2.1	7	134-135
7.3.3	7	134-135
7.3.4	7	134-135
7.3.5	7	134-135

Chapter 8: Exterior Live Fire Training Props

Objectives	Corresponding Chapter(s)	Corresponding Page(s)
8.1.1		
8.1.2	8	160, 170-172
8.1.3	8	167
8.1.4	8	160, 170-172
8.1.5	8	161
8.1.6	8	170-172
8.1.7	8	161
8.2.1	8	170-172
8.2.1.1	8	170-172
8.2.2	8	165
8.2.3	8	170-172

Chapter 9: Reports and Records

Objectives	Corresponding Chapter(s)	Corresponding Page(s)
9.1.1	5	95
9.1.2	6	103
9.1.3	6	121
9.1.3.1	6	121
9.1.4	9	183

Acquired Structure Live Fire Training Model SOP

The following model Standard Operating Procedure (SOP) should be used as a guide to create an agency-specific SOP for your department. Included here is a model SOP for live fire training in acquired structures. This SOP is a model and specific information will have to be added by the user as needed to reflect local requirements and procedures.

Purpose

This policy provides a standard procedure for the training of personnel engaged in structural firefighting operations under live fire conditions. This policy focuses on training for coordinated interior firefighting operations with minimum participant exposure risk. Live fire training fires present the same hazards as those encountered at actual fire ground incidents. The incident management system employed at actual fire incidents will be the standard operating procedure at *all* structural training fires.

Interior live fire training in a suitable acquired structure awaiting demolition is an excellent means of training fire fighters. While this type of training provides a high level of authenticity, it also carries with it most of the hazards of interior fire fighting at an actual emergency. Interior live fire training evolutions shall be planned with great care and supervised closely by instructional personnel. When conducting live fire training evolutions, this agency will comply with applicable state/provincial laws, NFPA 1403, *Standard on Live Fire Training Evolutions*, and additional requirements herein.

Initial Evaluation

Once an inquiry or request is received, an initial site visit will preliminarily determine the suitability of the acquired structure or indicate problems that might immediately rule out the structure, or require significant mitigation. The initial evaluation should be documented, specifying any of the concerns below:

- Proximity to other structures and other exposures (utilities, infrastructure, sheds, trees and heavy vegetation on and adjacent to property, etc.)
- Adjacent properties that could adversely be affected by the smoke or possible business disruptions.
- Transportation that such operations could interfere with (busy streets or any highways, railroads, airports, etc.)
- Access problems to the site, parking for apparatus and equipment.
- Obvious structural integrity (sagging roof or floors, cracks in brick or masonry walls).
- Prior damage due to fire, collapse, or other.
- Contamination to the building or site, due to past occupancy.
- Does the acquired structure provide for the fire department's training objectives?
- Does the interior configuration allow the fire department to conduct the desired evolutions in a safe manner?
- Are there any hazardous features or fixtures?

Procedure

The following procedure shall be used when an owner desires to dispose of property, and that property can be utilized for live fire training purposes.

1. Information such as location, owner, contact phone number, and type of property must be obtained and provided to the training division of the fire department. The owner may also contact the training division directly. The training division will complete the Acquired Structures Contact Information Form.

2. Once all pertinent information is obtained, a designee from the training division will physically inspect the property and its location. During this inspection, all aspects of NFPA 1403 will be considered to determine if the property and/or structure is acceptable for live fire training purposes, which may include either contained live fire training evolutions or a destructive live fire burn. During the site inspection, the member from the training division will take pictures of the structure to document the suitability of the property and/or structure for live fire training purposes.

3. Once a structure or property has been deemed acceptable, the training division will obtain all pertinent information and begin the documentation process. The documentation process includes the completion and tracking of forms specific to the legal process and NFPA 1403 requirements. All required forms shall be maintained by the fire department. The department will assure the building owner has completed, notarized, and returned the following required paperwork:

   a. Two original signed and notarized copies of the License Agreement Relating to Fire Rescue Training and Structure Burn/Proof of Insurance Cancellation or a signed statement of nonexistence of insurance
   b. Insurance Certificate Form
   c. Written Title Opinion issued by a licensed attorney or title company indicating clear title within the last 180 days
   d. Documentation as to application for a Demolition Permit from the appropriate government unit.

… APPENDIX C

4. Upon receipt of the items listed above, a member from the department will legally execute the above documents. This process may take 2–4 weeks to complete.
5. The training division will complete the Asbestos Project Request/Estimate/Authorization Form and forward it to the [appropriate government unit] Risk Management department to request an asbestos inspection. This process may take 2–6 weeks to complete. The property owner will be notified, in writing, as to the results of the asbestos inspection. The ability to utilize the acquired structure for training purposes will rely on the asbestos survey and abatement, if required. If the acquired structure is positive for asbestos, an asbestos abatement will be completed or the structure will not be used. Typically the expense of the asbestos abatement is the responsibility of the property owner. [Requirements vary by location—insert requirements by the authority having jurisdiction here.]
6. Once the approved paperwork process is completed, a Checklist for Live Fire Training Evolutions in Acquired Structures will be completed to identify what, if any, modifications or improvements must be made to make the acquired structure safe for live fire training evolutions. All live fire training evolutions must be approved by the fire chief.
7. No trespassing signs and hazard signs will be posted at the property and on the acquired structure. The training division will take pictures of the posted signs to document that this process has been completed. The signs must remain on the property throughout its scheduled use. Additional warning signs must be posted when a destructive live fire burn is actually conducted.
8. Prior to any training evolution that contains reduced visibility and/or smoke, a High Hazard Safety Checklist must be completed, the water source must be secured, and the identified Command Structure, Duties, and Assignments form must be completed.
9. Prior to any destructive live fire burn, the following must be completed:
   a. Checklist for Live Fire Training Evolutions in Acquired Structures
   b. Command Structure, Duties, and Assignments Form
   c. Approval by the fire chief
   d. Inspection by the Department of Health
   e. Required environmental forms
   f. Completed demolition permit from the [appropriate government unit]. (must be on-site at time of burn)
   g. Written documentation to the property owner stating the date of the burn, expected final condition of the acquired structure after the burn, and that they will be responsible for security of the site and any removal of debris from the property
10. Prior to any destructive live fire burn, notification to the following personnel/agencies must be completed:
    a. 911 and communications center(s) for Fire, emergency medical services (EMS) and law enforcement
    b. Fire chief of the affected battalion
    c. Public information officer
    d. Local emergency medical service provider.
    e. Appropriate helicopter ambulance service (with GPS to acceptable landing zone)
    f. [Appropriate government unit] Risk Management Department
    g. Locally required notifications depending on AHJ
    h. The owners of the property and affected neighbors
11. The instructor-in-charge is responsible for providing written documentation of the live burn as well as any additional required paperwork provided by the training division and for following NFPA 1403.
12. All records related to acquired structures used in training will be maintained in the training division.

Initial Preparation

In preparation for live fire training, an inspection of the acquired structure shall be made to determine that the floors, walls, stairs, and other structural components are able to withstand the weight of contents, participants and accumulated water. The Checklist for Live Fire Training Evolutions in Acquired Structures will be completed to identify what, if any, modifications or improvements must be made to make the structure safe for training drills. The Checklist for Live Fire Training Evolutions in Acquired Structures will be utilized in developing the preburn plan, and when preparing the building, the site, and ensuring overall preparedness.

Preburn Plan

Prior to any live fire evolution the Checklist for Live Fire Training Evolutions in Acquired Structures will be completed. This checklist will ensure that all components are examined and deemed to be safe for live fire training in accordance with NFPA 1403. A written preburn plan inclusive of objectives, order of operations, and emergency plans, will reviewed by all participants prior to initiating evolutions.

Fuel Materials

The fuels that are utilized in live fire training evolutions shall have known burning characteristics of such a nature to be as controllable as possible. Unidentified materials, such as debris found in or around the structure, materials of undetermined composition, which may burn in unanticipated ways, react violently, or create environmental or health hazards, shall not be used. Materials shall be used in only the amounts necessary to create the desired size fire. No flammable or combustible liquids of any type shall be used during interior structure training evolutions. *Note: Acceptable Class A materials include straw, wooden pallets, hay, pine excelsior and other ordinary combustibles. A reasonable effort should be made to ascertain that straw or hay, if used, has not been treated with pesticides or other harmful chemicals.*

The use of flammable or combustible liquids, as defined in NFPA 30, *Flammable and Combustible Liquids Code*, shall be prohibited for use in live fire training evolutions in acquired structures.

Water Supply

The minimum water supply and delivery for the live fire training evolutions shall meet the criteria identified in NFPA 1142, *Standard on Water Supplies for Suburban and Rural Fire Fighting*.

A minimum reserve of additional water in the amount of 50 percent of the fire flow demand shall be available to handle exposure protection or unforeseen situations.

Separate sources shall be identified for supply of attack lines and back up lines in order to preclude the loss of both water supply sources at the same time. *Note: Reliability should be considered when determining what constitutes a separate source.*

The intent of this section is to prevent the simultaneous loss of both attack lines and back-up lines in the event of a pump or water supply failure. If a public water supply system is used, two pumpers on two different hydrants, should be used. Two pumpers drafting from the same pond, river and/or folding tanks would also be appropriate if the source contains sufficient usable water. Two separate pumpers should be used to supply the attack and back-up lines.

Command Structure

The live fire evolution will be operated under the Incident Management System. The instructor-in-charge (incident commander) is responsible for completing any required paperwork provided by the training division and for following NFPA 1403. The Command Structure, Duties, and Assignments Form will be used to document assignments and responsibilities.

Note: NFPA 1403 refers to both an instructor-in-charge and an incident commander. This SOP refers to the two being one in the same to avoid conflict or confusion. This is an authority having jurisdiction (AHJ) decision that needs to be clear in this policy.

The assignments for any live fire evolution require the following positions to be filled:

- instructor-in-charge
- safety officer
- ignition officer/fire control team
- instructors

Instructor-in-Charge

A preburn plan shall be prepared for the acquired structure and shall be utilized in the preburn briefing. All interior rooms, hallways, exterior openings, and access and egress points shall be indicated on the plan. Prior to conducting any live fire training evolutions all participants shall conduct a walk-through of the acquired structure, to gain knowledge and familiarity with the layout of the building and the emergency evacuation signal, in the event an emergency evacuation becomes necessary.

The instructor-in-charge is responsible for establishing a method of communication between command, interior divisions, exterior divisions/sectors, the safety officer, instructors, and external requests for assistance.

The instructor-in-charge shall determine, prior to each specific live fire evolution, how many attack lines and back-up lines will be necessary. Each hose line shall be capable of delivering a minimum of 95 gpm (360 L/min). Sufficient back-up lines shall be provided to ensure adequate protection for personnel on training attack lines. *NOTE: A minimum flow rate of 95 gallons per minute is required in order to have adequate quantities of water/extinguishing agent available to handle the planned evolution plus a reserve for unanticipated emergencies. The appropriate quantity and exact flow rates that will be needed for fire control and extinguishment should be calculated in advance, and certain factors such as equipment, personnel, fire area, and topography should be taken into consideration. Knowledge of the hose line sizes, types of nozzles, what fire stream will be utilized, and the principles of fire attack and deployment will aid in determining the exact flow rates which will be necessary.*

The Instructor-In-Charge shall:

1. Assign one (1) instructor to each functional crew, which shall not exceed five (5) students,
2. Assign one (1) instructor to each back-up line(s),
3. Assign sufficient additional personnel to back-up lines to provide mobility

Prior to conducting actual live fire training evolutions in the acquired structure, the instructor-in-charge will ensure a preburn briefing is conducted for all participants. All evolutions to be conducted shall be discussed during this briefing and assignments shall be made for all crews participating in the training session.

The instructor-in-charge shall assess the selected fire room environment for factors that will affect the growth, development, and spread of fire. The instructor-in-charge, as a minimum, shall document fuel loading, including wall and floor coverings and ceiling materials; type of construction, including type of roof and combustible void spaces; and the dimensions of room.

The live fire training evolution shall be immediately stopped if the instructor-in-charge or safety officer determines through continuing assessments that the combustible nature of the environment represents a potential hazard. The evolution shall continue only when the appropriate actions have been taken to reduce the hazard.

Instructor-in-Charge Sample Checklist
Responsibilities

- Plan and coordinate all training activities
- Monitor activities to ensure safe practices

☐ Assign instructor positions
☐ Brief instructors on responsibilities
- Accounting for assigned students
- Assessing student performance
- Inspecting PPE and SCBA
- Monitoring safety
- Achieving tactical and training objectives
☐ Review and communicate the objectives of the live fire training evolution
☐ Assign coordinating personnel as needed:
- Emergency Medical Services
- Apparatus operators
- Water supply officer
- Rehab
- Public relations – media
- Additional safety officers – specific assignments
☐ Coordinate participant readiness with the safety officer
☐ Go/No-Go sequence
☐ Assure adherence to NFPA 1403 standard by all persons in the training area

Safety Officer

A safety officer shall be appointed for all live fire training evolutions. The safety officer shall have the authority, regardless of rank, to intervene and correct any aspect of the operations when, in their judgment, a potential or real danger, accident, or unsafe condition exists. The safety officer shall be responsible for the safety of all persons on the scene including students, instructors, visitors, and spectators.

The safety officer shall be established and an evacuation signal shall be demonstrated to all participants in the live fire training evolution. The evacuation signal identified in [Insert agency name] will be used: [Insert local evacuation signal here, i.e., horn siren 30 seconds on, 30 seconds off, repeat].

The safety officer shall not be assigned other duties inconsistent with safety responsibilities. Additional safety personnel, as deemed necessary by the safety officer, shall be strategically placed within the structure to react to any unplanned or threatening situation or condition.

Emergency medical services shall be available on site to handle any injuries with an ambulance with transport capability. Written reports shall be made on all injuries and on all medical aid rendered.

No person(s) shall be placed inside the building to play the role of victim. Rescue mannequins shall not be dressed in protective clothing similar in color or configuration to firefighter's protective clothing to avoid confusing a rescue mannequin and a "fire fighter down." A thorough search of the acquired structure shall be conducted to ensure that no unauthorized persons, animals or objects are in the structure immediately prior to ignition.

NOTE: *Participants involved in the live fire training evolution should be instructed to report to a predetermined location for a roll call should evacuation of the building be signaled. Instructors should report immediately to the instructor-in-charge any personnel not accounted for.*

Safety Officer Sample Checklist
Responsibilities

- Prevention of unsafe acts
- Elimination of unsafe conditions

☐ Establish safety zones
☐ Establish water supply and ensure reserve water supply is in place
☐ Ensure accountability system is in place
☐ Ensure/ignition hose line is in place and charged
☐ Ensure rapid intervention crew (RIC) is established and in place

- ☐ Communication check – perform radio check
- ☐ Evacuation signal demonstrated
- ☐ Fire set is checked for size and location – fuel materials are compliant
- ☐ Assure PPE is compliant for all participants
  - Personal protective clothing
  - SCBA
  - PASS alarms
- ☐ Ensure building is vacated prior to ignition
- ☐ Notify instructor-in-charge that evolution can begin

Ignition Officer/Fire Control Team

One member of the fire control team shall be designated as the ignition officer to control the materials being burned and to ignite the training fire in the presence of and under the direct supervision of the safety officer. This person shall wear full protective clothing including self-contained breathing apparatus (SCBA). A charged hose line shall accompany the ignition officer when igniting the fires. The decision to ignite the training fire shall be made by the ignition officer in coordination with the instructor-in-charge and the safety officer. Nobody, including the ignition officer, shall operate in the structure alone.

Awareness of weather conditions, wind velocity, and wind direction shall be maintained. In all cases, immediately before actual ignition, a final check shall be made for changes in weather conditions.

Ignition Officer Sample Checklist
Responsibilities

- Prepare scenario within the standards of NFPA 1403

- ☐ Confirm scenario objectives with lead instructor
- ☐ Confirm locations of fire sets
- ☐ Confirm location of props and obstacles for the evolution
- ☐ Ensure the safety/ignition hose line are in place and charged
- ☐ Prepare the fire set – ensure fuel materials are compliant
- ☐ Communication check – perform radio check
- ☐ Ensure the building is vacated prior to ignition
- ☐ Go/No-Go Sequence – Confirm with safety officer and instructor-in-charge that the evolution can begin
- ☐ Ignite fire – notify instructor-in-charge once the fire is set
- ☐ Notify instructor-in-charge when companies can begin evolution

Instructors

All Instructors shall be deemed qualified to deliver live fire training by the AHJ. [Insert local requirements here]

Instructors should meet the criteria outlined in NFPA 1041, *Fire Service Instructor Professional Qualifications*, and have training specific to training firefighters using live fire in acquired structures. The AHJ needs to specify what these requirements are.

The participating student-to-instructor ratio shall not be greater than five (5) to one (1). It is important that the participating student-to-instructor ratio be monitored so as to not exceed the span of control necessary to provide for the safe and proper supervision of trainees. Instructor trainees may be present and utilized but may not directly supervise students.

■ Participant Safety

Protective Clothing and Equipment

Each participant involved in live fire operations shall be equipped with full protective clothing and self-contained breathing apparatus (SCBA). All participants shall be inspected by the safety officer to ensure the protective clothing and SCBA are being properly worn prior to entry into a live fire training evolution.

Protective equipment shall meet the requirements of NFPA 1971, *Standard on Protective Ensemble for Structure Fire Fighting*, and must be inspected for integrity. Self-contained breathing apparatus (SCBA), shall meet the requirements of NFPA 1981, *Standard on Self-Contained Breathing Apparatus for Fire Fighters*. Protective footwear shall meet the requirements of NFPA 1971 and OSHA 29 CFR 1910.156 (e) (2) (ii) and (e) (2) (iii). Personnel are not allowed on site without proper footwear. Sneakers, soft sole shoes, and sandals, are prohibited. *NOTE: Clothing worn under protective clothing can degrade and cause injury to the user, even without damaging the protective clothing. All persons should be aware of the dangers of clothing that is made from certain all-synthetic materials melting, sticking to, and burning the user.*

Participants need to be briefed prior to the day of evolutions to wear proper clothing. Fire retardant fabrics and all natural fibers should be given consideration.

All students, instructors, safety personnel, and other personnel participating in any evolution or fire suppression operation during the live fire training evolution shall breathe from the SCBA air supply whenever one or more of the following conditions exist:

1. Operating in an atmosphere that is oxygen deficient or contaminated by products of combustion, or both.
2. Operating in an atmosphere that is suspected of being oxygen deficient or contaminated by products of combustion or both.

3. Operating in any atmosphere that may become oxygen deficient or contaminated, or both.
4. Operating below ground level or in a confined space as defined by OSHA.

Note: No one should be allowed to breathe smoke, toxic vapors or flames, products of combustion, or other deficient atmosphere.

Vehicle Parking/Staging

Adequate areas for staging, operating, and parking of fire apparatus that will be used in the live fire training evolution shall be designated. An area shall be designated to park fire apparatus and vehicles which are not a part of the evolution so as not to interfere with the fire-ground operations. If required, parking areas for police vehicles or for the news media shall be designated.

A parking area for an ambulance shall be designated. Consideration shall be given to designate this area for prompt response. Consideration shall be given to the designation and layout of entry/egress in order to assure their availability in the event of an emergency.

Chief's Approval

Dated signature blocks for the training chief and fire chief must be on the plan.

Live Burn Manual Forms

Sample Letter to Property Owner

[Date]
[Property Owner Name]
[Property Owner Address]

Dear [sir/madam]:

Thank you for your interest in making your structure at [Insert property address] available to [Insert agency name] for the purpose of live fire training.

To utilize your structure, we will need the attached paperwork completed, notarized, and returned to [Insert agency name and address], attention Training Division.

1. License Agreement Relating to Fire Rescue Training and Structure Burn
2. Insurance Certification Form
3. Written Title Opinion issued by a licensed attorney or title company, demonstrating clear title to the above-referenced
4. Proof of Insurance Cancellation or a signed statement of nonexistence of insurance
5. Documentation as to application for a Demolition Permit from the [Insert appropriate government unit].

Once we have received the properly completed documents, a member from [Insert agency name] will execute the documents. This process may take 2–4 weeks to complete. The training division will request an asbestos survey through the [Insert appropriate government unit] Risk Management Department. This process may take 2–6 weeks to complete. Our ability to utilize the structure for training purposes will rely on the asbestos survey and abatement, if required. The cost of any required asbestos abatement will be the responsibility of the property owner. A member from the training division will maintain documentation files as well as communication with you throughout the process.

Should you desire [Insert agency name] to perform a live fire burn which results in the destruction of the building, a demolition permit from the [insert unit of governance] must be obtained and a copy made available to [Insert agency name] ____ days prior to the live fire training being performed. At that time, the property and remaining structure, if applicable, will be returned to your control to secure the site and for further disposal according to local ordinances.

If you have any questions, please contact the training division at [Insert agency phone number].

Sincerely yours,
[Insert name and address]

Appendix C

■ Sample Acquired Structure Property Owner Contact Information Sheet

Name: _____

Address: _____

Phone: _____

Cell Phone: _____

Structure Availability Date: _____

How Long Is Structure Available: _____

Type of Structure: _____

Limitations: _____

Remarks: _____

_____

_____

■ Sample Insurance Certification Form

I (We) certify that any and all insurance on the aforesaid building located at [Insert property address] has been canceled or there is a non-existence of such insurance on the said structure, and that the existence of any insurance on the structure, of any kind, would constitute perjury and establish prima facia evidence of intent to defraud which offenses are punishable by a court of competent jurisdiction.

Dated this _____ day of _____, 20 _____.

WITNESSES:

_____   _____

_____   _____

_____   _____

Agent/Owner   Agent/Owner

Sworn to and subscribed before me this _____ day of _____, 20_____.
[insert Notary information appropriate for your state/province]

Sample Asbestos Letter to Property Owner

[Date]
[Insert property owner name]
[Insert property owner address]

As a result of your request to have the property located at [Insert property address] to be used in a live fire training evolution, an asbestos abatement survey has been completed. The results of that survey indicate [Insert survey results here].

In order to safely remove the asbestos for the purpose of live fire training, there will be an estimated cost of $[Insert estimated cost here]. Additionally, please feel free to contact me regarding assistance with this matter.

We greatly appreciate your time and consideration. Should you have any questions or require any further information, please do not hesitate to contact me at (xxx) xxx-xxxx.

Sincerely yours,
[Insert your name, agency, and agency address]

Sample Letter to Neighbors

[Insert Date]

Dear [Insert Neighborhood] residents:

[Insert agency name] plans to perform live fire training at the structure located at [Insert property address] on [Insert date and time of evolutions]. The purpose of this exercise is to provide training to your firefighters under realistic conditions. They have been practicing search and rescue skills along with fire attack, ventilation, and exposure protection activities. These exercises allow us to maintain our skills at an optimum level, helping to assure that we are ready when a real fire emergency occurs.

This exercise will be controlled. We have planned additional resources to help insure that this training is conducted safely; both for our personnel, and for the safety and property protection of others.

We will be constantly monitoring the wind conditions and the effect that this exercise may have on the residents in the area. We request that you keep your windows closed during the exercise, and to notify us if you feel that smoke is getting into your house. A fire department officer will be at the burn location until the fire is completely extinguished.

You are welcome to watch and take pictures on the day of the burn; there is an area specifically for those wishing to observe the training.

Should you have any questions or concerns regarding this exercise, please contact me at [Insert phone number]. I will be at [Insert property address] on [Insert day of week], and will be able to answer questions. The training is designed with you in mind. A live fire exercise is the best type of training that we can provide your firefighters.

Sincerely,
[Insert your name, title, and phone number]

Sample Checklist for Live Fire Evolutions in Acquired Structures

General Information
Property Address:

Initial Inspection
The acquired structure must be inspected prior to acceptance from the owner to determine if the structure is eligible for live fire evolutions or basic firefighter training.
Reasons:

Inspection completed by:

Date completed:

The acquired structure must also be inspected prior to live fire evolutions to ensure compliance with NFPA 1403 prior to the burn.
Reasons:

Inspection completed by:

Date completed:

Student Prerequisites

Yes/No	JPR	Compliance Requirement	Comments
	4.3.1	Ensure students have meet the Firefighter I training requirements	
	4.3.2	Students participating in a live fire training evolution who have received the required minimum training from other than the authority having jurisdiction (AHJ) shall not be permitted to participate in any live fire training evolution without first presenting prior written evidence of having successfully completed the prescribed minimum training to the levels specified in 4.3.1.	

Structures and Facilities

Yes/No	JPR	Compliance Requirement	Comments
	5.1.4	Application and receipt of permission from building owner, must be in writing.	
	5.1.4	Building ownership has been determined.	
	5.1.5	Evidence of clear title has been received and filed.	
	5.1.6	Written permission shall be secured from the owner of the structure in order for the fire department to conduct live fire training evolutions within the acquired structure.	
	5.1.6	A clear description of the anticipated condition of the acquired structure at the completion of the evolution and the method of returning the property back to the building owner has been put in writing and filed, and was acknowledged by the building owner.	
	5.1.8	Proof of insurance cancellation or a signed statement of nonexistence of insurance has been received and filed.	
	4.2.2, 5.1.9	Agencies requiring permits will be contacted.	
	5.2.2	All hazardous storage conditions shall be removed from the structure or neutralized in such a manner as to not present a safety problem during use of the structure for live fire training evolutions.	
	5.2.3	Closed containers and highly combustible materials shall be removed from the structure.	
	5.2.3.1	Oil tanks and similar closed vessels that cannot be removed shall be vented sufficiently to prevent an explosion or overpressure rupture.	
	5.2.3.2	Any hazardous or combustible atmosphere within the tank or vessel shall be rendered inert.	
	5.2.4	All hazardous structural conditions shall be removed or repaired so as to not present a safety problem during use of the structure for live fire training evolutions.	
	5.2.4.1	Floor openings shall be covered.	
	5.2.4.2	Missing stair treads and rails shall be repaired or replaced.	

(Continues)

(Structures and Facilities Continued)

Yes/No	JPR	Compliance Requirement	Comments
	5.2.4.3	Dangerous portions of any chimney shall be removed.	
	5.2.4.4	Holes in walls and ceilings shall be patched.	
	5.2.4.6	Low-density combustible fiberboard and other unconventional combustible interior finishes shall be removed.	
	5.2.4.7	Extraordinary weight above the training area shall be removed.	
	5.2.5	All hazardous environmental conditions shall be removed prior to conducting live fire training evolutions in the structure.	
	5.2.6	Debris creating or contributing to unsafe conditions shall be removed.	
	5.2.4.5	Roof ventilation openings that are normally closed but can be opened in the event of an emergency shall be permitted to be utilized.	
	5.3.1	Utilities shall be disconnected.	
	5.2.7	Any toxic weeds, insect hives, or vermin that could present a potential hazard shall be removed.	
	5.2.5.1	All forms of asbestos deemed hazardous to personnel shall be removed by an approved asbestos removal contractor.	
	5.4	Exits from the building shall be identified and evaluated prior to each training burn.	Doors and windows used for access or as emergency exits from the building shall have unimpeded clear access with no obstructions, brush, debris, or significant height differences that would preclude entry/egress. No room will be used without at least two separate means of egress or escape available. No room with limited access or long escape routes will be used.

(Structures and Facilities *Continued*)

Yes/No	JPR	Compliance Requirement	Comments
			Windows need to be easily accessed for emergency egress. Any covering must be removable from inside or out by hand without tools. All glass shall be removed from the windows along with any hardware, horizontal or vertical crosspieces. High window sills will be low enough for exiting if there are not sufficient means of egress for the evolution planned.
			Door locking mechanisms shall be disabled, door closers shall be removed. Remove any storm doors, screen doors and secondary security doors and any hardware on doors that may snag or catch on PPE.
			Clearly mark doors that cannot be used as exits with a spray painted "X".
			Mark primary exits and exit routes clearly with high visibility paint along the baseboard or floor.
			Sides of the building shall be clearly marked using the local protocol (i.e. "A" front of the building, then clockwise).
	5.4.1	Participants of the live fire training shall be made aware of exits from the building prior to each training burn.	
	5.1.1.1	Buildings that cannot be made safe as required by this chapter shall not be utilized for interior live fire training evolutions.	Each exit shall be clearly identified inside and out with a numerical designation and alpha for the side of the building.
	5.1.2	Adjacent buildings or property that might become involved shall be protected or removed.	They must also receive written notice of scheduled burn date.
	5.3.2	Utility services adjacent to the live burn site shall be removed or protected.	

(*Continues*)

(Structures and Facilities *Continued*)

Yes/No	JPR	Compliance Requirement	Comments
	5.2.8	Trees, brush, and surrounding vegetation that create a hazard to participants shall be removed.	
	5.2.9	Combustible materials, other than those intended for the live fire training evolution, shall be removed or stored in a protected area to preclude accidental ignition.	
	4.15.8	Property adjacent to the training site that could be affected by the smoke from the live fire training evolution, such as railroads, airports or heliports, and nursing homes, hospitals, or other similar facilities, shall be identified.	
	4.15.8.1	Adjacent properties shall be informed in writing of the date and time of the evolution.	
	4.15.9	Streets or highways in the vicinity of the training site shall be surveyed for potential effects from live fire training evolutions and safeguards shall be taken to eliminate possible hazards to motorists.	
	4.14.5	Fire lines shall be established to keep pedestrian traffic in the vicinity of the training site clear of the operations area of the live burn.	
	4.6.11	Awareness of weather conditions, wind velocity, and wind direction shall be maintained, including a final check for possible changes in weather conditions immediately before actual ignition.	
	4.11.1	The lead instructor and safety officer shall determine the rate and duration of water flow necessary for each individual live fire training evolution, including the water necessary for control and extinguishment of the training fire, the supply necessary for backup lines to protect personnel, and any water needed to protect exposed property.	Rate and duration of water flow _____
	4.11.4	The minimum water supply and delivery for live fire training evolutions shall meet the criteria identified in NFPA 1142, *Standard on Water Supplies for Suburban and Rural Fire Fighting*.	Minimum water supply _____
	4.11.5	A minimum reserve of additional water in the amount of 50 percent of the fire flow demand determined shall be available to handle exposure protection or unforeseen situations.	50% reserve flow _____

(Structures and Facilities *Continued*)

Yes/No	JPR	Compliance Requirement	Comments
	4.11.6	Separate sources shall be utilized for the supply of attack lines and backup lines in order to preclude the loss of both water supply sources at the same time.	Source 1 _____ Source 2 _____
	4.13.1	Areas for the staging, operating, and parking of fire apparatus that are used in the live fire training evolution shall be designated.	
	4.13.2	An area for parking fire apparatus and vehicles that are not a part of the evolution shall be designated so as not to interfere with fireground operations.	
	4.13.3	If any of the apparatus are in-service to respond to an emergency, the apparatus shall be located in an area to facilitate a prompt response.	
	4.13.4	Where required or necessary, parking areas for police vehicles or for the press shall be designated.	
	4.10.2	A parking area for an ambulance or an emergency medical services vehicle shall be designated and located where it will facilitate a prompt response in the event of personal injury to participants in the evolution.	
	4.13.5	Ingress/egress routes shall be designated, identified, and monitored during the training evolutions to ensure their availability in the event of an emergency.	
	4.15.2	Prior to conducting actual live fire training evolutions, a preburn briefing session shall be conducted for all participants, in which all facets of each evolution to be conducted are discussed and assignments made for all crews participating in the training session are given.	
	4.15.5	The location of simulated victims shall not be required to be disclosed, provided that the possibility of victims is discussed during the briefing.	Rescue mannequins shall not be dressed similar to firefighter's PPE to avoid confusing a rescue mannequin and a "firefighter down".
	4.15.1	A preburn plan shall be prepared and shall be utilized during the preburn briefing sessions.	

(*Continues*)

(Structures and Facilities Continued)

Yes/No	JPR	Compliance Requirement	Comments
	4.15.1.1	All features of the training areas and structure shall be indicated on the preburn plan.	
	4.15.7	Prior to conducting any live fire training, all participants shall be required to conduct a walk-through of the structure in order to have a knowledge of and familiarity with the layout of the building and to facilitate any necessary evacuation of the building.	
	4.14.1	All spectators shall be restricted to an area outside the operations area perimeter established by the safety officer.	
	4.14.2, 4.14.5	Control measures such as ropes, signs, and fire line markings shall be posted to indicate the perimeter of the operations area.	
	4.14.3	Visitors who are allowed within the operations area perimeter to observe operations shall be escorted at all times.	
	4.14.4	Visitors allowed within the operations area perimeter shall be equipped with and shall wear complete protective clothing.	
	4.12.11.9	All possible sources of ignition, other than those under the direct supervision of the person responsible for the start of the training fire, shall be removed from the operations area.	

■ Fuel Materials

Yes/No	NFPA	Compliance Requirement	Comments
	4.12	The fuels that are utilized in live fire training evolutions shall have known burning characteristics that are as a controllable as possible.	
	4.12.4	Unidentified materials, such as debris found around the structure that could burn in unanticipated react violently, or create environmental or health hazards shall not be used.	
	4.12.2	Pressure-treated wood, rubber, and plastic, and hay treated with pesticides or harmful chemicals shall not be used.	

(Fuel Materials *Continued*)

Yes/No	NFPA	Compliance Requirement	Comments
	4.12.6	Fuel materials shall be used only in the amounts necessary to create the desired fire size.	
	4.12.7	The fuel load shall be limited to avoid conditions could cause an uncontrolled flashover or backdraft.	
	4.12.3, 4.12.11.1	Flammable or combustible liquids, as defined in NFPA 30, *Flammable and Combustible Liquids Code* shall not be used in live fire training evolutions in acquired structures.	
	4.12.8	The instructor-in-charge shall assess the selected room environment for factors that can affect the growth, development, and spread of fire. The instructor-in-charge shall document fuel loading including all of the following: (1) Furnishings (2) Wall and floor coverings and ceiling materials (3) Type of construction of the structure including type of roof and combustible void spaces (4) Dimensions of the room	
	4.12.10	The training exercise shall be immediately stopped when the instructor-in-charge or safety officer determines through ongoing assessment that the combustible nature of the environment represents a potential hazard.	
	4.12.10.1	An exercise stopped as a result of an assessed hazard shall continue only when actions have been taken to reduce the hazard. [And after the "Go/No Go" sequence]	

Safety

Yes/No	NFPA	Compliance Requirement	Comments
	4.4.1	A safety officer shall be appointed for all live fire training evolutions.	
	4.4.3	The safety officer shall have the authority, regardless of rank, to intervene and control any aspect of the operations when, in his or her judgment, a potential or actual danger, accident, or unsafe condition exists.	

(*Continues*)

(Safety Continued)

Yes/No	NFPA	Compliance Requirement	Comments
	4.4.4	The responsibilities of the safety officer shall include, but shall not be limited to, the following: (1) Prevention of unsafe acts (2) Elimination of unsafe conditions	
	4.4.5	The safety officer shall provide for the safety of all persons on the scene including students, instructors, visitors, and spectators.	
	4.4.6	The safety officer shall not be assigned other duties that interfere with safety responsibilities.	
	4.11.1	The safety officer and instructor-in-charge of the live fire training evolutions shall determine, prior to each specific evolution, the number of training attack lines and backup lines that are necessary.	
	4.11.2	Each hoseline shall be capable of delivering a minimum of 95 gpm (360 L/min).	
	4.11.3	Backup lines shall be provided to ensure protection for personnel on training attack lines.	
	4.6.4	The instructor-in-charge then shall assign the following personnel: (1) One instructor to each functional crew, which shall not exceed five students (2) One instructor to each backup line (3) Additional personnel to backup lines to provide mobility (4) One additional instructor for each additional functional assignment	
	4.6.7	Additional safety personnel, as deemed necessary by the safety officer, shall be located strategically within the structure to react to any unplanned or threatening situation or condition.	
	4.9.1	A method of fireground communications shall be established to enable coordination among the incident commander, the interior, the exterior, the safety officer, and external requests for assistance.	
	4.6.5	The instructor-in-charge shall provide for rest and rehabilitation of participants operating at the scene, including any necessary medical evaluation and treatment, food and fluid replenishment, and relief from climatic conditions. (See Annex D.)	

(Safety *Continued*)

Yes/No	NFPA	Compliance Requirement	Comments
	4.6.5.1	Instructors shall be rotated through duty assignments.	
	4.6.6	All instructors shall be qualified by the AHJ to deliver live fire training.	
	4.6.9	Instructors shall take a personal accountability report (PAR) when entering and exiting the structure or prop during an actual attack evolution conducted in accordance with this standard.	
	4.9.2	A building evacuation plan shall be established, including an evacuation signal to be demonstrated to all participants in an interior live fire training evolution.	A "Meet at" location to conduct a personnel accountability report (PAR) outside of the structure shall be at _____.
	4.10, 4.10.1	Emergency medical services shall be available on site to handle injuries with transport capabilities.	
	4.10.3	Written reports shall be filled out and submitted on all injuries and on all medical aid rendered.	
	5.1.10	A search of the structure shall be conducted to ensure that no unauthorized persons, animals, or objects are in the building immediately prior to ignition.	
	5.1.11	No person(s) shall play the role of a victim inside the building.	
	5.1.12	Only one fire at a time shall be permitted within an acquired structure.	
	4.5	The training session shall be curtailed, postponed, or canceled, as necessary, to reduce the risk of injury or illness caused by extreme weather conditions.	
	4.8.1	Each participant shall be equipped with full protective clothing and self-contained breathing apparatus (SCBA).	
	4.8.2	All participants shall be inspected by the safety officer prior to entry into a live fire training evolution to ensure that the protective clothing and SCBA are being worn properly and are in serviceable condition.	
	4.8.3	Protective coats, trousers, hoods, footwear, helmets, and gloves shall have been manufactured to meet the requirements of NFPA 1971, *Standard on Protective Ensemble for Structural Fire Fighting*.	

(*Continues*)

(Safety Continued)

Yes/No	NFPA	Compliance Requirement	Comments
	4.8.4	Self-contained breathing apparatus (SCBA) shall have been manufactured to meet the requirements of NFPA 1981, *Standard on Open-Circuit Self-Contained Breathing Apparatus for the Fire Service*.	
	4.8.5	Where station or work uniforms are worn by any participant, the station or work uniform shall have been manufactured to meet the requirements of NFPA 1975, *Standard on Station/Work Uniforms for Fire and Emergency Services*.	
	4.8.6	Personal alarm devices shall have been manufactured to meet the requirements of NFPA 1982, *Standard on Personal Alert Safety Systems (PASS)*.	
	4.8.2	All students, instructors, safety personnel, and other personnel shall wear all protective clothing and equipment whenever they are involved in any evolution or fire suppression operation during the live fire training evolution.	The "hot zone" area shall be marked to identify areas that PPE and SCBA are required.
	4.8.7	All students, instructors, safety personnel, and other personnel participating in any evolution or operation of fire suppression during the live fire training evolution shall breathe from an SCBA air supply whenever operating under one or more of the following conditions: (1) In an atmosphere that is oxygen deficient or contaminated by products of combustion, or both (2) In an atmosphere that is suspected of being oxygen deficient or contaminated by products of combustion, or both (3) In any atmosphere that can become oxygen deficient or contaminated, or both (4) Below ground level	
	4.7.11	One person who is not a student shall be designated as the "ignition officer" to control the materials being burned and will be a member of the fire control team.	
	4.7.4	The fire control team shall wear full protective clothing, including self-contained breathing apparatus (SCBA) when performing this control function.	
	4.7.5	A charged hoseline shall accompany the ignition officer when he or she is igniting any fire.	

(Safety *Continued*)

Yes/No	NFPA	Compliance Requirement	Comments
	4.7.2	The decision to ignite the training fire shall be made by the instructor-in-charge in coordination with the safety officer and incident commander.	
	4.7.3	The fire shall be ignited by the ignition officer/fire control team.	
		No one shall operate inside the structure alone including the ignition officer or safety personnel.	

Instructors

Yes/No	NFPA	Compliance Requirement	Comments
	4.6.6	All instructors shall be qualified to deliver fire fighter training per AHJ requirements.	
	4.6.4	The participating student-to-instructor ratio shall not be greater than 5 to 1.	
	4.6.7	Additional instructors shall be designated when factors such as extreme temperatures or large groups are present, and classes of long duration are planned.	
	4.6.2	The instructor-in-charge shall be responsible for full compliance with this standard.	
	4.6.8	Prior to the ignition of any fire, instructors shall ensure that all protective clothing and equipment specified in this chapter are being worn according to manufacturer's instructions.	Prior to the ignition of any fire, the instructor-in-charge shall conduct a "Go" in the "Go/No Go" sequence with all primary positions to insure readiness.
	4.6.9	Instructors shall take a head count when entering and exiting the building during an actual attack evolution.	
	4.6.10	Instructors shall monitor and supervise all assigned students during the live fire training evolution.	
	4.6.5	The instructor-in-charge shall provide for rest and rehabilitation of members operating at the scene, including necessary medical evaluation and treatment, food, and fluid replacement, and relief from climatic conditions.	

(Continues)

(Instructors *Continued*)

Yes/No	NFPA	Compliance Requirement	Comments
	4.6.3	Prior to the "final burn down" the building will be checked for personnel, equipment and conditions, PAR will be taken, and an audible "burn down" signal other than the emergency evacuation shall be sounded before ignition.	
	4.15.4	Assignments shall be made for all crews participating in the training session.	

Preburn Notification Contact Sheet

Property Location: _____

_____

Burn date: _____

Agency	Phone	Contact Name	Date Contacted
Property Owner			
Health Dept.			
Fire Chief/Local District Chief			
Affected Law Enforcement Agency			
Affected Neighbors/Owners			
Risk Management			
Public Information Officer (PIO)			
Hydrant Use			
Environmental Regulator			

Sample Participant Duties and Assignments

Instructor-in-Charge

[*Note: NFPA 1403 refers to both an instructor-in-charge and an incident commander. This sample plan refers to them both as instructor-in-charge to avoid conflict or confusion. This is a decision that needs to be made by the AHJ.*]

Name: _____

Duties:
- Assist and advise the safety officer.
- Identify the method of fireground communications between the instructor-in-charge, the interior, the exterior, the safety officer, and external requests for assistance.
- Assist and advise the instructional staff.
- Assist and advise the local or host fire department administration.
- Monitor and document weather conditions to ensure safe operations.
- Confirm the completion of all forms and records prior to training.
- Plan and coordinate all training activities.
- Assess the selected room environment for factors that can affect the growth, development, and spread of fire.
- Ensure the following personnel are made:
  - One instructor to each functional crew, which shall not exceed five students
  - One instructor to each backup line
- Ensure the ignition of the training fire is made by the instructor-in-charge in coordination with the safety officer and incident commander.
- Determine the number of training attack lines and backup lines that are necessary.
- Determine the rate and duration of water flow necessary for each individual live fire training evolution, including the water necessary for control and extinguishment of the training fire, the supply necessary for backup lines to protect personnel, and any water needed to protect exposed property.
- Determine a minimum reserve of additional water in the amount of 50 percent of the fire flow demand determined shall be available to handle exposure protection or unforeseen situations.
- Provide for rest and rehabilitation of members operating at the scene, including necessary medical evaluation and treatment, food, and fluid replacement, and relief from climatic conditions.
- Document fuel loading including all of the following:
  - Furnishings
  - Wall and floor coverings and ceiling materials
  - Type of construction of the structure including type of roof and combustible void spaces
  - Dimensions of the room
- Monitor activities to ensure safe practices.
- Inspect building prior to each fire evolution.
- Assign burn instructors and instructor trainees.
- Brief instructional staff on their responsibilities.
- Assign coordinating personnel as needed.
- Ensure that this document is followed by all in the training area.

Safety Officer

Name: _____

Deputy safety officer name: _____

Duties:
- Prevent all unsafe acts.
- Eliminate all unsafe conditions.
- Intervene and terminate, if needed, any unsafe acts or condition.
- The safety officer shall not be assigned other duties that interfere with safety responsibilities.
- Locate additional safety personnel within the structure to react to any unplanned or threatening situation or condition.
- Supervise additional safety officer personnel as needed.
- Conduct a search of the structure to ensure that no unauthorized persons, animals, or objects are in the building immediately prior to ignition.
- Ensure the fire is ignited by the ignition officer in the presence of and under the direct supervision of the safety officer.
- Inspect all participants prior to entry into a live fire training evolution to ensure that the protective clothing and SCBA are being worn properly and are in serviceable condition.
- Ensure all spectators shall be restricted to an area outside the operations area perimeter.
- Ensure an accurate accounting at all times of all participants.
- Ensure for the safety of all persons on the scene including students, instructors, visitors, and spectators.
- Intervene and control any aspect of the operations where a potential or actual danger, accident, or unsafe condition exists.

Ignition Officer/Fire Control Team

NAME: _____   NAME: _____
NAME: _____   NAME: _____

Duties:
- Confirm the completion of all forms and records of this document prior to training.
- Ensure water supply is sufficient for expected fire involvement.
- Provide ignition and monitor fire involvement during entire training drill.
- Ensure there is only one fire at a time within an acquired structure.
- Light the fire in the presence of and under the direct supervision of the safety officer.
- Wear full protective clothing, including self-contained breathing apparatus (SCBA).
- Ensure a charged hose line accompanies the ignition officer when he or she is igniting any fire.

Instructors

NAME: _____   NAME: _____
NAME: _____   NAME: _____
NAME: _____   NAME: _____
NAME: _____   NAME: _____

Duties:
- Monitor and supervise assigned students (no more than 5 at a time).
- Account for assigned students before and after evolution.
- Familiarize students with building layout.
- Familiarize students with location of safe zone and critique area.
- Take a head count when entering and exiting the building during an actual attack evolution.
- Eliminate all unsafe conditions.
- Report all possible unsafe conditions to safety officer.
- Inspect students protective clothing and equipment.
- Monitor and supervise all assigned students during the live fire training evolution.
- Ensure all students have portable radio.
- Inform safety officer when ready for ignition of materials.
- Ensure that all protective clothing and equipment specified in this chapter are being worn according to manufacturer's instructions.
- Instruct students on tactical and assign objectives.
- Critique students after each team evolution.

Instructor Trainees

NAME: _____   NAME: _____
NAME: _____   NAME: _____
NAME: _____   NAME: _____
NAME: _____   NAME: _____

Duties:
- Monitor for the purposes of learning, and assist the instructor.
- Eliminate all unsafe conditions.
- Report all possible unsafe conditions to your assigned instructor or safety officer.
- Inspect participants' protective clothing for safety.
- Monitor, for the purposes of learning, during the critique of students after each team evolution.

Students

Duties:
- Acquire prerequisite training.
- Provide documentation on prerequisite training if from outside local participating departments.
- Report all possible unsafe conditions to safety officer.
- Familiarize yourself with structure and building layout.
- Wear fully-approved protective clothing.
- Use portable radios.
- Wear approved self contained breathing apparatus.
- Obey all instructions from your instructor and safety rules of the training area.

Sample Preburn Plan

Location

Intended day(s) of training exercises, specify types of exercises for each day (forcible entry, search and rescue with smoke, etc.) and what (days) live fire exercises will occur.

1. Specific objectives for the training exercise
   - No new objectives or evolutions without briefing of all instructors and approval by instructor-in-charge and safety officer, and only within parameters approved by the fire chief.
2. Water supply requirements: identify flow, secondary source etc.
3. Apparatus needs: locate on plan what pumpers, tenders or other units are needed and their positioning. Be prepared for the wind to be different on the day of the fire.
4. The building plan
   - Show dimensions of the building, all rooms, windows and doors with all features of the training areas inclusive of primary and emergency ingress/egress needs.

- Number each room that will have burn evolutions and in what sequence they will be burned on the form and in the structure itself.
- Indicate fuel loading.
- Mark the sides of the actual building per local protocol (Side "A, B, etc.") and number each door and window with the exit letter/number designation. (A-1 is side A window 1). Make sure actual marking inside and outside the building are easily seen and match the plan.

5. The site plan: Show area measurements, the structure, command post, rehab/medical area, operations area, staging area (see further below), placement of engines, parking area, etc.
6. Parking and areas of operations
7. Emergency plans
   - RIC: staffing, equipment, and deployment
   - Mayday procedure
   - Lost communication with interior
   - Evacuation signal/personnel accountability report (PAR)/Meet at location
   - Loss of one of the two water supplies
8. Any time the instructor-in-charge determines that fuel, fire or any other condition represents a potential hazard, the training exercise shall be stopped immediately. If an exercise is stopped, it should only be restarted once the hazard identified has been resolved. and after the "Go/No Go" sequence.
9. Non-exercise emergencies
   - Plan for on-site emergencies not exercise related
   - Plan for emergency cessation of evolution for off-site response of personnel
10. Weather: acceptable parameters, conditions requiring temporary cessation (i.e. lightening).
11. List of the evolutions
12. Order of operations: Any changes must be included in the briefing to all of the instructors and participants.
    - Initial briefing of activities and safety brief for preignition activities and assignments.
    - Complete set up for operations, rehab, etc.
    - Initial PPE inspection
    - Instructor briefing and walk-through by the instructor-in-charge and the safety officer with clear objectives tied to specific evolutions.
    - Participants briefing and walk-through by the instructor-in-charge and the safety officer with clear objectives tied to specific evolutions.
    - The preburn briefing shall include the location of the burn sets, sequence of burns and which order the rooms will be burned with the rooms numbered, the exit markings, ventilation points and the primary and secondary means of egress.
    - All participants need to know the operations area where all PPE is to be worn, and the fireground (warm zone) where a minimum of a helmet may only be required.
    - Assignments given to personnel reporting to their positions.
    - Flow all lines with proper pressures.
    - Instructors and students in place for a Go/No Go sequence.
    - Ignition
13. The emergency medical plan
    - On scene resources
    - Calling for assistance
    - Landing zone location and procedures for helicopter ambulance
14. Communications plan
15. Staffing and organization: ICS chart, instructors and assignments, participant teams, etc.
    - Staff and participant rotation
    - Instructor to participant ratio
    - Accountability
16. Demobilization plan

Preburn Site Plan

[Insert property address]
- Mark North on plan
- Identify sides of building
- Mark exposures
- Show: fence lines, power lines
- Hazards
- Hydrants

[Insert prepared by]
- Locate command, RIC, rehab, primary and secondary pumper, water supply, hot zone air cylinder refill, EMS unit, staging, parking for apparatus, attendees, etc.
- Take extra copies without above in case of weather caused changes

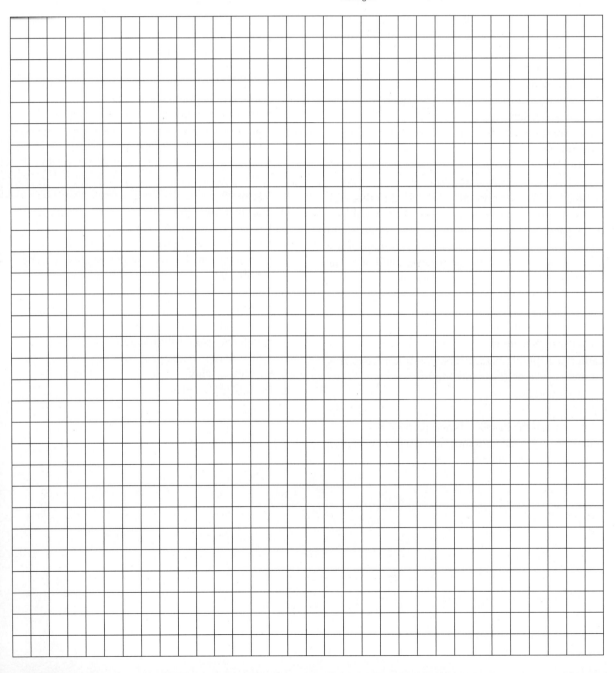

Appendix C

Preburn Floor Plan

[Insert property address]
- Mark North on plan
- Identify sides of building
- Dimensions of all rooms & building

- Show and number all doors, windows means of egress
- Number each room that will have burn evolutions, show sequence

[Insert prepared by]
- Location of fire sets
- Hazards
- Fuel loading

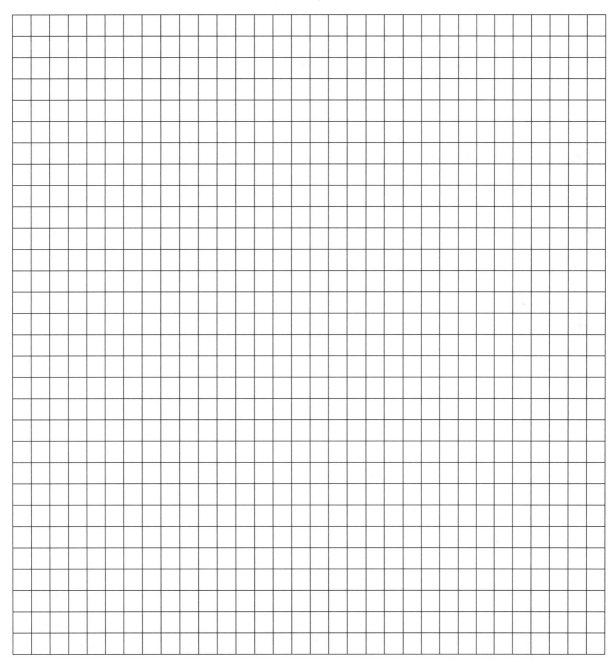

Permanent Structure Live Fire Training Model SOP

NOTE to users: Italic type signifies notations or comments for agencies utilizing information from this sample for their own agency use. This sample policy covers both class A- and B-fueled structural props. Utilize the information that is appropriate to your local facilities. When information is specific to a type of prop either by fuel used or construction, it so noted in [square brackets] or a text box so the local user can choose the language appropriate to their prop(s).

Purpose

This standard operating procedure is for the training of personnel engaged in structural firefighting operations under live fire conditions using the purpose-built structural prop at the _____ facility. This policy focuses on training for aggressive, coordinated interior firefighting operations with a minimum exposure of risk for the participants. Training fires present some of the same hazards as those encountered at actual fire-ground incidents and compliance to this policy and NFPA 1403 will reduce the danger to participants during high-hazard training. The Incident Management System (IMS) employed at actual fire incidents will be the Standard Operating Procedure at *ALL* structural training fires. All live fire-training evolutions shall be planned with great care and supervised closely by instructional personnel. All departmental policies shall be followed, including the use of Command and Accountability Tactical Boards, forms, job aids, and adopted live fire training–specific forms and checklists. (*See sample Burn plan for forms and Job Aids.*)

Preburn Information

Training Structures/Props

The permanent live fire training structural props utilized by _____ are [*engineered props that have been designed specifically for the purpose of repeated live fire training evolutions and include safeguards which only become unacceptable hazards through misapplication of use or improper maintenance. or locally built props utilizing shipping containers that do pose many of the hazards of an acquired structure*]. The fuel utilized to create the fire in the structural prop is [*propane, compressed natural gas, class A (agency-approved wooden pallets, straw, hay, or excelsior)*]

[*All props that use pressure to move fuel to the fire shall be equipped with remote fuel shutoffs. The remote fuel shutoff shall be within sight of the prop and the entire field of attack for the prop, but shall be outside of the safety perimeter. During the entire time the prop is in use, the remote shutoff shall be continuously attended by safety personnel trained in its operation.*]

Preparation of Training Center Burn Buildings

A thorough inspection should be conducted by the Instructor-in-Charge and the Safety Officer before each use. All doors, windows, roof scuttles [*and automatic ventilators, mechanical equipment, lighting, manual or automatic sprinklers, and standpipes necessary*] for the live fire training evolutions shall be checked [*and operated, where appropriate,*] prior to any live fire training evolution to ensure correct operation. Access and egress paths, along with all interior floors, need to be checked for debris that could injure participants or impede the operation.

[*Instructors responsible for conducting live fire training evolutions with a gas-fueled training system shall be trained properly in the complete operation of the system. The training of instructors shall be performed by an individual authorized by the system manufacturer.*]

Provisions for participant and instructor rehab shall be verified.

[*For Class A props, the following five items and the Burn Sequence chart must be defined:*

1. Fuel loads for each evolution conducted in the room (see "Sequential Live Fire
2. Burn Evolutions," NFPA 1403.7.3. Sequential evolutions will heat the room or building and the amount of fuel will need to be reduced).
3. The maximum number of evolutions for the room in a given training period before allowing the room to cool off (see "Sequential Live Fire Burn Evolutions," NFPA 1403.7.3).
4. Placement and configuration of fuel (are burn racks used or can the fuel be stacked on the floor, or leaned against walls? AHJ must clearly define beyond just fire loading).
5. Ignition procedures (devices and materials used, specify safe removal if LP "torches" are used—more follows in "Ignition Officer" section).
6. Ventilation techniques and/or issues (define to what degree the fire room is ventilated during ignition and operations).

APPENDIX D

7. Burn Sequence Chart shall be developed specific to your prop and followed (see "Sequential Live Fire Burn Evolutions," NFPA 1403.7.3. Room designations may follow the sides of the building designations or use clearly identified alpha/numerical designations. Sample is a basic four-room prop with a center hallway). The amount of fuel used will need to be determined by experimentation.

Policy should clearly allow for <u>less</u> fuel loading based on weather or interior heating, but should also as clearly state loading location, quantity or content shall not exceed the policy as developed.]

Burn Sequence Chart for _____ Training Center Structural Live Fire Prop
Maximum Fire Loads and Maximum Number of Evolutions Per Room

	Room Alpha/Bravo	Room Bravo/Charlie	Room Bravo/Charlie	Room Alpha/Delta
Burn Sequence 1	3 pallets 1 bale	3 pallets 1 bale	2 pallets 1 bale	3 pallets 1½ bales
Burn Sequence 2	2 pallets 1 bale	2 pallets 1 bale	1 pallet ½ bale	3 pallets 1 bale
Burn Sequence 3	1 pallet ½ bale	2 pallets 1 bale	End use or allow 30-minute cool down	2 pallets 1 bale
Burn Sequence 4	End use or allow 30-minute cool down	End use or allow 30-minute cool down		1 pallet ½ bale
				End use or allow 30-minute cool down

- Specify pallet and bale size and bale material (hay/straw or excelsior)
- Estimate time intervals between each room being used
- NOTE: In the above example, the rooms are not the same size, hence some will need to be cooled sooner than others

Class A Burn Sequence Information

Burn Sequences: The following facts will affect conditions encountered in a burn room during a live fire evolution:

(1) Larger burn rooms and rooms with higher ceilings will have more cubic feet of air than smaller burn rooms.
(2) Generally, with a given quantity of fuel, the lower the cubic footage in a room, the higher the temperatures and more rigorous the environment will be.
(3) As the number of openings in a burn room increase, the available ventilation area increases, resulting in typically lower temperatures and less severe environments.
(4) The construction of the burn room will affect how much energy the room will retain with each successive evolution. All burn rooms will retain a level of heat with each burn. The temperature and radiant heat in the burn room will increase with each additional evolution. At some point, every room will become too hot to safely conduct further training. Outside environmental conditions might also affect this.

Preparation for Evolutions

A written burn plan shall be prepared that clearly states the evolution objectives, emergency plans (including evacuation signal, emergency medical attention, communications, emergency cessation of operations due to unrelated issue {weather, off-site emergency, etc.}), the communications plan, accountability, the list of the evolutions, and the Order of Operations.

Any room used for fire suppression should have two means of egress. Extreme caution must be used with any rooms not having two means of egress.

All instructors and support staff shall be fully briefed on the objectives, plan, and their assignments. All instructors, support staff, and participants shall participate in the briefings and walkthroughs.

The evacuation signal will be demonstrated and the rally point identified. (NOTE: NFPA 1403 does not require a rally point; however it is an easy safety step to institute in case of an evacuation, where all personnel should gather.) Participants will be assigned to crews for accountability and advised of their assignments.

Order of Operations

The "Order of Operations" is the steps in proper sequence to conduct the live fire evolution and is major portion of the Burn Plan. It serves as a script for the exercises. It is determined after the list of the evolutions is drawn up. See LFTPP Chapter 6 for more detail. On the day of the evolution, such an order includes:

a. Setup according to the site plan, unless conditions dictate a change. <u>Any changes must be included in the briefing to all of the instructors and students.</u>
b. Briefing of all participants (the instructors first, followed by the students), with clear objectives and a consensus understanding. The location of the burn sets, sequence of burns with the rooms numbered, exit markings, ventilation points, and the primary and secondary means of egress should be discussed. Review emergency plans.

Fuel Loading

Class A Fuels

The fuels that are utilized in live fire training evolutions shall be of known burning characteristics of such a nature to be as controllable as possible. Unidentified materials, such as debris found in or around the structure or prop, or any materials of undetermined composition, which may burn in unanticipated ways, react violently, or create environmental or health hazards, shall not be used. Materials shall be used in only the amounts necessary to create the desired-size fire. No flammable or combustible liquids of any type shall be used during interior structure training evolutions.
NOTE: Acceptable Class A materials include straw, wooden pallets, hay, pine excelsior, and other ordinary combustibles. A reasonable effort should be made to ascertain that straw or hay, if used, has not been treated with pesticides or other harmful chemicals.
The use of flammable or combustible liquids, as defined in NFPA 30, *Flammable and Combustible Liquids Code*, shall be prohibited in live interior structure fire training evolutions.

Follow the manufacturer's directions for the operation of Class B props. Do not add Class A materials.

Water Supply

[NFPA 1403 no longer requires gas-fueled props to meet NFPA 1142, *Standard on Water Supplies for Suburban and Rural Fire Fighting*. Individual attack hose lines need to be capable of flowing 95 gpm.]

The minimum water supply and delivery for the live fire training evolutions shall meet the criteria identified in NFPA 1142, *Standard on Water Supplies for Suburban and Rural Fire Fighting*. [Currently 1403 does not differentiate between acquired structures and a permanent live fire training structural prop that utilizes class A fuel. As of this writing, there is consideration by the 1403 Committee to reduce the fire flow requirements to only the fuel itself. If that happens, a Technical Interpretation to that effect will be posted and the standard revised during its normal cycle.]

A minimum reserve of additional water in the amount of 50 percent of the fire flow demand shall be available to handle exposure protection or unforeseen situations.

Separate sources shall be identified for supply of attack lines and backup lines in order to preclude the loss of both water supply sources at the same time [per 4.11.6.1, *a single water source is acceptable if a backup power source or backup pump is provided to ensure an uninterrupted supply of water*].

NOTE: Reliability should be considered when determining what constitutes a separate source. The intent of this section is to prevent the simultaneous loss of both attack lines and backup lines in the event of a pump or water supply failure.

Initiation of Evolutions

1. The water supply shall be verified prior to ignition. All hose lines shall be tested for flow and pressure, and to ensure that they are in proper position and staffed.
2. Communications onscene and to the local dispatch center shall be tested.
3. Rehab shall be in place.
4. Final walk through by the Safety Officer and the Instructor-in-Charge shall be completed.
5. The _____ Training Center shall utilize the Go/No-Go sequence to initiate fire exercises for each ignition. This rapid process provides each operational crew/functional unit a formal step to advise they are ready and allows for each unit to cross check other units they have visual contact with.
    a. By radio, the Instructor-in-Charge says: "All personnel stand by for a Go/No-Go." Start with the support group and then move on to the attack group.

From the Instructor-in-Charge	Reply by Group
Medical: Go/No-Go?	Medical: Go
Rehab: Go/No-Go?	Rehab: Go
Engine 1: Go/No-Go?	Engine 1: Go
Engine 2: Go/No-Go?	Engine 2: Go
Entry: Go/No-Go?	Entry: Go
RIT: Go/No-Go?	RIT: Go
Backup: Go/No-Go?	Backup: Go
Attack: Go/No-Go?	Attack: Go
Ignition? Go/No-Go?	Ignition: Go
Safety: Go/No-Go?	Safety: Go

    b. Prior to the Safety Officer giving a "Go," he or she must inspect the structure to make sure it is clear of any occupants and that personnel are ready.
        (1) Personnel (students and instructors) are ready. This includes protective clothing and breathing apparatus inspection, covered in detail in Chapter 10 of this text (protective clothing, self-contained breathing apparatus, PASS, inspections, uniform apparel).
        (2) Another example of a "No-Go" would be the safety officer observing students still donning their equipment. Even if all of the positions had advised "Go," the safety officer stops the process
    c. All positions share the responsibility and can stop the process. Any position seeing less than 100 percent preparedness needs to give a "No-Go" report.
    d. Any time operations are shut down, a "Go/No-Go" sequence should be repeated before continuing the live fire evolution.
    e. After the Safety Officer advises a "Go," the Instructor-in-Charge would declare "We have a Go for ignition."
    f. The "STOP DRILL" order by the Instructor-in-Charge or Safety Officer shall be repeated three times consecutively to stop the evolutions. This is <u>not</u> the same as the evacuation order.

Emergency Plan

In case of an emergency involving interior operations, normal fire-ground procedures will be utilized to deploy the Rapid Intervention Team and backup line if appropriate.

Likewise, the order to evacuate the building (Prop) will follow normal procedures, with a PAR taken upon the evacuation of personnel.

Should there be a non-fire-related onsite emergency, the Instructor-in-Charge will determine whether or not suspend drill operations. The safety of all participants is always the priority.

Should weather conditions change, the Instructor-in-Charge, in consultation with the Safety Officer, will determine whether or not to suspend drill operations.

If units that are committed to training are needed for coverage or response, the Instructor-in-Charge, in consultation with the Safety Officer, will determine whether to continue with fewer participants or to suspend drill operations. All primary positions at drills must remain staffed. The Instructor-in-Charge will make the determination to extinguish the fires or what normal demobilization processes will be completed.

Command Structure, Duties, and Assignments

The live fire evolution will be operated under the Incident Management System. The Instructor-in-Charge serves as the Incident Commander and is responsible for completing any required paperwork and for following NFPA 1403. *NOTE: This may vary locally, with some agencies having an Incident Commander, who has overall responsiblity for all facets of the training exercise, and an Instructor-in-Charge, who directly oversees the training component of the exercises and reports to the Incident Commander.* The Command Structure, Duties, and Assignments Form will be used to document assignments and responsibilities.

The assignments for any live fire evolution <u>require</u> the following positions to be filled:
- Instructor-in-Charge
- Safety Officer
- Ignition Officer/Fire Control Team
- Instructors

Instructor-in-Charge

The Instructor-in-Charge utilizes the training center's normal Burn Plan with regard to emergency plans, communications plan, fuel loading, rehab, evacuation signal, etc. Fuel loading, rotations, etc., shall not exceed this policy.

The Instructor-in-Charge will ensure a preburn briefing and walkthroughs are conducted. All evolutions to be conducted and their objectives shall be discussed and assignments shall be made for all crews participating in the training session.

The training exercise shall be immediately stopped if the Safety Officer or Instructor-in-Charge determines, through continuing assessments, that the combustible nature of the environment represents a potential hazard. The exercise shall continue only when the appropriate actions have been taken to reduce the hazard. The Instructor-in-Charge s is responsible for establishing the method to communicate between command, interior divisions, exterior divisions, the Safety Officer, and external requests for assistance.

The Instructor-in-Charge will get a PAR as each crew enters the live fire training structural prop and upon exiting the same.

The Instructor-in-Charge shall then:

1. Assign one (1) instructor to each functional crew, which shall not exceed five (5) students.
2. Assign one (1) instructor to each a backup line(s).
3. Assign sufficient additional personnel to backup lines to provide mobility.

Safety Officer

A Safety Officer shall be appointed for all live fire training evolutions. The Safety Officer will be a qualified Burn Instructor. The Safety Officer shall have the authority, regardless of rank, to intervene and correct any aspect of the operations when, in their judgment, a potential or real danger, accident, or unsafe condition exists. The Safety Officer shall be responsible for the safety of all persons on the scene, including students, instructors, visitors, and spectators.

The Safety Officer shall be knowledgeable in the operation and location of safety features available within the burn building, such as emergency shutoff switches, gas shutoff valves, and evacuation alarms. The Safety Officer shall not be assigned other duties inconsistent with safety responsibilities.

Additional safety personnel, as deemed necessary by the Safety Officer, shall be strategically placed within the structure to react to any unplanned or threatening situation or condition.

Emergency medical care shall be available on site to handle any injuries. Written reports shall be made on all injuries and on all medical aid rendered.

<u>No person(s) shall be placed inside the building or prop to play the role of victim.</u> No "rescue dummy" shall have bunker gear on it.

Ignition Officer/Fire Control Team

One person shall be designated as the Ignition Officer to control the materials being burned and to ignite the training fire in the presence of and under the direct supervision of the Safety Officer. This person shall be a member of the Fire Control Team (added NFPA 1403, 4.7.1.1.1, 2012 ed.) and wear full protective clothing, including self-contained breathing apparatus. A charged hose line shall accompany the Ignition Officer when igniting fires. The decision to ignite the training fire shall be made by Ignition Officer in coordination with the Safety Officer. The Fire Control Team shall monitor the fire and take actions as necessary to control the fire and protect the participants.

Awareness of weather conditions, wind velocity, and wind direction shall be maintained at all time. In all cases, immediately before actual ignition, a final check shall be made for changes in weather conditions.

Instructors

All Instructors shall be deemed qualified to deliver structural firefighting training by the Training Center staff (*and applicable state/municipal/provincial requirements*). Instructors shall verify that all participants are accounted for through a PAR, both when entering and exiting an actual attack evolution conducted in accordance with this document.

NOTE: Instructors should meet the criteria outlined in NFPA 1041, *Fire Service Instructor Professional Qualifications*, for Level II Instructor or higher.

The participating student–instructor ratio shall not be greater than five (5) to one (1). It is important that the participating student–instructor ratio be monitored so as to not exceed the span of control necessary to provide for the safe and proper supervision of trainees.

In addition, Instructor Trainees may be present and utilized, but may not directly supervise students.

Protective Clothing and Equipment

Each participant involved in live fire operations shall be equipped with full protective clothing and self-contained breathing apparatus (SCBA). All participants shall be inspected by the Safety Officer to ensure the protective clothing and SCBA are being properly worn prior to entry into a live fire training evolution.

All protective equipment shall meet the requirements of the current NFPA standards. Clothing worn under PPE should meet the requirements of NFPA 1975, *Standard on Station/Work Uniforms for Fire Fighters* or at least be selected for the fabrics' ability to resist ignition or melting. Fire retardant fabrics and all natural fibers should be given consideration. Synthetic materials should not be worn.

Personal alarm devices shall be used by all participants, and the devices shall meet the requirements of NFPA 1982, *Standard on Personal Alert Safety System (PASS) for Fire Fighters*. and OSHA regulations.

All students, instructors, safety personnel, and other personnel shall wear all protective clothing and equipment whenever these persons are involved in any training evolution or fire suppression operation.

All students, instructors, safety personnel, and other personnel participating in any evolution or fire suppression operation during the live fire training evolution shall breathe from the SCBA air supply whenever one or more of the following conditions exist:

- Operating in an atmosphere that is oxygen-deficient, contaminated by products of combustion, or both.
- Operating in an atmosphere that is suspected of being oxygen-deficient, contaminated by products of combustion, or both.
- Operating in any atmosphere that may become oxygen-deficient, contaminated, or both.
- Operating below ground level or in a confined space, as defined by OSHA.

No one should be allowed to breathe smoke, toxic vapors or flames, products of combustion, or any other deficient atmosphere.

INSTRUCTOR-IN-CHARGE
NAME: _____

- Confirm the presence and completion of all forms and records prior to training.
- Plan and coordinate all training activities.
- Ensure the following personnel are in place:
  - One instructor to each functional crew, which shall not exceed five students.
  - One instructor to each backup line.
  - Additional personnel to backup lines to provide mobility.
  - One additional instructor for each additional functional assignment.
- Ensure the ignition of the training fire is made in coordination with the Safety Officer.
- Determine the number of training attack lines and backup lines that are necessary.
- Determine the rate and duration of water flow necessary for each individual live fire training evolution, including the water necessary for control and extinguishment of the training fire, the supply necessary for backup lines to protect personnel, and any water needed to protect exposed property.
- Determine a minimum reserve of additional water in the amount of 50 percent of the fire flow demand determined shall be available to handle exposure protection or unforeseen situations. *(Not required for Class B fueled props)*
- Provide for rest and rehabilitation of members operating at the scene, including necessary medical evaluation and treatment, food, fluid replacement, and relief from climatic conditions.
- Monitor activities to ensure safe practices.
- Inspect building prior to each fire evolution.
- Assign Burn Instructors and Instructor Trainees.
- Brief instructional staff on their responsibilities.
- Assign coordinating personnel as needed.
- Ensure that this policy is followed by all in the training area.

SAFETY OFFICER(S)
NAME: _____
NAME: _____

- Monitors operations for unsafe acts and takes necessary steps to correct any noted.
- Monitors conditions.
- Intervenes and terminates, if needed, any unsafe acts or conditions.
- The safety officer shall not be assigned other duties that interfere with safety responsibilities.
- Locate additional safety personnel within the structure to react to any unplanned or threatening situation or condition.
- Supervise additional safety officer personnel as needed.
- Conduct a search of the structure to ensure that no unauthorized persons, animals, or objects are in the building immediately prior to ignition.
- Supervise the fire ignition by the Ignition Officer/Team, or ensure that the Ignition Officer/Team is under the direct supervision of the designated deputy Safety Officer, with a hose line present for protection *(hose line for Class A props only)*.
- Inspect all participants prior to entry into a live fire training evolution to ensure that the protective clothing and SCBA are being worn properly and are in serviceable condition.
- Ensure all spectators are restricted to an area outside the operations area perimeter.
- Ensure an accurate accounting at all times of all participants.
- Ensure for the safety of all persons on the scene, including students, instructors, visitors, and spectators
- Intervene and control any aspect of the operations where a potential or actual danger, accident, or unsafe condition exists

<u>Takes Direction From:</u>
Instructor-in-Charge

IGNITION OFFICER/TEAM
NAME: _____
NAME: _____
NAME: _____
NAME: _____

- Ensures water supply is sufficient for expected fire involvement.

- Provide ignition for and monitors fire involvement during entire training drill.
- Lights the fire in the presence of and under the direct supervision of the Safety Officer or his/her designated deputy Safety Officer.
- Wears full protective clothing, including self-contained breathing apparatus (SCBA)
- Ensures a charged hose line accompanies the Ignition Officer when he or she is igniting any fire.

Takes Direction From:
Instructor-in-Charge, Safety Officer

INSTRUCTORS
NAME: _____
NAME: _____
NAME: _____
NAME: _____
NAME: _____
NAME: _____
NAME: _____
NAME: _____

- Monitor and supervise assigned students (no more than five at a time).
- Account for assigned students before and after evolution.
- Familiarize students with building layout.
- Familiarize students with location of safe zone and critique area.
- Take a head count when entering and exiting the building during an actual attack evolution.
- Eliminate all unsafe conditions.
- Report all possible unsafe conditions to Safety Officer.
- Inspect students for proper donning of PPE and SCBA.
- Monitor and supervise all assigned students during the live fire training evolution.
- Ensure all students have portable radio.
- Inform Safety Officer when ready for ignition of materials.
- Ensure that all protective clothing and equipment specified in this chapter are being worn according to manufacturer's instructions.
- Instruct students on tactical and assign objectives.
- Critique students after each team evolution.

Takes Direction From:
Instructor-in-Charge, Safety Officer, Ignition Officer/Team

INSTRUCTOR TRAINEE(S)
NAME: _____
NAME: _____
NAME: _____
NAME: _____
NAME: _____
NAME: _____
NAME: _____
NAME: _____

- Monitor for the purposes of learning, and assist your assigned instructor.
- Assist fire sets under the direction of the Ignition Officer/Team.
- Eliminate all unsafe conditions.
- Report all possible unsafe conditions to your assigned Instructor or the Safety Officer.
- Inspect students for proper donning of PPE and SCBA.
- Monitor, for the purposes of learning, during the critique of students after each team evolution.

Takes Direction From:
Instructor-in-Charge, Safety Officer, Ignition Officer/Team, Instructors

STUDENTS/PARTICIPANTS
- Acquire prerequisite training.
- Provide documentation on prerequisite training if from an outside department.
- Report all possible unsafe conditions to the Safety Officer.
- Be fully familiarized with structure and building layout.
- Properly wear approved protective clothing and self-contained breathing apparatus.
- Obey all instructions from your instructor and the safety rules of the training area.

Takes Direction From:
Instructor-in-Charge, Safety Officer, Ignition Officer/Team, Instructors

APPENDIX E

PERMANENT STRUCTURE LIVE FIRE TRAINING BURN PLAN FOR _____ FIRE TRAINING CENTER

Procedures and specifics must be revised for local use to reflect local procedures, facilities, etc. Much of the form's contents can be reused or simply revised each time to reflect changes in participants, objectives, etc. Content in **blue** indicates sample content for local completion of that section—it is not part of the form.

Date:	Time:	Prop:
Participants (agency/company/class); attach roster		
Instructor-in-Charge/Incident Commander:		Safety Officer:
Additional Safety Personnel:		
Fire Control Team: (this may change)		
NOTE: Standard accountability and tactical boards will be utilized to maintain assignments and personnel accountability.		

OBJECTIVES FOR EVOLUTIONS:	**Participant Level:** In-service fire companies

1. Refresher for Accountability system.
2. Visualize airflow from front door to burn room utilizing a thermal imager (TI).
3. Refresher for TI use and analysis of readings. Obtain surface temperatures in burn room and hallway, differentiate heat levels.
4. Visualize changes in fire conditions caused by ventilation.
5. Demonstrate hose handling, stream control.
6. Suppression with direct attack, cooling temperatures above thermal balance toward ceiling.

Safety Plan

Signify each of the below has been completed and documented
- ❏ Instructor-in-Charge and Safety Officer have inspected the site and building:
- ❏ Weather conditions checked (copy of forecast from Internet attached):
- ❏ Prop/Grounds Inspection by Instructor-in-Charge and Safety Officer:
- ❏ Assignments given:
- ❏ Assignments given to Deputy Safety Officers:
- ❏ Briefing given:
- ❏ Walkthrough:
- ❏ Safety Plan approved and briefed by Safety Officer:
- ❏ RIT staffed:
- ❏ Backup hose line in place, readied, and charged (minimum 95 gpm), flow checked:
- ❏ Attack team hose lines in place, readied, and charged, flows checked:

Communications Plan

- ❏ Check radio communications inside/outside prop, confirm with communications center.
- ❏ All drill ground communications shall be on channel ___.
- ❏ Emergency channel: ____.
- ❏ All instructors shall be issued radios. A minimum of one radio per team should be issued if there are not enough radios for all participants.
- In case of emergency at the drill ground, Dispatch shall be notified (specify if by radio or, for Training Centers with a fire department, by cell phone). One radio shall monitor Dispatch or the channel assigned for responding units.

MEDICAL PLAN	Medical Aid Station Location	
Level of Care ❏ ALS ❏ BLS		**Transport Provider**

Medical Emergency Procedures
1. Participants shall have their blood pressure and pulse checked and documented prior to operations.
2. Care for minor injuries will be provided by the onsite medical team with proper documentation.
3. For injuries or medical emergencies requiring transport: initial care will be provided by onsite medical team, EMS will be requested.

Landing Zone location: (address and national grid coordinates)

Fire Dept standby required at LZ?   ❏ yes   ❏ no

REHAB PLAN	Rehab Station Location

Rehab Procedures:
1. Participants and instructors shall remove their bunker coats and pants after exiting the prop and hot zone during Rehab.
2. Participants shall rehydrate and cool down.
3. Rehab shall check participant blood pressure and heart rates, monitor for heat-related issues.
4. Rehab shall notify Medical of participants who are hypertensive, have high heart rates that do not resume normal rates, or exhibit signs of heat emergencies or other medical concerns/injuries.

ORDER OF OPERATIONS

Order of Operations can be revised by Instructor-in-Charge with acknowledgement by all instructional and support personnel.

1. Instructor-in-Charge and Safety Officer shall inspect site and building.
2. All participant and staff PPE will be inspected.
3. After initial briefing, attack, backup, and supply hose lines will be set up and checked. Medical, Rehab, and support positions (SCBA refill, staging, etc.) will be readied.
4. NOTE: All personnel will utilize (specify your system) accountability system during interior operations. Each crew (Attack, Ignition, etc.) will advise PAR with number of personnel when entering and upon exiting, once outside.
5. Following participant briefing and crew assignments, Attack Team 1 and 2 will be assigned, comprised of 3 fire fighters and an Instructor each. Attack Teams 3 and 4 will be assigned and stand by without PPE donned. RIT will be assigned *(if participants are students and not experienced fire fighters, RIT should be staffed by experienced fire fighters and led by an officer, depending upon prop)*. Fire Control Team/Ignition Team will have already been assigned and, with the Safety Officer, will check the prop. Once the Attack and RIT teams advise Command, a Go/No-Go sequence will be initiated. Once a Go is given:
6. Evolution 1: Fire in room Bravo/Charlie. (Follow sequence chart for fuel loading.) Designated attack team will enter through door on Alpha side. A fire fighter will remain at the outside door for door control. The attack team will proceed into the hallway and close burn room door. Perform controlling airflow at the outside and burn room doors; visualize the difference with a thermal imager, with door both open and closed, of the hallway airflow and of fire conditions and heat flow from room.
7. After closing exterior door, engage fire, perform partial knock down, reposition to hallway and allow fire to build back, and change fire fighters on nozzle. Try to get all three on rotation.
8. Evolution 2: Restock same room per sequential chart (2 pallets, half bale of hay), repeat for other Attack Team.
9. Evolution 3: Attack Team 3 and 4, same as Evolution 1 and 2 in Charlie (center/rear).
10. Evolution 4: Restock same room per sequential chart (2 pallets, half bale of hay), repeat for other Attack Team.
11. Cool building, extinguish all fires. Utilize PPV to cool interior atmosphere, open all doors and windows in PPV sequence for 15 minutes.
12. List additional evolutions, specify if different from first round
13. Upon the last evolution, extinguish all fires and thoroughly ventilate prop. Do not cool down walls or any interior service with hose lines. Remove smoldering fuel to (specify location).
14. Rehab personnel (including medical surveillance).
15. Complete demobilization procedure (check all SCBA are cleaned, filled, and ready; likewise, ensure fire apparatus is ready for use).
16. Conduct Post-Incident Critique/debrief
17. Advise Communications and any units on standby that operations are complete.
18. Secure facility.

PARTICIPANT BRIEFING (all students, instructors, and support personnel will participate)

Review the following information with all participants:
- ❏ PPE inspection
- ❏ Objectives of drill
- ❏ Order of Operations, Burn Plan
- ❏ Emergency plans and procedures. (fire, nonfire, offsite)
- ❏ Medical and Rehab plans and procedures.
- ❏ Go/No-Go procedure (all crew members check each other, each crew visualizes other crews)
- ❏ Communication plan
- ❏ Evacuation signal demonstration
- ❏ Rally point
- ❏ Burn Building orientation tour
- ❏ Primary entrances/exits
- ❏ Location of planned exercises, entrances to be used
- ❏ "Victim" scenario: no live "victims," no "victims" in PPE, and possibility of simulated victims (if used)
- ❏ Emergency egress
- ❏ Potential hazards that could be encountered
- ❏ Identify Instructor-In-Charge *(and Incident Commander if a separate position is used)*
- ❏ Assignments given
- ❏ Accountability implemented

STRUCTURAL LIVE FIRE TRAINING PROPSITE PLAN

Actual diagram in plan needs to show all primary positions (CP, Rehab, apparatus locations, hose lines, etc.).

Appendix E

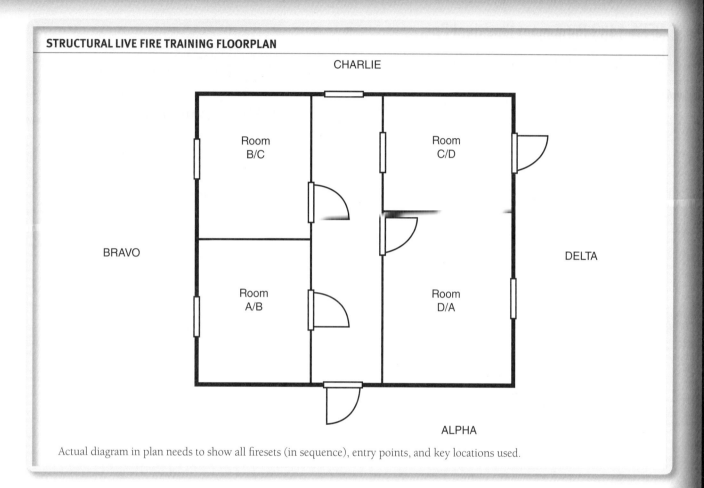

STRUCTURAL LIVE FIRE TRAINING FLOORPLAN

Actual diagram in plan needs to show all firesets (in sequence), entry points, and key locations used.

GO/NO-GO PROMPT	
Instructor-in-Charge/Command: "All personnel stand by for a Go/No-Go." (Utilize local terms, as not all positions may be necessary and some that are used in your department may not be listed.)	
From the Instructor-in-Charge/Command	Reply (repeats assignment)
1. Fire Control (Ignition), Go/No-Go?	Fire Control GO
2. Staging, Go/No-Go?	Staging GO
3. Rehab, Go/No-Go?	Rehab GO
4. Attack Team 1, Go/No-Go?	Attack Team 1 GO
5. Backup Team, Go/No-Go?	Backup Team GO
6. RIT, Go/No-Go?	RIT GO
7. Safety: Go/No-Go?	Safety GO
8. "We have a Go for ignition."	Ignition copies GO for ignition

NFPA 1403 Permanent Structural Prop Instructor-in-Charge Check Sheet

Confirm By ✓	Standard	2012 Edition
	Students meet prerequisites	4.3
	Instructor-in-Charge shall be trained to meet the minimum JPRs for Instructor I in NFPA 1041	4.6.1
	Visually inspected prior to use	6.2.1/7.21
	Damage documented	6.2.2/7.2.1.1
	If structural damage exists and impacts student safety, cannot use structure	6.2.2/7.2.2.
	All openings, automatic ventilators, mechanical equipment, sprinklers, and standpipes checked and operated prior to live fire evolution	6.2.3/7.2.3
	All safety devices checked prior to training	6.2.4 / 7.2.4
	Run gas-fueled systems prior to exposing students	6.2.5
	Structure left in safe condition after training	6.1.2 / 7.1.2
	Remove debris in access or egress paths	6.1.3 / 7.1.3
	Awareness of weather conditions prior to ignition	4.6.11
	Instructor-in-Charge and Safety Officer determine rate and duration of water flow to handle attack and backup lines and protect exposures	4.11.1
	Minimum water supply per NFPA 1142	4.11.4
	50 percent of fire flow demand held in reserve	4.11.5
	Separate water sources for attack and backup lines, unless training center has adequate volumes and backup power source, pumps, or both	4.11.6 4.11.6.1
	Designated areas for staging, operating, and parking of fire apparatus	4.13.1
	Area for apparatus not involved in live fire training designated	4.13.2
	If apparatus not involved in fire training is in service, locate in area for prompt response	4.13.3
	Parking for ambulance/EMS vehicle provided for easy access/egress	4.10.2
	Preburn briefing conducted prior to live fire training	4.15.2
	During the briefing, reinforce that any "victims" (manikins) will not be in PPE	Not in current 1403
	Preburn plan prepared and utilized during preburn briefing	4.15.1
	Preburn plan includes all features of area and structure	4.15.1.1
	Walkthrough of structure prior to live fire training	4.15.7
	Spectators restricted outside operations area perimeter	4.14.1
	Control measures posted to indicate perimeter	4.14.2
	Control measures established to keep pedestrians outside of operations area	4.14.5
	Visitors inside operations perimeter must be escorted	4.14.3
	Visitors inside operations perimeter shall wear complete protective clothing	4.14.
	Ignition sources other than identified are to be removed	4.12.11.9
	Ensure room around all props for attack and backup lines	4.11.7
	Fuel materials	4.12
	Only wood products for fuel	4.12.1

Appendix E

Confirm By ✓	Standard	2012 Edition
	Debris/unidentified materials that could burn not used	4.12.4
	Fuel-fired buildings and props permitted to use appropriate fuels	4.12.1.1
	Flammable or combustible liquids shall not be used	4.12.3
	Flammable gas fires shall not be ignited manually	6.1.4
	Instructor-in-Charge and Safety Officer assess room	4.12.8
	Instructor-in-Charge or Safety Officer stops training if potential hazard exists	4.12.10
	If exercise stopped due to potential combustible nature hazard, must reduce hazards before resuming training	4.12.10.1
	Safety	1.1
	Safety officer must be appointed	4.4.1
	Safety Officers shall be trained on the application of the requirements of this standard	4.4.2
	Safety Officer has authority to intervene and control unsafe conditions	4.4.3
	Safety Officer responsibilities include prevention of unsafe acts and elimination of unsafe conditions	4.4.4
	Safety Officer responsible for all persons, including students, instructors, visitors, and spectators	4.4.5
	Safety Officer cannot be assigned other duties	4.4.6
	Safety Officer must be knowledgeable in operation and location of safety features	4.4.7
	Instructor-in-Charge determines number of attack and backup lines prior to each evolution	
	Each hose line must be able to deliver minimum of 95 gpm	4.11.2
	Backup lines provide protection for attack line personnel	4.11.3
	Instructor-in-Charge assigns instructor to each functional crew (1:5 instructor:student ratio), one instructor to each backup line, additional personnel to backup lines as needed, and one additional instructor for each additional functional assignment	4.6.4
	Safety officer can have additional safety personnel as needed within the structure	4.4.8
	Fire-ground communications established	4.9.1
	Building evacuation plan established to include evacuation signal, which is demonstrated to all personnel	4.9.2
	EMS available on site	4.10.1
	Injuries documented and submitted	4.10.3
	Search of structure prior to ignition to assure no people, animals, or objects are in the structure	5.1.10
	Victims cannot be real people	5.1.11
	Cancellation of training if extreme weather conditions exist	4.5
	Participants must have full protective clothing and SCBA	4.8.1
	Safety Officer checks PPE prior to entry into evolution	4.8.2
	All personnel must wear protective clothing and equipment if involved in any evolution	4.8.1
	All personnel must use SCBA if in oxygen-deficient environment, an environment suspected of being oxygen-deficient or contaminated by products of combustion, any environment that may become oxygen-deficient, or below ground level	4.8.7
	Fire Control Team consist of minimum of two personnel	4.7.1
	Ignition Officer cannot be a student and is in charge of igniting, maintaining, and controlling burning materials	4.7.1.1

Confirm By ✓	Standard	2012 Edition
	Ignition Officer member is a of the Fire Control Team	4.7.1.1.1
	Fire Control Team member is in the area to observe the Ignition Officer ignite and maintain fire and monitor and report adverse conditions	4.7.1.2
	Instructor-in-Charge coordinates with Safety Officer to ignite training fire	4.7.2
	Fire shall be ignited by Ignition Officer (previous edition of NFPA 1403 does not identify Ignition Officer)	4.7.4
	Charged hose line available when Fire Control Team is igniting or tending fire	4.7.5
	Fire ignited only if Instructor visually confirms that flame area clear of personnel	4.7.6
	Fuels utilized in Class A live fire training shall only be wood products	4.12.1
	Fuel-fired buildings and props are permitted to use appropriate fuels as designed	4.12.1.1
	Lighters, fuses, matches, and similar devices can be used to ignite training fires if removed immediately after ignition	4.12.5
	RIC, trained to NFPA 1407, shall be provided during live fire training	5.5
	Training of Instructors provided by individual authorized by gas-fueled training system manufacturer or others qualified to perform type of training	4.6.12.2
	Instructor-in-Charge clearly identified	Not in current 1403
	Instructor-in-Charge responsible for coordination of overall live fire training activities for safety	4.6.3
	Instructor-in-Charge follows Sequential Live Burn Chart developed for the prop, specific to sequential evolutions in each room, fuel load, etc.	7.3
	Assign participants into crews, give crew assignments/functions	4.15.4

APPENDIX F

Annotated Changes and Additions to NFPA 1403, 2012 Edition

The new standard has moved the repetitive standards such as student prerequisites, instructor and safety qualifications and duties, and water supply from the individual sections (Acquired Structures, Gas-Fired Live Fire Training Structures, Non-Gas-Fired Live Fire Training Structures and Exterior Live Fire Training Props) to Chapter 4.

Maritime props are now within 1403's purview, the Ignition Officer is now part of a Fire Control Team of at least two trained (not student) members, a new Sequential Burn Matrix is required, and the training requirements for Live Fire Trainers, Rapid Intervention Crews and other positions is clarified. Eleven new terms (Chapter 3 Definitions) have been added relating to fire behavior.

Other changes include:

1.3.1	Live fire training must comply with current edition of NFPA 1403
1.3.2	Includes marine structures or vessels
2.2	Adds NFPA 1041 - Instructor Qualifications, 1407 - Training Fire Service Rapid Intervention Crews
2.4	Adds NFPA 1001 – STANDARD FOR FIRE FIGHTER PROFESSIONAL QUALIFICATIONS
3.2 & 3.3	Adds significantly to NFPA and Official and General Definitions
4.1	Live fire training shall comply with this chapter 4 (1403) and appropriate chapter for type of training
4.11.1	Adds Safety Officer with Instructor-in-Charge to determine rate and duration of water flow to handle attack and backup lines and protect exposures
4.2	Safety Officers shall be trained on the application of the requirements of this standard
4.2.3	Runoff from live fire shall comply with environmental requirements for site
4.6.1	IIC shall be trained to meet the minimum JPRs for Instructor I in NFPA 1041
4.6.5.1	Instructors rotated through duty assignments
4.6.6	All instructors must be qualified to delivery live fire training by the AHJ
4.6.9	Instructors shall take a personal accountability report (PAR) when entering and exiting the structure or prop during an actual attack evolution
4.7	Fire Control Team
4.7.1	A fire control team shall consist of minimum of two personnel
4.7.1.1	Ignition officer cannot be student or the safety officer and to ignite, maintain and control burning materials

4.7.1.1.1	Ignition Officer member of fire control team
4.7.1.2	Fire control team member in area to observe ignition officer ignite and maintain fire and monitor and report adverse conditions
4.7.3	Fire shall be ignited by ignition officer– previous edition outside prop not identified as ignition officer
4.7.4	The fire control team shall wear full personal protective clothing, including SCBA, when performing this control function.
4.7.5	Charged hose line available when fire control igniting or tending fire
4.10.1.1	For acquired structures, Basic Life Support Emergency Medical Service (EMS) on scene with transport capabilities
4.11.1	Instructor-in-Charge and Safety Officer determines rate and duration of water flow to handle attack and backup lines and protect exposures (new text)
4.11.5.1	Minimum reserve of additional water does not apply to permanently sited gas-fueled training systems
4.12.1	Fuels utilized in live fire training shall only be wood products
4.12.1.1	Fuel-fired buildings and props are permitted to use appropriate fuels as designed
4.12.5	Lighters, fusees, matches and similar devices can be used to ignite training fires if removed immediately after ignition
4.12.8	The instructor-in-charge and the safety officer shall assess the selected fire room environment for factors that can affect the growth, development, and spread of fire. (new text)
4.12.9	The instructor-in-charge and the safety officer shall document fuel loading, including all of the following: (1) Fuel material; (2) Wall and floor coverings and ceiling materials; (3) Type of construction of the structure, including type of roof and combustible void spaces; (4) Dimensions of the room. (new text)
4.12.10	The training exercise shall be stopped immediately when the instructor-in-charge or the safety officer determines through ongoing assessment that the combustible nature of the environment represents a potential hazard. (new text)
4.14.2	Control measures shall be posted to indicate the perimeter of the operations area.
4.14.5	Control measures established to keep pedestrian traffic outside operations
4.15.5	The location of the manikin shall not be required to be disclosed, provided that the possibility of victims is discussed in the preburn briefing.
4.7.4	Fire control team (includes ignition officer) shall wear protective clothing and SCBA
4.8	2007 Standard 7.4.17.2(1) allowed personnel that were not engaged in firefighting or exposed to hazards of structural firefighting to wear OSHA approved helmets. This was removed in the 2012 version.
4.12.1.1	Fuel-fired buildings and props permitted to use appropriate fuels designated by the manufacturer
4.12.11	The use of flammable gas, such as propane and natural gas, shall be permitted only in live fire training structures specifically designed for their use.
4.12.11.1	Liquefied versions of the gases specified in 4.12.11 shall not be permitted inside the live fire training structure
5.1.11	No person (s) shall play the role of a victim inside the acquired structure
5.5	RIC trained to NFPA 1407 shall be provided during live fire training

6.1.1	Section of standard pertains to all interior spaces where gas-fired live fire training occurs.
6.2.6	Structural integrity evaluated and documented annually by AHJ (2007 ed 5.2.2.2 required periodic evaluation and documentation by a licensed professional engineer)
6.2.6.1	If visible structural defects found, owner shall have follow-up evaluation by licensed professional engineer with live fire training structure experience
6.2.7	At least every 10 years structural integrity evaluated/documented by licensed professional engineer or another competent individual identified by AHJ.(2007 ed 5.2.2.3 (1)(2)(3)(4) had every one to three years)
6.2.8	Structures constructed with calcium aluminate refractory structure concrete inspected by structural engineer every 3 years.
6.2.8.1	Inspection includes removal of concrete core samples
6.2.9	At least once every 10 years, removal and reinstallation of representative area of thermal linings for inspection (2007 ed had every five years)
7.3	Sequential Live Burn Evolutions
7.3.1	AHJ develops and utilizes action plan when multiple sequential burn evolutions conducted per day in each burn room
7.3.2	Burn sequence matrix chart shall be developed for burn rooms in live fire training structure
7.3.3	Burn sequence defines maximum fuel load for first and each successive burn
7.3.4	Burn sequence matrix specifies maximum number of sequential evolutions in each room not to be exceeded
7.3.5	Fuel loads per evolution and maximum number of sequential burns cannot be exceeded

Glossary

Acquired prop A piece of equipment that was not designed for burning but is used for live fire training.

Acquired structure A building or structure acquired by the authority having jurisdiction from a property owner for the purpose of conducting live fire training evolutions.

Authority having jurisdiction (AHJ) An organization, office, or individual responsible for enforcing the requirements of a code or standard, or for approving equipment, materials, an installation, or a procedure.

Backdraft A deflagration resulting from the sudden introduction of air into a confined space containing oxygen deficient products of incomplete combustion.

British thermal units (Btu) A unit of measurement indicating the amount of heat required to heat one pound of water 1° Fahrenheit (F) at sea level.

Burn building A common nontechnical term for a permanent live fire training structure.

Combustible Capable of burning, generally in air under normal conditions of ambient temperature and pressure, unless otherwise specified. Combustion can occur in cases where an oxidizer other than oxygen in air is present (e.g., chlorine, fluorine, or chemicals containing oxygen in their structure).

Conduction Heat transfer to another body or within a body by direct contact.

Construction Classification A set of predetermined factors that are used to help determine the minimum water supply based on the type of building construction of an acquired structure.

Convection Heat transfer by circulation within a medium such as a gas or a liquid.

Core temperature The body's internal temperature.

Deflagration Propagation of a combustion zone at a velocity that is less than the speed of sound in the unreacted medium.

Demonstration The act of showing a skill.

Dry run Used during initial live fire training, an attack on a prop with no fire and no water flowing to give students confidence in hose handling techniques. A dry run is typically followed by second evolution flowing water, and the third evolution introducing fire.

Evolution A set of prescribed actions that result in effective fireground activity.

Exterior prop A nonstructural, outdoor live fire training prop.

Fire control team Comprised of a minimum of two members including the ignition officer or assigned ignition person and utilized any time a fire is ignited. Observes the ignition officer, ignites and maintains the fire, and watches for adverse conditions.

Fire flow rate The amount of water pumped per minute (gallons per minute or liters per minute) for a fire. There are several different formulas that are commonly used to calculate this.

Flameover (rollover) The condition in which unburned fuel (pyrolysate) from the originating fire has accumulated in the ceiling layer to a sufficient concentration (i.e., at or above the lower flammable limit) that it ignites and burns. Flameover can occur without ignition of or prior to the ignition of other fuels separate from the origin.

Flashover A transition phase in the development of a compartment fire in which surfaces exposed to thermal radiation reach ignition temperature more or less simultaneously and fire spreads rapidly throughout the space, resulting in full room involvement or total involvement of the compartment or enclosed space.

Fuel load The total quantity of combustible contents of a building, space, or fire area, including interior finish and trim, expressed in heat units or the equivalent weight in wood.

Gas-fired live fire training structure A permanent live fire training structure where the fires are fueled primarily by propane or liquefied natural gas.

Go/No Go sequence A verbal confirmation via radio communication that each and every participant is ready for action in the live fire training environment.

Heat release rate The amount of heat energy released by a material over time in a fire.

Incident scene rehabilitation A function on the fireground that cares for the well being of the fire fighters. It includes physical assessment, revitalization, medical evaluation and treatment, and regular monitoring of vital signs.

Instructor An individual qualified by the authority having jurisdiction (AHJ) to deliver fire fighter training, who has the training and experience to supervise students during live fire training evolutions.

Instructor-in-charge An individual qualified as an instructor and designated by the authority having jurisdiction to be in charge of the live fire training evolution.

Live fire Any unconfined open flame or device that can propagate fire to a building, structure, or other combustible materials.

Manufactured prop A type of exterior prop that is built to resemble an actual emergency for the purposes of live fire training.

Minimum water supply requirement The total amount of water, not flow, required for a given structure, based on its size, construction, and proximity to other structures or properties that could be damaged (exposures).

Non-gas-fired live fire training structure A permanent live fire training structure where the fires are fueled by Class A materials such as excelsior, hay, and pallets.

Occupancy Hazard Classification A set of predetermined factors that are used to help determine the minimum water supply based on the hazard levels of certain combustible materials.

Order of operations The sequence of steps followed to conduct a procedure. In this context, the steps are in proper sequence to conduct the live fire evolution, but order of operations could also refer to any emergency scene operation. Most commonly refers to the sequence a mathematical equation is solved.

Participant Any student, instructor, safety officer, visitor, or other person who is involved in the live fire training evolution within the operations area.

Personnel accountability report (PAR) A verification by person in charge of each crew or team that all of their assigned personnel are accounted for.

Preburn plan A briefing session conducted for all participants of live fire training in which all facets of each evolution to be conducted are discussed and assignments for all crews participating in the training sessions are given.

Pyrolysate Product of decomposition through heat; a product of a chemical change caused by heating.

Pyrolysis The process of decomposition of a material into other molecules when it is heated.

Radiation Heat transfer by way of electromagnetic energy.

Safety officer An individual appointed by the authority having jurisdiction as qualified to maintain a safe working environment at all live fire training evolutions.

Sequential live burn evolutions matrix A chart developed when multiple sequential burn evolutions are conducted. Must include fuel loading per evolution, number of evolutions conducted per room, and maximums that cannot be exceeded.

Stroke volume The amount of blood pumped with each contraction of the heart.

Thermal gradient The rate of temperature change with distance.

Thermal tolerance The body's ability to cope with high heat conditions.

Thermoregulation The process by which the body regulates body temperature.

Ventilation-controlled fire A fire in which the heat release rate or growth is controlled by the amount of air available to the fire.

Vent point ignition Smoke is at or above its ignition temperature and is lacking oxygen. The smoke will ignite as it exits the opening and falls within the flammable range.

Index

A

Accountability, 48–49, 90
Acquired prop, 159, 160, 173
Acquired structure, 102, 122
  emergency plans for, 103–104
  evolution preparation
    equipments, 115
    for ignition, 115–116
  exterior preparation
    asbestos, 108
    asphalt, 108
    chimneys, 109
    hazards, 109
    sides of building, 108
    utilities, 108
    ventilation, 108–109
  final controlled burn, 121
  initial preparation
    access, 106
    entry and egress routes, 106–108
    equipments, 106, 109
  interior preparation
    attics, 112
    burn locations, 113
    burn set, 113
    ceilings, 111
    doors, 111–112
    exits, 113
    flooring, 111
    fuel loads, 113
    furniture, 110–111
    hazards, 109–110
    kitchens, 112
    line the room, 114
    oil tanks, 112
    walls, 111
    windows, 111
  neighborhood preparation, 114–115
  operations
    fire behavior considerations, 117
    fire ignition, 117
    Go/No Go, 116–117
    preburn briefing, 116
    setup, 116
    walk-throughs, 116
  post-incident analysis, 120
  preparation
    adjacent properties, 102
    AHJ, 103
    fire department, 103
    preburn plan for, 103
    water supply, 104–105
Adams, Donald, 29, 32
AFFF (aqueous film forming foam), 164
Agency notification checklist, 94
AHJ (authority having jurisdiction), 6, 8, 21, 40, 88, 103, 134, 160
Ambient temperature, 61
Apparatus, fire
  areas for operating and parking of, 85
  need for, 84

Aqueous film forming foam (AFFF), 164
Asbestos riding shingles, 108
Asphalt shingles, 108
Attack
  group, 89, 90
  instructor, 90
Attic fire evolution, 182
Authority having jurisdiction (AHJ), 6, 8, 21, 40, 88, 103, 134, 160
Avoidance, 33–34

B

Backup instructor, 90
Barrel drill, barrels arrangement for, 20
Barrett, Rusty, 20
Blood pressure, 71
Body temperature, 61, 71
British thermal units (Btu), 43, 115–116, 122
Building
  construction, 19
  evacuation plan, 143
  plan, 84–85, 142
Burn
  buildings, 130, 150
  characteristics of, 110, 115
  Class A, 136
  scenarios, 133
    locations, 135
    types, 134
  sequence, 135, 229

C

Cardiac emergencies, prevention of, 69–70
Cardiovascular
  disease, 72
    risk factors, 69
  fitness, 59
  strain
    factors affecting, 58–60
    of firefighting, 56–57
Car props, 137
CC (construction classification), 104
Celsius to Fahrenheit formula, 43
Chimneys, 109
CISD (critical incident stress debriefing), 27
Civil litigation, 33
Class A burns, 136
Class A props, 91
  ignition of, 171
  safety considerations for, 161
  types of, 160–161
Class B props, 161–165
  ignition of, 171–172
Code requirements, 130
Command staff, 89, 90
Communications, 167
  plan, 88, 142
Computer-controlled systems, 133
Concrete live fire training structure, 131
Construction classification (CC), 104, 122
Cooling techniques, 70
Core temperature, 57, 73

Critical incident stress debriefing (CISD), 27

D

Death plan, 26–27
Dehydration, 60
Demobilization plan, 94, 142
Demonstration, 9, 21
Drillground, 145
Dry run, 172, 173
Ducted smoke distribution, 141

E

EH (exposure hazard), 104
Electrolyte replacement, 70–71
Emergency
  cardiac, prevention of, 69–70
  evacuation, sound systems for, 137
  medical plan, 88, 142
  operations center, 27
  plans, 85–87, 142–143
    for acquired structure, 103
Emergency Incident Rehabilitation, 60
Emergency medical services (EMS), 9, 71, 85, 106
Employment, 33
EMS (emergency medical services), 9, 71, 85, 106
Engineering requirements, 144
Entry officer, 90
Environmental conditions, physiological response to firefighting, 58–59
Environmental Protection Agency (EPA), 165
Evaporative cooling, 57, 61
Evolutions, 178, 185
Excess body fat, 59
Exposure hazard (EH), 104
Exterior props, 159, 173
  types
    Class A, 160–161
    Class B, 161–165
External reviews, 29

F

Fahrenheit to Celsius formula, 43
Farago, Joe, 29
FFFIPP (Fire Fighter Fatality Investigation and Prevention Program), 27
Final burn down signal, 121
Fire alarm, sound systems for, 137
Fire behavior, 17, 41, 43–45
  drill, 180
Firebox, 114
Fire dynamics in structure, 43–45
Fire extinguishers, portable, 17, 19
Fire Fighter Fatality Investigation and Prevention Program (FFFIPP), 27
Fire fighters
  fatality, aftermath of
    civil litigation, 33
    employment, 33
    internal strife, 32–33
    news media scrutiny and public opinion, 33

Index

regulatory and criminal charges and prosecution, 32
fitness level of, 59
medical conditions of, 59
qualifications, professional, 181
safety, 41
style, 45
training, 42
Firefighting
cardiovascular and thermal strain of, 56–57
physiological responses to, 58–60
thermal strain of, 56–57
Fire flow rate, 122
Fireground noises, sound systems for, 137
Fire ignition, 117
Fire lighting, 148–149, 171–172
Fire locations and fuel, 134
Fire service, 41
recruits, 179
Fitness level of fire fighters, 59
Flammable liquid props, 163
Flashover, 121
live fire training structures, 136–137
Florida's live fire training laws, 49–50
Foam, 164
Forcible entry, 19, 41
Fuel
and fire locations, 134
load, 146, 160
materials, 134–135
and oil separator, 165

G

Gas-fired live fire training structures, 131, 133–134, 150
preburn plans for
communications plan, 142
emergency plans, 142–143
on-site facilities, 143–144
spectators, media, and visitors, 143
preparation and inspection phase of, 145–146
Gas-fired systems, ignition of, 148–149
Go/No Go sequence, 87–89, 96, 116–117, 146, 148, 149

H

Heart attacks, 69, 70
Hearths, fire resistance for, 114
Heat
cramps, 61
emergencies, 60–65
exchange, effectiveness of, 61
exhaustion, 62
illnesses, 60, 61
classifications, 62
prevention, 64–65
risk factor, 63–64
release rate, 43–44, 52
of common materials, 43
stress prevention guidelines, 65
Heatstroke, 62–63
High temperature linings, 135
HUMIDEX chart, 64

Hydration, 60
urine color and, 60

I

ICS (incident command system), 41
IDLH (immediately dangerous to life and health), 8, 183
Ignition officer, 90, 94
Immediately dangerous to life and health (IDLH), 8, 183
Incident command organization for structural live fire exercises, 89, 90
Incident command system (ICS), 41
Incident reports, 10–15
Baltimore, Maryland - 2007, 110–112
Boulder, Colorado - 1982, 10–11
Green Cove Springs, Florida - 1990, 92–93
Greenwood, Delaware - 2000, 184
Hollandale, Minnesota - 1987, 12–13
Lairdsville, New York - 2001, 30–31
Milford, Michigan - 1987, 14–15
Parsippany, New Jersey - 1992, 168–169
Pennsylvania State Fire Academy - 2006, 46–47
Poinciana, Florida - 2002, 66–68
Port Everglades, Florida - 2003, 138–140
Incident scene rehabilitation, 70–73
Instructional technique, 172
Instructor, 40, 52
Instructor-in-charge, 8, 21, 87–90
acquired structures
burn locations, 113
emergency plans for, 103–104
fire ignition, 117
ignition preparation, 115–116
preburn briefing, 116
preburn plan, 103
and safety officer, 113, 116–117
water supply, 104, 105
of live fire training, 16
Insulation panels, 132, 133
Insuring care, 27
Interior safety crew, 179
International Society of Fire Service Instructors (ISFSI), 50
Investigation, 27, 35
ISFSI (International Society of Fire Service Instructors), 50

K

Knoff, Richard, 92

L

Ladders, 19
Law enforcement, 27
LFAT (Live Fire Adjunct Trainer), 49
LFTI. *See* Live Fire Training Instructor
Lightning detectors, 87
Line method, fire resistance, 114
Line-of-duty significant injury, 26–27
Liquefied petroleum gas (LPG), 163
props, 170

Liquid-propane (LP) props, 7
Live fire, 8, 21
Live Fire Adjunct Trainer (LFAT), 49
Live Fire Master Instructor, 49
Live fire training, 56
evolutions, 7–9
experienced students, 182
history of, 5–6
incidents, 6
last session before, 19
learning objectives, 178
operations, 183
postevolution debriefing, 183
recruit students, 179–182
student and, 16–17
impact of NFPA 1403 on, 7
incident scene rehabilitation for, 70–71
instructor-in-charge of, 16
ISFSI program, 50
laws, 49
principles, 107–108
structural, 132
student prerequisites for, 9
Live Fire Training Instructor (LFTI), 49, 181
qualities of, 40–41
requirements for, 40–43, 49–50
training for, 43–49
LPG. *See* Liquefied petroleum gas
LP (liquid-propane) props, 7

M

Malo, Walt, 32, 33
Manufactured prop, 159, 173
Media, 115, 143
news (*See* News media)
Medical conditions of fire fighters, 59
Medical monitoring, 71
Miami-Dade's Fire Rescue, 141
Minimum water supply (MWS), 104, 122
Mobile live fire training structure, 131, 132
Multiple fires, 147
Multistory props, 136
MWS (minimum water supply), 104, 122

N

National Fire Protection Association. *See* NFPA
National Incident Management System (NIMS) model, 89
National Institute for Occupational Safety and Health (NIOSH), 68, 117, 149
investigations, 27–28, 35
National Institute of Standards and Technology (NIST), 47, 68
National Wildfire Coordinating Group (NWCG), 50–51
News media, 29, 32, 35
scrutiny and public opinion, 33
NFPA 30, *Flammable and Combustible Liquids Code*, 7
NFPA 58, *Liquefied Petroleum Gas Code*, 7, 170
NFPA 59, *Utility LP-Gas Plant Code*, 7, 170
NFPA 1001, *Standard for Fire Fighter Professional Qualifications*, 8, 17, 41, 52, 181

NFPA 1041, *Fire Service Instructor Professional Qualifications*, 7, 40, 52
NFPA 1142, *Standard on Water Supplies for Suburban and Rural Fire Fighting*, 8, 104, 105
NFPA 1402, *Guide to Building Fire Service Training Centers*, 130
NFPA 1403, *Standard on Live Fire Training Evolutions in Structures*, 3–7, 16, 25, 39, 40, 49, 55, 56, 64, 77–81, 83, 84, 86, 88, 91, 94, 95, 97, 99–101, 106, 125–129, 137, 153–158, 160, 161, 177
  application of, 7
  impact of, 7
  purpose of, 5, 21
  of structural evaluation, 144
  using, 8–9
NFPA 1406, *Standard on Outside Live Fire Training Evolutions*, 6
NFPA 1582, *Standard on Comprehensive Occupational Medical Programs for Fire Departments*, 59
NFPA 1583, *Standard on Health-Related Fitness Programs for Fire Department Members*, 59
NFPA 1584, *Standard on the Rehabilitation Process for Members During Emergency Operations and Training Exercises*, 70
NFPA 1971, *Standard on Protective Ensembles for Structural Fire Fighting and Proximity Fire Fighting*, 8, 91, 167
NFPA 1975, *Standard on Station/Work Uniforms for Fire and Emergency Service*, 8, 91, 167
NFPA 1981, *Standard on Open-Circuit Self-Contained Breathing Apparatus (SCBA) for Emergency Services*, 8, 91, 167
NFPA 1982, *Standard on Personal Alert Safety Systems (PASS)*, 8, 91, 167
NIOSH. *See* National Institute for Occupational Safety and Health
NIST (National Institute of Standards and Technology), 47, 68
Nonexercise emergencies, 87
Non-gas-fired live fire training structures, 131, 134–135, 150
  preburn plans for, 141
    communications plan, 142
    emergency plans, 142–143
    management, 144
    on-site facilities, 143–144
    spectators, media, and visitors, 143
    preparation and inspection phase of, 146–147
Non-gas-fired systems, ignition of, 149
Nonstructural elements, evaluating, 144
NWCG (National Wildfire Coordinating Group), 50–51

O

Occupancy Hazard Classification (OHC), 104, 122
Occupational Safety and Health Administration (OSHA), 6, 27, 130
OHC (Occupancy Hazard Classification), 104, 122
On-site facilities, 143–144
OPB (Orientated Strand Board), 144

Order of operations, 87–88, 96
Orientated Strand Board (OPB), 144
Orientation drills, 9, 20
OSHA. *See* Occupational Safety and Health Administration
Overhaul, 19, 121

P

PAR. *See* Personnel accountability report
Participants, 8, 21
  in live fire training, 84
  rotation during live fire training, 94
Participant safety, 167
PASS (personal alert safety system), 8, 46, 47
Pennsylvania State Fire Academy (PSFA), 50
Pennsylvania Suppression Instructor Development Program (ZFID), 46, 50
Perflurochemicals (PFCs), 164
Personal alert safety system (PASS), 8, 46, 47
Personal protective equipment (PPE), 9, 48, 56, 58–59, 142, 167, 181, 182
  design of, 19
  level of, determine, 65, 180
  use of, 17, 94, 163
Personnel accountability report (PAR), 48, 52, 67, 86, 96, 121, 143
  drills, 48–49
PFCs (perflurochemicals), 164
Physiological responses to firefighting, 58–60
Portable fire extinguishers, 17, 19
Position task books (PTB), 50, 51
Positive pressure ventilation (PPV), 121
Postevolution debriefing, 183
Post-incident analysis, 93, 120
  Baltimore, Maryland - 2007, 120
  Boulder, Colorado - 1982, 11
  Green Cove Springs, Florida - 1990, 93
  Greenwood, Delaware - 2000, 184
  Hollandale, Minnesota - 1987, 13
  Lairdsville, New York - 2001, 31
  Milford, Michigan - 1987, 15
  Parsippany, New Jersey - 1992, 169
  Pennsylvania State Fire Academy - 2006, 47
  Poinciana, Florida - 2002, 68
  Port Everglades, Florida - 2003, 140
PPE. *See* Personal protective equipment
PPV (positive pressure ventilation), 121
Preburn briefing, 87, 94, 95, 116
  session, 170–171
Preburn plan, 83, 96, 103
  for acquired structure, 103
  developing, 83
  emergency medical services, 166
  environmental concerns, 165
  for gas-fired live fire training structures, 141–144
  list of evolutions, 87
  maintenance, 170
  neighbors, 165
  for non-gas-fired live fire training structures, 141–144

  operations area, 166
  parking areas, 166
  safety officer, 166–167
  student prerequisites, 165
  using, 94–95
  water supply, 166
  weather, 166
Preparation phase, 145
  drillground, 145
  and inspection phase
    of gas-fired live fire training structures, 145–146
    of non-gas-fired live fire training structures, 146–147
Prerequisite training for live fire, 17
Primary engine/pumper, 90
Propane torch, 117
Protective clothing, 90, 91
PSFA (Pennsylvania State Fire Academy), 50
PTB (position task books), 50, 51
Pyrolysis, 43, 52

R

Rapid intervention crew (RIC), 67, 86, 90, 107, 180
  deployment, 30
Recruit
  fire fighters, 180
  to live fire training, 181
Rehab/medical officer, 90
Rehydration, 70–71
RIC. *See* Rapid intervention crew
Road flares, 117
Roof vent props, 141

S

Safety
  during evolutions, 172
  fire fighter, 41
  for live fire, 17
  and practice, 48–49
Safety officer, 9, 21, 30, 116–117
  responsibilities of, 89–91
SAR (search and rescue) team, 67
SCBA (self-contained breathing apparatus), 6, 8, 40, 46, 58, 90, 91, 94, 142, 167, 171, 183
Search and rescue (SAR) team, 67
Secondary engine/pumper, 90
Self-contained breathing apparatus (SCBA), 6, 8, 40, 46, 58, 90, 91, 94, 142, 167, 171, 183
Sequential live fire burn evolutions, 135, 229
Shipping containers, 136–137
Site plan, 85, 87, 142
Smoke, 44–45
SOG (standard operating guidelines), 7
SOP (standard operating procedure), 7, 83, 179, 182
Spatzer, Sam, 33, 34
Spectators, 114–115, 143
Split level cell, 136
Staffing
  and organization, 88–89, 148

requirements, 134, 135
rotation during live fire training, 94
Staging/air supply officer, 90
Standard and legal considerations, 6–9
Standard operating guidelines (SOG), 7
Standard operating procedure (SOP), 7, 83, 179, 182
Statter, Dave, 29, 30
Stroke volume, 57
Structural fire dynamics, 43–45
Structural fire training prop, 149
Structural live fire training, 132
Student
  and live fire training evolutions, 16–17
  prerequisites for live training, 0
  psychology, 45
  training, 17
    building construction, 19
    fire behavior, 17
    fire hose, appliances, and streams, 19
    forcible entry, 19
    ladders, 19
    overhaul, 19
    personal protective equipment, 19
    portable fire extinguishers, 17, 19
    safety, 17, 21
    ventilation, 19
    water supply, 19
Student-to-instructor ratio, 84
Support group, 89, 90
Swedish fire service training system, 136

T

Temperature monitoring, 135
Thermal balance, 61
Thermal gradient, 61, 73
Thermal imaging camera (TIC), 44, 48
  drill, 180
Thermal shock, 132
Thermal strain
  factors affecting, 58–60
  of firefighting, 56–57
Thermal tolerance, 60, 73
Thermoregulation, 61, 73
TIC. *See* Thermal imaging camera
Training. *See also* Live fire training
  firefighting, 57
  for LFTI, 43–49
  objectives, 84
  site, initial evaluation of, 83
Type X Gypsum wall board, 114

U

United States Fire Administration (USFA), 28
Urine color and hydration, 60
USFA (United States Fire Administration), 28

V

Ventilation, 19, 41, 108–109, 162
  PPV, 12z1
Vent point ignition, 44, 49, 52
Visitors, 143

W

Water supply, 19, 41, 104–105, 143, 166
  MWS, 104, 122
  need for, 84
Weather, 87
  parameters, 142

Z

ZFID (Pennsylvania Suppression Instructor Development Program), 46, 50

Credits

Live Fire Training Instructor in Action Courtesy of Christopher L. Dyer, BravePhoto, LLC; **Background charred wood image** © Josiah Garber/Dreamstime.com

Chapter 1
1-1 Courtesy of Frank Young, City of Boulder Fire Department; **1-3** Courtesy of Chris Dilley, Clay County Fire Rescue; **Boulder IR Figures A and B** Courtesy of Dave Demers, Demers Associates; **Hollandale IR Figure A** Reproduced with permission of the Minnesota Department of Labor and Industry, Minnesota OSHA Compliance; **Hollandale IR Figure B** Courtesy of the Minnesota Department of Labor and Industry, Minnesota OSHA Compliance; **Milford IR Figure A** Courtesy of the U.S. Fire Administration; **Milford IR Figure B** Reproduced with permission from NFPA, Copyright © 1987, National Fire Protection Association.

Chapter 2
Lairdsville IR Figure A From the NIOSH Fire Fighter Fatality Investigation and Prevention Program, Investigation Report F2001-38 NY.

Chapter 3
3-1 Courtesy of Chris Dilley, Clay County Fire Rescue; **3-5, 3-6, 3-7, 3-8** Courtesy of Joshua Bauer, Seminole Tribe of Florida Fire Rescue; **Pennsylvania State Fire Academy IR Figures A and B** From the NIOSH Fire Fighter Fatality Investigation and Prevention Program, Investigation Report F2005-31.

Chapter 4
4-3 Courtesy of the USFA; **Poinciana IR Figures A and B** Courtesy of the Florida State Fire Marshal; **Poinciana IR Figure C** From the NIOSH Fire Fighter Fatality Investigation and Prevention Program, Investigation Report F2002-34.

Chapter 5
5-1 Courtesy of Shawn Morgan, Corvallis Fire Department

Chapter 6
6-13 Courtesy of Kriss Garcia; **6-16** Courtesy of the USFA Media Production Center; **6-17** Courtesy of Christopher L. Dyer, BravePhoto, LLC; **Baltimore IR Figure A** Courtesy of the USFA Media Production Center

Chapter 7
Opener Courtesy of Fireblast 451, Inc.; **7-1** Courtesy of the Loudoun County Department of Fire, Rescue & Emergency Management; **7-4, 7-6** Courtesy of Fireblast 451, Inc.; **7-8, 7-9** Courtesy of Kidde Fire Trainers, Inc.; **7-14** Courtesy of American Fire Training Systems, Inc.; **7-15** Courtesy of Ed Hartin, CFBT-US LLC; **7-16** Courtesy of Kidde Fire Trainers, Inc.; **7-20** Courtesy of Robert Hernandez, Miami-Dade Fire Rescue; **7-21** Courtesy of Kidde Fire Trainers, Inc.; **7-23** Courtesy of Stephen Carter, University of Maryland University College; **7-24** Courtesy of WHP Trainingtowers; **7-28** Courtesy of Ed Hartin, CFBT-US LLC

Chapter 8
8-1 Courtesy of Kidde Fire Trainers, Inc.; **8-2** Courtesy of Walt Malo; **8-5A, 8-5B, 8-8** Courtesy of Christopher L. Dyer, BravePhoto, LLC

Chapter 9
Greenwood IR Figure A From the NIOSH Fire Fighter Fatality Investigation and Prevention Program, Investigation Report F2000-27.

Unless otherwise indicated, all photographs and illustrations are under copyright of Jones & Bartlett Learning or have been provided by the authors.

Reproduced with permission from NFPA 1403, *Standard on Live Fire Training Evolutions*, Copyright © 2007, National Fire Protection Association. This reprinted material is not the complete and official position of the NFPA on the referenced subject, which is represented only by the standard in its entirety.

Reproduced with permission from NFPA 1584, *Standard on the Rehabilitation Process for Members During Emergency Operations and Training Exercises*, Copyright © 2008, National Fire Protection Association. This reprinted material is not the complete and official position of the NFPA on the referenced subject, which is represented only by the standard in its entirety.